Advances in Nanoporous Materials

纳孔材料进展

Stefan Ernst

科 学 出 版 社

北 京

图字:01-2011-1543 号

This is a annotated version of
Advances in Nanoporous Materials
by **Stefan Ernst**

ISBN: 9780444531797

Authorized English language reprint edition published by the Proprietor.
ISBN 13:9789812728944

Elsevier(Singapore)Pte Ltd.
3 Killiney Road
08 01 Winsland House 1
Singapore 239519
Tel:(65)6349-0200
Fax:(65)6733-1817

First Published 2011
<2011>年初版

图书在版编目(CIP)数据

纳孔材料进展=Advances in Nanoporous Materials:英文/(德)恩斯特(Ernst,S.)主编.—北京:科学出版社,2011
ISBN 978-7-03-030570-1

Ⅰ.①纳… Ⅱ.①恩… Ⅲ.①纳米材料:多孔性材料-简介-英文 Ⅳ.①TB383

中国版本图书馆 CIP 数据核字(2011)第 043959 号

责任编辑:霍志国/责任印制:钱玉芬/封面设计:耕者设计工作室

科学出版社 出版
北京东黄城根北街 16 号
邮政编码:100717
http://www.sciencep.com

双青印刷厂 印刷

科学出版社发行 各地新华书店经销

*

2011 年 4 月第 一 版 开本:B5(720×1000)
2011 年 4 月第一次印刷 印张:21 3/4
印数:1—1 800 字数:438 000

定价:99.00 元

(如有印装质量问题,我社负责调换)

导　读

纳孔（nanoporous）材料是指孔尺寸为纳米（nanometer）级的固体孔材料。提到纳孔材料，广大读者自然而然地会联想到纳米材料。两者既有联系，又有区别。纳米材料已经是公共术语，在过去的二十年，特别是近十年，纳米科技是最受关注的科学技术领域。现在，恐怕不知道“纳米”的人很少，但了解“纳孔”的，也许不是很多，纳孔材料远远没有纳米材料那样为公众所熟知。在阅读本书之前，有必要对二者的确切含义给予简要说明。

纳米材料和纳孔材料这两个术语中的“纳”字是英文 nano 的音译，1 纳米（nm）等于 10^{-9} 米（m），即十亿分之一米，相当于大约 4 倍铯原子（225pm）的尺度，或大约 40 倍氦原子（32pm）的尺度。它是一种尺度的概念，像厘米、毫米一样，其本身并没有特殊的物理内涵。人们关注纳米材料，是因为从 20 世纪 70 年代开始，科学家观察到当物质尺度达到纳米级以后，大约是在 1～100nm 这个范围内，物质有时会出现特殊性能。这种具有既不同于原子、分子，也不同于宏观物质的特殊性能材料，即为纳米材料。过去，人们只注意原子、分子或常见的宏观尺寸，常常忽略这个中间领域，而这个领域实际上大量存在于自然界中，只是以前没有认识到物质在这个尺度范围的特殊性能。80 年代中期，人们就正式把这类材料命名为纳米材料。随着纳米材料科技的进步和广泛应用，人们倾向于把尺度在纳米级的材料都称为纳米材料，多孔固体材料就是其中之一。多孔固体材料的颗粒尺寸可以是纳米级的，但更多是微米级或毫米级，甚至更大。然而，多孔固体材料中的孔道的尺寸和结构，的确是纳米级的，被称为纳米结构。这种纳米结构的内涵包括孔道与窗口的尺寸和形状，孔道的维数和走向，孔壁的组成与性质等。这种具有纳米结构的纳孔材料可以被视作纳米材料的一种，从而扩展了纳米材料的范畴。

在描述多孔材料的各项特性指标中，孔道的大小尺寸是最重要的特征。根据国际纯粹和应用化学联合会（International Union of Pure and Applied Chemistry，IUPAC）的定义，多孔固体材料可以按其孔径分为以下 3 种：小于 2 nm 为微孔（micropores）；介于 2～50 nm 之间为介孔（mesopores），有时也被称为中孔；大于 50 nm 为大孔（macropores）。与其相对应的固体孔材料就自然而然地称为微孔材料（microporous materials）、介孔材料（mesoporous materials）和大孔材料（macroporous materials）。介孔材料从结构特点和合成规律等方面均不完全相同于微孔材料，但二者又有很多共同的规律和特征。有些学者将上述 3 种固体多孔材料都称为纳孔材料，但本书的主编给出的定义是小于

50nm 的多孔材料属于纳孔材料，明确地将大于 50nm 的多孔固体排除在纳孔材料的范畴之外。由此可见，将多孔固体称为纳孔材料，并不是因为这种材料的表观尺寸，而是因为其中的孔的孔道尺寸。当这种孔道尺寸减小到一定程度后，其吸附、扩散性能就会发生显著地变化。人们认识到这种孔材料特性要远远早于对纳米材料特性的认识。早在 1756 年，人们就注意到天然沸石具有可逆的脱水作用。1773 年，Scheele 就观察到气体可以被木炭吸附。气 固吸附能产生的原因是多孔固体大的比表面积和微小的孔径尺寸。这种狭小孔道内由于孔壁吸附势能的叠加效应使得气体吸附能力进一步增强。比表面积主要产生于微孔，以活性碳的数千平方米的比表面积为例，至少 90% 的比表面积与小于 2nm 的微孔有关，不到 10% 来自于介孔，而大孔对比表面积几乎没有贡献。所以，本书作者将纳孔材料的范畴定义为微孔材料和介孔材料，而将大孔材料排除在外是有科学道理的。需要指出的是，固体中的孔道结构是多种多样的，具有不同的尺寸和形状。常用的孔径概念是用来描述圆柱形孔道的，对于楔形孔道，常用孔宽来描述。除了 BET 比表面、孔径这些参数外，孔体积也是常用评价纳孔材料特性的指标。通常用 77K 条件下的 N_2 吸附来确定总孔体积，用 273K 条件下的 CO_2 吸附来确定微孔体积。另外，纳孔这一术语并不是 IUPAC 推荐使用的，但它已经是科技界的热门词汇，所以，本书标题中的纳孔材料这种表述本身就是本书的一个亮点。

本丛书的主编是德国凯泽斯劳腾（Kaiserslautern）大学化学系的著名教授斯特凡·恩斯特（Stefan Ernst）博士。凯泽斯劳腾大学始建于 1970 年，是一所强调应用技术的研究型大学。斯特凡·恩斯特博士领导的化工技术科研小组致力于纳孔固体材料作为催化剂和吸附剂的多相催化和分离/纯化研究。在固体酸和双功能沸石分子筛择型催化、胺基催化中心在沸石分子筛中的构建、用纳孔吸附剂脱除 CO_2 和 H_2S 等研究领域做出了很多开创性的研究成果。本书编辑委员会的其他成员也是本领域的著名学者，如加利福尼亚（California）大学的 Yaghi 教授，是当今世界上金属有机骨架化合物（Metal Organic Frameworks，MOFs）合成、表征和应用方面最著名的科学家。

由于纳孔材料丰富的内涵，使得属于其范畴的固体孔材料种类非常多，如硅铝沸石分子筛及其种类繁多的同系物、全硅分子筛、磷酸分子筛及其同系物、活性碳、碳分子筛、碳纳米管、氧化硅、非硅孔材料、金属孔材料、金属有机框架化合物（MOFs）等。其中有些材料在化学催化、吸附、分离、膜分离、膜反应、离子交换、传感器等方面已经起到了极其重要的作用，有些则有广泛的应用前景。具体的应用领域如石油加工、石油化工、煤化工、精细化工等领域中的工业催化剂，分离、净化、干燥领域中的吸附剂，洗涤剂工业、矿厂、放射性废料处理等领域的离子交换材料，医学领域的药物载体、骨材料等。在如此广泛和众多的纳孔材料及其应用的题材中，本丛书的编辑选择了沸石分子筛膜、烃类芳构化催化剂、介孔非硅材料和微孔磷酸铝催化剂这四个方面的综述构成了本丛书第一卷的内容。选题跨度大、内容宽泛是本卷的一个特点。此外，本卷所收录的每个专题在注重学术方面综述的同时，都介绍了相应孔材料的实际应用或潜在的实用领域，构成了该书的另一个特点。

第 1 章综述了沸石分子筛膜的现状和前景，是本卷的重点内容。过去人们对沸石分子筛的印象是固体颗粒，本章综述的则是沸石分子筛膜，是一个新的领域，学术价值很高。颗粒状固体分子筛已经广泛用于催化和吸附分离工业领域。当被用作吸附剂时，强调其孔结构、比表面和吸附容量，其操作过程是吸附和脱附交替的间歇操作过程。当被用作催化剂时，除强调上述特性外，还强调其表面化学特性。而对膜状分子筛来讲，强调的则是分子选择性地透过膜材料，因而膜通量和选择性是其最重要的表征和评价指标。将沸石分子筛膜化，或换言之，开发沸石分子筛膜的研究始于 80 年代末，由已故的新西兰籍英国科学家、英国皇家科学院院士 R. M. Barrer（1910—1996）率先开始，现在是学术界和工业界的热门研发领域。沸石分子筛膜是现阶段有工业应用前景或已经工业化的 3 种无机膜材料之一，另外两种是用于氢气分离的钯基金属膜和用于氧气分离的钙钛矿型膜。与有机聚合物膜相比，无机膜具有热稳定性好，抗有机溶剂性强，机械强度大，寿命长等优点。与传统的分离方法相比，膜分离的最大优势是降低分离成本和连续操作过程。本章在介绍膜分离概念、沸石膜的制备、评价和表征的同时，用了很多篇幅综述了沸石分子筛的应用实例。如氢气分离，二氧化碳分离，亲水沸石分子筛膜用于乙醇和丙醇脱水，憎水沸石分子筛膜用于发酵过程中乙醇的移除等。对沸石分子筛膜应用领域的详细总结是本章的亮点之一，读者们应该给予关注。值得一提的是，作者在本章中还简要介绍了金属有机框架化合物分子筛膜的研究现状。金属有机框架化合物作为纳孔材料的研究是近十年才刚刚兴起的，其膜化、表征及应用是近几年才开始的，所以这部分新内容非常值得读者给予关注。还有，本章中沸石分子筛膜反应器的最新研究进展和应用也值得读者们阅读。另外，本章的作者们引用了近 400 篇相关文献，最新的文献是 2009 年发表的研究论文。读者在阅读本书内容时，要充分利用好这一丰富的文献资源。

第 2 章介绍了沸石分子筛催化剂在芳构化过程中的应用进展。沸石分子筛在石油炼制过程和石油化工过程中有广泛的应用，占据技术核心的地位。全世界沸石分子筛催化剂的 85% 被用在了石油炼制过程中的流化催化裂化工艺过程，用于生产汽油和柴油馏分。人们日常使用汽油中的每个分子，大都出自于沸石分子筛催化剂的孔道。石油炼制过程中的催化重整工艺也是非常重要的烃类加工过程，目的是将烷烃通过芳构化过程生产苯、甲苯和二甲苯，用到的催化剂是含有贵金属的沸石分子筛催化剂。金属中心起脱氢作用，沸石分子筛的酸性中心提供裂化、异构化功能。这两种功能的协同作用，实现了烃类分子的芳构化过程，用来生产高辛烷值汽油组分和石油化工基本原料。随着环保要求的提高，汽油中苯含量的上限由原来的3%变更到1%，总的芳烃限制量由36%调整到25%。这种调整使得芳构化过程更主要用来生产基本化工原料。读者可以参考第99页的图表，了解苯、甲苯和二甲苯在石油化工过程中的重要基础原料地位，以及与此相关的各种化学反应，如烷基化、烷基转移、异构化、歧化等。这些反应都基于酸性催化反应，本章的主要内容就是综述了这些相关的化学过程和用到的纳孔沸石分子筛催化剂。对从事基础化学研究的读者，本章呈现的石油化工知识可能有些陌生。同样，对化工领域的读者，可能对沸石分子筛这种纳孔材料的相关基础知识了解的不够深入。所以，本章的内容对化学和化工领域的科研工作者都是非常好的参考，可以扩展化学科学工作者的化工视野，可以加强化工从业人员的化学基础，是一种“双功能”的学习材料。

第3章的内容是介孔非硅材料及其功能的综述。含硅孔材料如MCM-41因其典型的介孔特性为广大科研工作者所熟知。近些年大量出现的介孔材料，如过渡金属氧化物、介孔碳、介孔聚合物、氮化物、介孔半导体等，由于其不含硅，被统称为非硅介孔材料。这些材料除了可以用作催化剂和吸附剂，还可以用来制造纳米器件、传感器、微电子器件等，已经引起了广泛关注。本章系统地综述了这些材料的制备、表征和应用。本章图文并茂，内容全面，条理非常清楚。其中介绍了很多最新的介孔材料，如碳纳米笼等。在本章的最后，作者列出了一个简短的专业术语表，方便读者学习和掌握相关内容。

第4章介绍了基于磷酸铝分子筛（$AlPO_4$）和磷酸硅铝分子筛（SAPO）的催化剂的催化作用，包括金属磷铝分子筛催化剂（MeAPOs）和金属磷酸硅铝分子筛催化剂（MeAPSOs)。磷铝分子筛由Wilson和Flanigen等于1982年率先合成成功，是继沸石分子筛（硅酸铝分子筛）之后又一个里程碑式的重要进展。这种新型纳孔分子筛的最大特点是可以将多达20种元素，包括主族金属与过渡金属以及非金属元素，引入其纳孔骨架，从而形成数百种不同结构的纳孔化合物，在催化领域有广泛的应用前景。

本书为研究生及相关科研人员提供了纳孔材料领域很好的最新进展资料。在阅读本书、学习纳孔材料相关内容的过程中，最好是将该书通读一遍。如果时间有限，或只对某一领域或部分章节比较熟悉和感兴趣，也可以将本书作为一本参考书。该书最后的“主题词索引”（Subject Index）是非常有效的工具。读者通过这个索引，可以查找一些术语的确切含义，查阅某一主题的详细内容，并且可以通过正文引用的参考文献，了解相关主题的研究进展。如果读者对本书内容比较生疏，可以先阅读徐如人院士的《分子筛与多孔材料化学》一书，在中文语境中了解相关背景知识，再阅读这本英文原版综述集，会收到更好的效果。

本书是《纳孔材料进展》丛书的第一卷，仅仅包括了纳孔材料领域众多专题中的四项内容。由于纳孔材料的科学研究和实际应用范围广阔，进展迅速，让我们共同期待本丛书的后续综述文集早日与读者见面。

赵学波
中国科学院能源科学与技术研究中心
中国科学院青岛生物能源与过程研究所

目　　录

（赵学波　译）

前　言

《纳孔材料进展》是专门针对各种纳孔固体材料科学和应用方面的新系列丛书。本书是该丛书的第一卷，出版目的是向读者介绍由著名专家撰写的纳孔材料领域有价值的全面综述，范围涵盖纳孔固体材料的所有方面，包括材料的制备和结构、后期合成修饰、表征方法以及其在催化和吸附/分离方面的应用。此外，纳孔材料的潜在应用，如膜、主体/客体化学、环境保护、电化学、传感器、光学器件以及纳孔材料的理论处理和模拟等，也是本系列丛书的内容。

纳孔材料这一术语是指孔径尺寸在大约 0.2～50nm，并且不论其化学组成、来源（天然的或合成的）和结构属性（无定型或晶体）的所有固体孔材料。典型的例子包括沸石及众多化学性质相似的分子筛，介孔氧化物如二氧化硅、硅酸铝等，金属有机框架化合物，层柱状粘土，多孔玻璃，多孔碳及相关材料。

本丛书的第一卷，即本书由 J. Caro 和 M. Noack 开篇，全面评述了沸石分子筛膜的制备、吸附/分离特性以及其在工业上的应用。此外，还讨论了新的合成概念和沸石分子筛膜的前景。

第 2 章中，C. Perego 和 P. Pollesel 综述了沸石分子筛在芳构化过程中的工业应用。他们的综述显示了在这个石油化学分支里，沸石分子筛在生产过程经济优化和系统优化方面起到的极其重要的作用。

第 3 章由 A. Vinu 执笔，介绍了介孔非硅分子筛的制备和性质，并且呈现了这种材料在传统应用领域（催化和分离）和电化学、传感器等领域的多种多样的潜在应用前景。

最后，M. Hartmann 和 S. P. Elangovan 贡献了一个全面的、有关含过渡金属微孔磷酸铝和磷酸硅铝晶体做催化剂的催化反应的文献综述。他们强调了这些材料在烃类酸和（或）双功能转化、选择性氧化、烯烃二聚/齐聚反应方面的潜在应用。

收录在本卷里的这四章综述很明显地反映出纳孔材料领域的多样性和宽泛内容，也表明《纳孔材料进展》这套丛书的内容十分丰富。计划将来每年再多出版 1～2 卷。

最后，主编衷心感谢本书的作者，他们付出了大量的时间和辛勤的努力来完成各自的章节，也衷心地感谢本书的编辑出版人员的付出。愿《纳孔材料进展》这套丛书能够成为高级研究者经常参考的有价值的研究综述文集和新读者了解精彩纷呈的纳孔材料领域的实用指导工具。

Stenfan Ernst

（赵学波　译）

CONTENTS

LIST OF CONTRIBUTORS

Juergen Caro
Leibniz University of Hannover, Institute of Physical Chemistry and Electrochemistry, Callinstr. 3-3A, D-30167 Hannover, Germany

S.P. Elangovan
Nippon Chemical Industry Co. Ltd., 9-11-1 Kameido, Koto-ku, Tokyo 136-8515, Japan

Martin Hartmann
Advanced Materials Science, University of Augsburg, Universitätsstr.1, D-86159 Augsburg, Germany and Erlangen Catalysis Resource Center, Friedrich-Alexander-Universität Erlangen-Nürnberg, 91058 Erlangen, Germany

Manfred Noack
Leibniz Institute for Catalysis at the University Rostock, Berlin Branch (former ACA), Richard–Willstätter-Str.12, D-12489 Berlin, Germany

C. Perego
Instituto Eni Donegani, Eni S.p.A, Via Fauser 4, 28100 Novara, Italy

P. Pollesel
Eni S.p.A, Refining & Marketing Division, Via Maritano 26, 20097 San Donato Milanese, Italy

Ajayan Vinu
International Center for Materials Nanoarchitectonics, World Premier International (WPI) Research Center, National Institute for Materials Science, 1-1, Namiki, Tsukuba, 305-0044, Japan

PREFACE

This is the first volume of *Advances in Nanoporous Materials*, a new book series devoted to the science and application of all kinds of nanoporous solids. Its intention is to publish comprehensive reviews of lasting value in the field of nanoporous materials written by renowned experts. Its scope covers all aspects of nanoporous solids, including their preparation and structure, their post-synthetic modification, methods for their characterization as well as their application in catalysis and adsorption/separation. Also, their fields of potential applications, for example, membranes, host/guest chemistry, environmental protection, electrochemistry, sensors, optical devices etc., and theoretical treatment and modeling of nanoporous materials are within the scope of the new series.

The term *nanoporous materials* is understood to comprise all kinds of porous solids that possess pore sizes in the range from ca. 0.2 up to ca. 50 nm, irrespective of their chemical composition, their origin (natural or synthetic) and their amorphous or crystalline nature. Typical examples are zeolites and chemically related molecular sieves, mesoporous oxides like silica, silica–alumina etc., metal organic frameworks, pillared clays, porous glasses and carbons and related materials.

This first volume starts with a review authored by *J. Caro* and *M. Noack* that gives an extensive overview on the preparation of zeolite membranes, their properties in adsorption/separation and their use in industrial applications. Moreover, novel synthesis concepts and the future of zeolitic membranes are discussed.

In the second chapter of this volume, *C. Perego* and *P. Pollesel* review the industrial applications of zeolites in aromatics processing. They show that in this branch of petrochemistry zeolite catalysts have gained tremendous importance with respect to optimizing economy and ecology of the production processes.

Chapter 3 by *A. Vinu* addresses the preparation and properties of non-siliceous mesoporous molecular sieves and presents an outlook on the manifold potential uses of these materials in classical applications (e.g., catalysis and separation) and in electrochemistry, sensing etc.

Finally, *M. Hartmann* and *S. P. Elangovan* present a thorough overview over the literature on reactions catalyzed by crystalline microporous aluminophosphates and silicoaluminophosphates containing transition metals. They highlight their possible usefulness in the acid and/or bifunctional conversion of hydrocarbons, in selective oxidations and in olefin dimerization/oligomerization reactions.

The four chapters presented in this volume impressively reflect the large diversity and the manifold aspects of nanoporous materials and, hence, the broad scope of *Advances in Nanoporous Materials*. For the future, it is planned to publish an additional one or two volumes annually.

Finally, the editor would like to express his sincere gratitude to the authors who spent much time and great efforts on their chapters and to the members of the editorial board for their input. It is hoped that *Advances in Nanoporous Materials* turns out to be both a valuable collection of reviews regularly consulted by the advanced researchers and a useful guide for newcomers to the fascinating and multifaceted world of nanoporous materials.

Stefan Ernst
Editor

CHAPTER 1

Zeolite Membranes – Status and Prospective

Juergen Caro[1,*] *and* **Manfred Noack**[2]
[1]Leibniz University of Hannover, Institute of Physical Chemistry and Electrochemistry, Callinstr. 3-3A, D-30167 Hannover, Germany
[2]Leibniz Institute for Catalysis at the University Rostock, Berlin Branch (former ACA), Richard-Willstätter-Str. 12, D-12489 Berlin, Germany

Contents

*Corresponding author. Tel.: +49 511 762 3175; Fax: +49 511 762 19121
E-mail address: caro@pci.uni-hannover.de

Advances in Nanoporous Materials, Volume 1 ISSN 1878-7959, DOI 10.1016/S1878-7959(09)00101-7

Abstract

The introduction of industrial membrane-based separation technologies can dramatically reduce the separation costs in comparison with thermally based separation technologies. In addition, membrane technologies allow the energy effective use and recovery of both valuable raw materials and the separation of wastes. Organic polymer membranes are increasingly used, but they suffer from stability at elevated temperatures and toward attack of organic solvents. Therefore, much effort is put into the development of temperature stable and solvent resistant inorganic membranes. Pd-based metal membranes for hydrogen separation, perovskite-type membranes for oxygen separation and zeolite-type molecular sieve membranes are on the jump into the industrial practice. This increasing application of inorganic membranes in gas separation – and on a later timescale in chemical membrane reactors – is a slow process. Because of the high investment costs, many companies prefer to play the role of an "observer." In this contribution, we reflect the state of the art of zeolite membranes. We report the first industrial application of zeolite membranes in bio-ethanol de-watering and parallel ongoing fundamental research on improving the thin zeolite layer crystallization on porous asymmetric supports following new synthesis concepts and the development of novel diagnostics. In this chapter, we also treat the molecular understanding of zeolite membrane separations since this knowledge is crucial for the proper use of zeolite membranes and for the exploration of new application fields.

1. INTRODUCTION: SETTING THE SCENE

Intelligent membrane engineering can help to realize the process intensification strategy. Integrated membrane separations and new membrane operations such as catalytic membrane reactors and membrane contactors will play a crucial role in future technologies. However, so far no inorganic membrane is used in large-scale industrial gas separation. The increase of the ^{235}U isotope concentration in a $^{238}U/^{235}U$ mixture from 0.7% in natural uranium to approximately 3.5% for nuclear fuel applications by separation of $^{235}UF_6$ and $^{238}UF_6$ on porous ceramic membranes according to the Knudsen mechanism[1] with an ideal separation factor of

[1]The Knudsen separation factor SF_{Kn} of a binary mixture of components m_1 and m_2 with the molecular mass M_1 and M_2 is $SF_{Kn}(m_1/m_2) = \sqrt{M_2/M_1}$.

1.0043 is an exception. However, nowadays exclusively gas centrifuges are used for uranium isotope separation. Membrane reactor technology has a huge potential in the development of processes that are more compact, less capital intensive, giving higher conversions and selectivities in both thermodynamically and kinetically controlled reactions, respectively. Membrane reactors are expected to save energy and costs of feed/product separation [1]. So far, no high-temperature membrane reactor for chemical reactions is in industrial operation. The use of porous ceramic filter membranes in biotechnology is an exception.

Inorganic membranes such as ceramics, metals, and glass show promising properties different from the organic ones. They can be backwashed frequently without damaging the separation layer. Inorganic membranes are highly resistant to cleaning chemicals, they can be sterilized and autoclaved repeatedly at 130–180 °C and can withstand temperatures up to at least 500 °C. These properties recommend them for biotechnological processes as well. Inorganic membranes should have longer life spans than organic ones. The life span of a typical hydrophilic organic membrane is approximately 1 year, of a hydrophobic membrane 2 years, and of fluoropolymers up to 4 years [2]. Inorganic membranes are, however, much more expensive than polymeric ones, and they are brittle.

Three types of inorganic membranes are near to a commercialization: Pd-based membranes in H_2 separation, perovskites in O_2 separation, and zeolite membranes in shape-selective separations. The regular pore structure of a zeolite molecular sieve suggests that a thin supported zeolite membrane layer can discriminate between molecules of different size and shape. The pore diameter of the separating zeolite layer is in the range of the kinetic diameter of the molecules to be separated to force molecular sieving as the determining diffusional regime. Furthermore, beside the molecular exclusion effect, due to the interplay of mixture adsorption and mixture diffusion, reasonable separation effects on zeolite membranes can be expected according to specific adsorptive interactions and/or differences in the molecular mobilities. The rapidly growing progress in the field of zeolite membranes is reflected by some recent reviews [3–10]. It is, therefore, not the aim of this contribution to give a comprehensive picture of zeolite membranes and to present all the fundamentals in detail, but to highlight and evaluate recent developments.

By the end of the 1980s, the idea was born to develop zeolite membranes and the first attempts to prepare them were reported, the first patents were claimed. With some pioneering papers, R.M. Barrer triggered the experimental work on zeolite membranes [11,12]. In parallel, he contributed to the theoretical understanding of mixture permeation through porous membranes [13,14]. The first one and the last one of Barrers altogether 407 publications were dealing with membranes [15,16]. The unit *Barrer* of gas permeability (flux in moles per time and area through a membrane of a given thickness and pressure difference) honours R.M. Barrer (Section 3.1).

Today, LTA (Linde Type A) membranes in the de-watering of alcohol by steam permeation or pervaporation have reached the commercial state. For shape-selective separations, other zeolite membranes with structure types such as MFI and DDR (deca-dodecasil 3R) are already in the technicum scale [8,17,18]. Further molecular sieve structures are tested as membranes (Table 1). Most progress in the development of molecular sieve membranes was achieved for MFI-type membranes (silicalite-1 and ZSM-5) since their preparation is relatively easy. They can be synthesized highly siliceous, which provides chemical stability and allows for oxidative regeneration [7]. Therefore, this contribution will mainly deal with MFI-type membranes.

Table 1 Claimed structures and common modifications of zeolite membranes [20]

Structures	Modifications
MFI – silicalite-1/ZSM-5	Isomorphous substitution
LTA – A-type	Ion-exchange
T-type	Defect healing
P-type	Heat treatment
FAU jasite – Y-type, X-type	Chemical treatment
MOR denite	
FER ricrite	
DDR	
SOD alite	
CHA basite – SAPO-34	
ANA lcime	
ETS	
AFI – $AlPO_4$-5	
BEA	
MEL-ZSM-11	

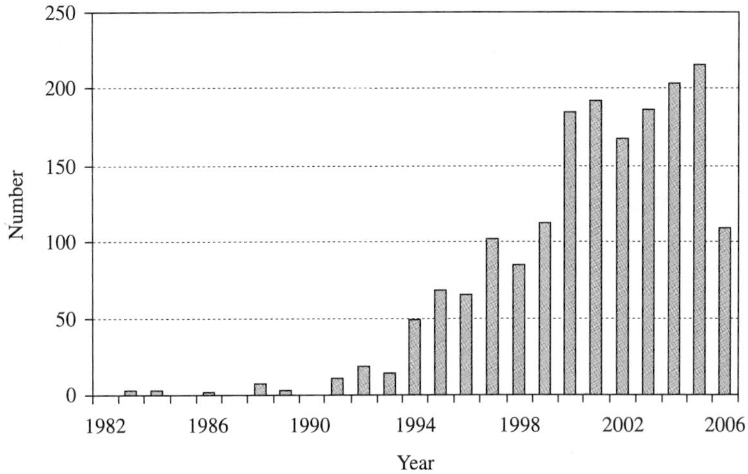

Figure 1 Development of the number of articles on zeolite/molecular sieve membranes/coatings/films in the open literature with searches on (zeolit* OR molecular sieve) AND (membrane* OR coating OR film) [20].

New ways of synthesis, improved permeation tests, and proper applications shall improve the zeolite membranes for their technical use. Increasing R&D activities are reflected by increasing publication activities (Fig. 1). It is the aim of this contribution to summarize the state of R&D on zeolite membranes as a relative young branch of the inorganic membrane family, 250 years after the discovery of zeolites by Cronsted [19]. It will be shown that the application of a crystalline molecular sieve as a zeolite membrane layer offers huge promises but it is still a challenge in itself.

2. PREPARATION OF ZEOLITE MEMBRANES

2.1 Peculiarities of zeolite membrane crystallization

As it will be described in more detail in Section 3.1, for high fluxes and a proper handling of zeolite membranes, a thin zeolite layer with a thickness of 1–20 μm is crystallized on a mechanically stable support. However, the chemical compositions of the crystallization solutions and their handling for zeolite membrane preparation as a thin supported layer differ from the standard recipes for a zeolite powder crystallization [21–23].

The following points are characteristic of the zeolite layer crystallization on supports [7]:

- At sufficient supersaturation, heterogeneous nucleation takes place on both the geometric outer surface of the support and inside the pores of the support. If externally prepared seed crystals are attached to the surface of the support, primarily the crystal growth of the seeds takes place but the simultaneous secondary nucleation at the surface of the seed crystallites and in/on the support cannot be suppressed completely. Therefore, diluted crystallization solutions are used to prevent the formation of new seeds and to have only growth of the attached seeds to a continuous layer.

- In the beginning of the growth of the seeds, the surface-to-volume ratio increases like in the case of the crystallization in the free solution. This is based on the effect that in the beginning of crystal growth, usually a parallel nucleation takes place, which results in a surface enlargement. In the subsequent process of crystal intergrowth, the individual crystals grow together to a continuous layer and the surface-to-volume ratio decreases drastically.
- The diffusion of the precursor species in the solution is not rate limiting. Since crystal growth is controlled by a first-order surface process, the growth rate decreases with the reduction of accessible surface.
- For the crystal intergrowth that is important for the sealing of voids between crystals, the viscosity of the synthesis solution should be low to enable mass transport in narrow slits. The driving force of the diffusion process is the concentration gradient. Therefore, the low viscosity should be realized rather by higher temperatures than by dilution. Another way to decrease the viscosity consists in an increase of the pH, which results in a higher concentration of low-connected silica species.
- During the crystal intergrowth of isolated crystals to a continuous layer, a large slit surface is in contact with a small volume of synthesis solution. Therefore, besides crystal growth, a strong heterogeneous secondary nucleation inside the slit can occur, which can lead to a closure of the macroscopic slit pore by many small crystals with intercrystalline transport pores between them. A post-synthesis thermal or hydrothermal treatment can result in a reorganization of these domains with improved membrane properties.
- The starting chemicals for the preparation of the synthesis batch should be selected with the aim to have low salt concentrations in the solution. Whereas these salts are not disturbing in the formation of the free crystals, the incorporation of neutral salt species – especially in multi-crystal layer formation – can be disturbing since defect pores are formed by their thermal decomposition (e.g., NH_4NO_3 and carbonate decomposition).
- It was found in a large number of studies that it is *de facto* impossible to crystallize defect-free Al-containing zeolite layers. Because of the strong negative surface charge (zeta potential) of Al-containing zeolites, the intergrowth of the crystals in the membrane layer is poor, and the grain boundaries represent defect pores in the mesopore region. This holds true for both the *in situ*-growth and the secondary growth with seeds. By using Intergrowth Supporting Substances (ISS), the crystal intergrowth in the zeolite membrane layer can be improved (Section 5.2).

For the most often prepared membrane types MFI, LTA, and FAU (faujasite), suitable synthesis batch compositions are given in the following. For MFI-type membrane crystallization, a recommended synthesis batch composition using tetrapropyl ammonium hydroxide or its salt as structure-directing agent (SDA) for the synthesis of MFI membranes is SiO_2:TPAOH:H_2O = 100:8.9:2220 for silicalite-1 and SiO_2:TPAOH:TPABr:NaOH:Al_2O_3:H_2O = 100:3.33:3.33:6.67:0.52:2000 for ZSM-5 with a molar Si/Al-ratio of 96.

The synthesis of LTA- and FAU-type crystals is very similar; in both cases, no organic template is used, the pH-value is above 13, the synthesis temperature is below 100 °C. For the membrane preparation of these membrane types, the use of a seeded support turns out to be advantageous. In many cases, multi-layer membranes were prepared with enhanced separation properties [24]. The use of microwave heaters operating at 2.45 GHz can accelerate the crystallization from the hour into the minute time scale (Section 5.1). This shortened synthesis times are beneficial since the formation of foreign phases is suppressed, and the dissolution and phase transformation of the zeolite layer already formed at long synthesis time is reduced [25,26]. These processes can lead to enlarged intercrystal slits between the crystals in the membrane layer as defects. The chemical composition for the LTA synthesis varies only slightly for the homogeneous gel synthesis: Na_2O/SiO_2 from 1.0 to 1.7, SiO_2/Al_2O_3 = 2, H_2O/Na_2O from 40 to 67 [24,27]. On the contrary, for a homogeneous solution synthesis the composition for the LTA synthesis is rather fixed: Na_2O/SiO_2 = 10, SiO_2/Al_2O_3 = 5, H_2O/Na_2O = 20 [25].

In the chemical composition for the FAU synthesis, we have to distinguish between the gel route and the clear solution route for FAU syntheses with different Si/Al-ratio: type X with a Si/Al-ratio of 1.2–1.3 and type Y with a Si/Al-ratio up to 2.5. In the gel route, the following molar ratios are used: SiO_2/Al_2O_3 near 10, Na_2O/SiO_2 from 1.2 to 1.4, H_2O/Na_2O from 50 to 60 [28–30]. In the clear solution route, the molar ratios are SiO_2/Al_2O_3 from 0.1 to 1, Na_2O/SiO_2 = 9, H_2O/Na_2O from 60 to 70 [31,32]. For a NaX membrane with a Si/Al-ratio of 1.3, the composition of the synthesis solution is SiO_2/Al_2O_3 = 3.6, Na_2O/SiO_2 = 1.4, H_2O/Na_2O = 50 [28]. For a NaY membrane with a Si/Al-ratio of 1.9 to 2.1, the composition of the synthesis solution is SiO_2/Al_2O_3 = 25, Na_2O/SiO_2 = 0.9, H_2O/Na_2O = 45 [28].

However, a completely different synthesis composition is used for the preparation of very thin supported LTA and FAU layers that are not used as membranes but as sensor devices. In these cases, the template tetramethyl ammonium hydroxide $(TMA)_2O$ and a high dilution are used: 10 SiO_2:1.6 Al_2O_3:0.267 Na_2O:11.9 $(TMA)_2O$:623 H_2O for LTA crystallization [33] and 10 SiO_2:2.3 Al_2O_3:0.15 Na_2O:5.5 $(TMA)_2O$:570 H_2O for type Y crystallization in colloidal solutions [34].

2.2 Direct *in situ* crystallization on supports

The direct hydrothermal *in situ* crystallization is the most widely applied technique to prepare layered zeolite membranes on a porous ceramic or stainless steel support [9]. In this route, the support is immersed in the zeolite synthesis solution, which can be a clear solution or an aqueous gel contained in an autoclave for crystallization temperatures >100 °C (e.g., 180 °C for MFI-type membranes) or in polymer bottles for syntheses <100 °C (e.g., FAU or LTA membrane preparations between 80 and 100 °C).

A gel layer on the surface of the support is formed by precipitation of the silica sol particles under certain concentration ranges and at given temperatures [35–39]. In the case of the MFI synthesis, the tetrapropyl ammonium ions (TPA^+) are found only in the solution but not in the precipitated gel layer. It is concluded, therefore, that MFI crystallization starts at the phase boundary between the liquid phase (as TPA^+ source) and the gel layer (as Si source). The crystals grow into the gel layer consuming the Si gel until the growing MFI crystals have reached the support. Using this synthesis route, different orientations of the MFI channel system relative to the support surface can be found. Often a *b*-orientation with straight channels perpendicular to the support can be obtained, which is a favorable orientation from the point of view of the anisotropy of mass transport in the MFI structure [40] (see Section 5.3 and cf. Figs. 5 and 39). This *b*-orientation can be explained as follows: Those nuclei that are oriented in parallel with the interface of the two nutrient pools in their fastest growth directions *c* and *a* show the largest growth rate and dominate the crystal orientation in the layer.

To get pinhole-free zeolite membranes, the crystal size in the zeolite layer should be less than 1 μm [41] since large well-faceted polyhedral crystals do not grow together to a defect-free layer. Another important issue for obtaining high-quality membranes by *in situ* crystallization on supports is (i) a smooth support surface and (ii) a hydrophilic support surface with good wettability [42]. Rougher support surfaces provide larger contact angles (like the Lotus effect) and the mass transfer in the *in situ* crystallization is reduced. Using porous titania supports (rutile), their wettability can be improved by UV radiation, which results in a better membrane quality.

The deposited gel can either form a surface layer or it can be soaked into the pore system of the support forming zeolite plugs [43]. In the latter case, the plugged zeolite membranes exhibit an improved mechanical stability but relative low fluxes. Whereas in some papers methods are described to grow the zeolite membrane layer for stability reasons within the pores of the support rather than on its surface [44–46], other authors propose to seal the support pores by adsorbed species to prohibit the penetration of the gel into the pores of the support thus having a high-flux membrane as a supported thin film. Using the latter technique to get higher fluxes, MFI membranes were prepared by *in situ* crystallization on porous α-Al_2O_3 disks that contained a diffusion barrier to limit the excessive penetration of siliceous species into the support pores [47]. The barrier was introduced into the ceramic pores by polymerizing a previously adsorbed mixture of furfuryl alcohol and tetraethyl orthosilicate followed by carbonization and a partial carbon burn-off to generate a carbon-free region for chemical bonding of the MFI layer to the support. The resulting MFI membrane had a smaller thickness and showed increased flows.

A two-step growth of MFI membranes was proposed by Vroon et al. [48]. In a first-step seed crystallites of 275–700 nm at relatively low temperature and high concentration of the crystallization batch are directly deposited on the support surface. Upon repeating the crystallization with fresh sol at elevated temperatures, a continuous zeolite layer with a thickness of 2–7 μm forms, which shows a separation factor for n-/i-butane of 50 at 25 °C. Further repetitions of the crystallization step did not give any improvements. On the contrary, the oxidative decomposition of the template resulted in a crack formation of the thick zeolite membrane layer.

Although the direct *in situ* crystallization can provide membranes of proven quality in gas separation, it has limitations [9,44]. There is little scope for the control of the microstructure of the final films since synthesis conditions have to be optimized for nucleation and growth [49].

2.3 Secondary crystallization using seeded supports

Two principal methods are used to suppress the effect of a homogeneous nucleation. One route is the so-called dry gel conversion [50,51], either as vapor-phase transport method when the SDA is in the vapor but not in the dry parent gel [52], or a steam-assisted crystallization with a dry gel containing the SDA [53].

The other strategy that has been established during the last few years for the controlled preparation of supported zeolite membranes is the seeding technique (secondary growth) using externally prepared seeds. The secondary growth of a supported seed layer is an effective approach for the formation of consolidated supported membranes and films with good quality and reproducibility [54,55]. By decoupling the nucleation step (at high supersaturation) from crystal growth (at low supersaturation), the seeds can grow in low concentrated solutions under suppression of the secondary nucleation. Unlike the X-ray amorphous metal oxide membranes, a polycrystalline zeolite layer prepared by hydrothermal synthesis from seed crystals deposited before on the support surface brings about the necessity to control the crystal intergrowth, so that pores between the individual zeolite crystals are avoided. This technique also requires a certain minimum membrane thickness or special techniques to achieve an orientation of the MFI zeolite crystals. Like in the direct *in situ* crystallization of zeolite membranes, masking techniques can be used to avoid the penetration of seeds and synthesis gel into the support pores. By a sophisticated polymethyl-methacrylate (PMMA)–polyethylene wax treatment in a laminar flow bench at high temperatures under vacuum, the support pores were filled by the wax which has a melting point above the synthesis temperature of 100 °C. By this procedure, the pores of the support were protected from the synthesis solution. Using colloidal nucleation seeds followed by hydrothermal growth at 100 °C, high-flux membranes with a thickness of approximately 0.5 μm could be prepared [56]. The power of the seeding technique was demonstrated by Matsukata et al. [57] showing that mordenite- or ZSM-5-type membranes could be prepared from identical organic-free aluminosilicate solutions under the same hydrothermal conditions by using either mordenite or MFI seeds.

To the authors' knowledge, the first patent to seed a support surface was submitted in 1994 [58], the first paper on seeding appeared in 1993 [59]. The use of seed crystals facilitates the formation of zeolite membranes since a seeded support grows to a pure-phase zeolite membrane more easily even when the crystallization conditions and the chemical batch composition are not optimum. There are three main ways to attach the seeds to the support: (i) variation of the pH to achieve that seeds and support have opposite surface charges (zeta potentials) for an electrostatic attachment, (ii) adsorption of positively charged polymers to re-charge the surface as condition for the following electrostatic attachment of negatively charged seeds, and (iii) immersion of the dried support into a seed solution. Tsapatsis et al. [60–62] change the pH of the solution to adjust different zeta potentials between the ceramic support (e.g., α-Al_2O_3) and the zeolite nanocrystals to be attached (e.g., silicalite-1 as pure SiO_2). Sterte et al. [63–65] adsorb cationic polymers on the support to create a positive surface charge and the

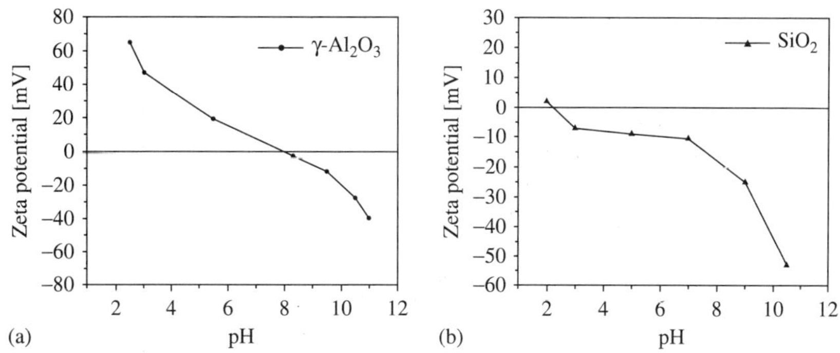

Figure 2 Zeta potential of alumina (a) and silica (b) nanoparticles suspended in 0.01 m KCl as electrolyte as a function of pH at 25 °C.

negatively charged zeolite seeds such as silicalite-1 become attached. Later, this method, which was first developed for coating Si wafers, was successfully transformed for seeding porous ceramic supports for membrane preparation [66]. As described in Section 5.3, the use of seeded supports usually results in a *c*-orientation of the MFI layer but under certain conditions also for secondary growth the desired *b*-orientation can be obtained [54].

1. *Charging the support surface by pH control*: It is looked for a pH where the zeta potentials of the support and the seed crystals show different signs to get an electrostatic attraction between support and seeds. As an example, negatively charged silicalite-1 seeds as pure SiO_2 phase become attracted by the positively charged alumina support in a wide pH range (Fig. 2)[2]. To avoid the formation of acid sites in the silicalite-1 layer by dissolving traces of aluminium species from the alumina support during the hydrothermal silicalite-1 membrane crystallization, porous titania and zirconia supports can be used. Additionally, by using titania, their wettability can be improved by UV radiation (Chapter 2.2). However, when titania supports are used, the zeta potential is negative over a wide pH range and SiO_2 nanoparticles such as the silicalite-1 seeds can not be attached electrostatically (Fig. 3). In this case, a positive surface charge can be generated by the adsorption of positively charged macromolecules as decribed in the following.

[2]Another consequence of the negative zeta potential of silicates and aluminosilicates in membrane preparation is that the intergrowth of neighboring seed crystals to a continuous layer is hampered. If two growing seed crystals form a narrow slit, then the negative surface charges can overlap and block the diffusive transport of the negatively charged silicate species of the synthesis solution to the surface of the crystals, thus forming a defect site. It will be shown in Chapter 5.2 that the use of so-called ISS can solve this problem.

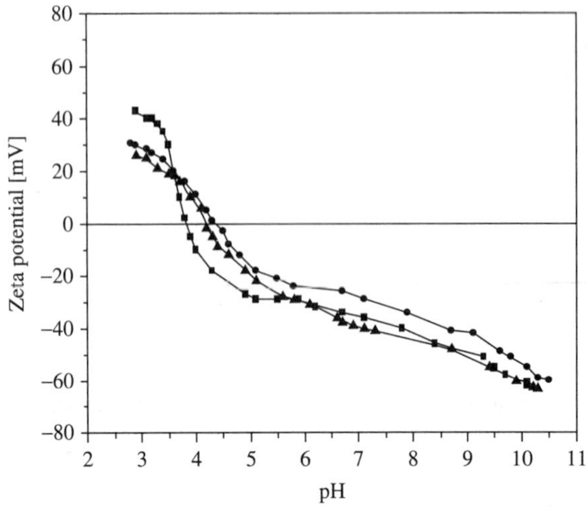

Figure 3 Zeta potential of suspended titania particles of different size as a function of pH in 0.01 m KCl as electrolyte at 25 °C.

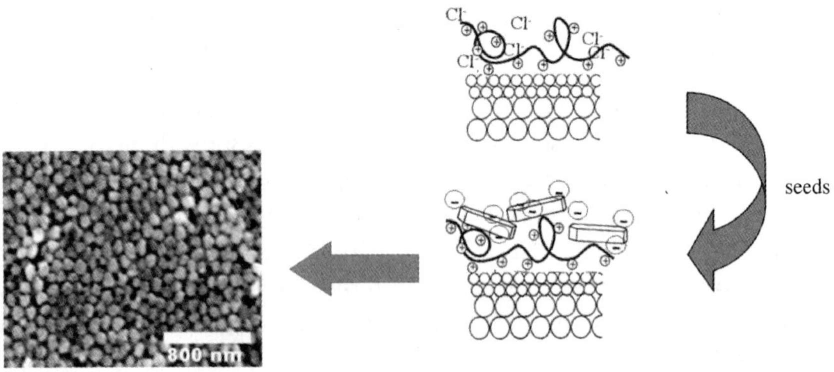

Figure 4 Scheme of the use of positively charged polymers for charging the support surface (after Ref. [66]).

2. *Charging the support surface by adsorption*: By the adsorption of positively charged polymers such as poly-DADMAC[3] or Redifloc[4] at the support surface, a positive surface charge can be generated. The counter ions of the ammonium polymer are usually chlorides which go into the solution, and negatively charged silica nanoparticles are attracted (Fig. 4).

[3]Poly-diallyl dimethyl ammonium chloride with a molecular mass of about 100,000.

[4]Trade name of EKA Chemicals, a polyamine.

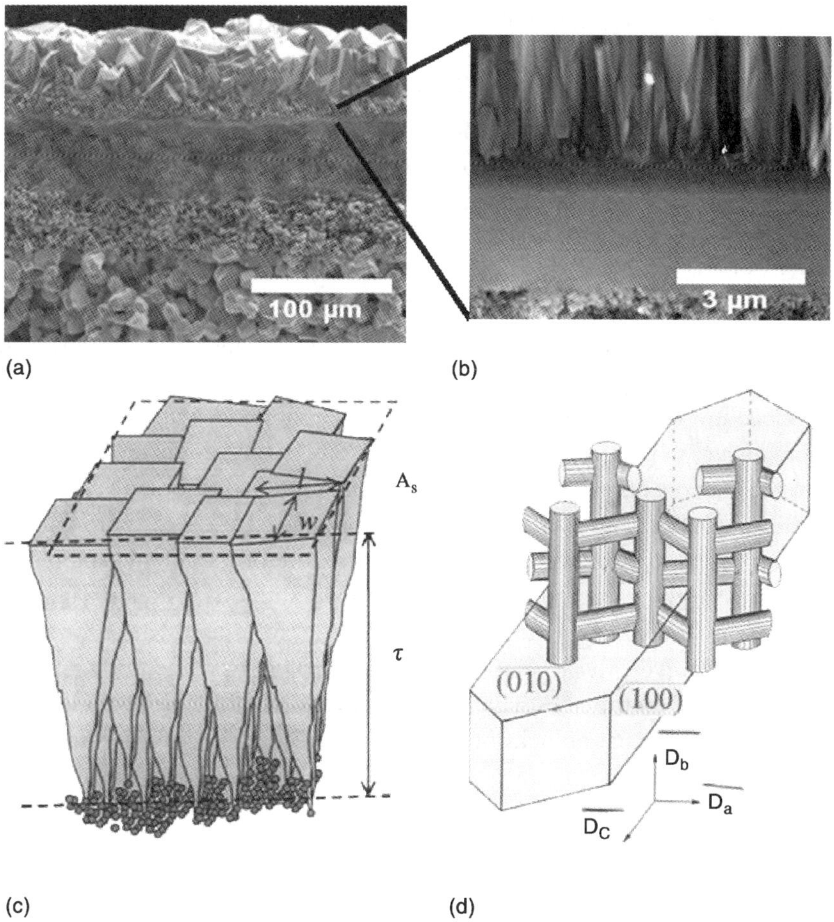

Figure 5 Typical polycrystalline MFI (silicalite-1) layer on the surface of a 1-cm tubular alumina support with asymmetric structure prepared by seeded crystallization [67]. (a) The silicalite-1 crystals start to grow from the seed layer resulting in a columnar growth structure, (b) scheme of the development of the crystallographic *c*-orientation by the evolutionary growth selection model [68], and (c) crystallographic orientation of the channel geometry of the MFI structure relative to the crystal shape (d). Reproduced from Ref. [18], reprinted with permission.

The membranes obtained show a typical columnar crystal structure in the cross section of the membrane (Fig. 5).

2.4 Use of silica nanoblocks as precursor

The use of zeolite nanoblocks is believed to trigger a new generation of extremely thin high-flux zeolite membranes [69]. The concept is based on coating a porous support with recently developed silica nanoblocks [70]. Silica polymerization in the presence of organic template molecules can yield identical rectangular silicalite-1 nanoblocks with the size of a few nanometer, for example, $4 \times 4 \times 1.4\,\mathrm{nm}^3$. These nanoblocks can be isolated and applied for membrane preparation using the self-assembly properties of these nanoblocks supported by surfactants. The silica nanoblocks are negatively charged (zeta potential) like the silicalite-1 seed crystals (Fig. 2). Therefore, the nanoblocks can be organized by cationic surfactants. Surfactants and organic templates can be removed by calcinations, and it should be noted that the surfactant molecule should not be too large since it can cause the formation of mesopores upon thermal decomposition.

Coatings were made on porous alumina supports with pore sizes of approximately 100 nm and 50–60 nm. To prevent intrusion of nanoblocks into the support pores, the supports were first coated with one or two intermediate colloidal titania sol-gel layers decreasing the pore size to approximately 2–3 nm [71]. Coatings were made using a mixture of silicalite-1 nanoblocks and surfactants by dip-coating flat or tubular supports. After drying and calcinations, an extremely thin supported silicalite-1 membrane is obtained (Fig. 6). However, the separation behavior of these new membranes is still poor, the best membranes have cut offs of 250 Dalton[5]. Nevertheless, the use of small-scale nanoblocks opens new perspective for the preparation of ultra-thin defect-free membranes. The challenges consist in achieving a perfect stacking with sufficient adhesion to the porous support after removal of the surfactant molecule and in the intergrowth of the nanoblocks.

3. SEPARATION BEHAVIOR OF MOLECULAR SIEVE MEMBRANES

3.1 Apparatus and definitions

In the overwhelming R&D on zeolite membranes, a thin zeolite layer is crystallized on a porous support. This support can be a porous ceramic, sinter metal, or carbon. Sintered metal supports are relative easy to mount to gas tight modules, whereas the ceramic supports show similar thermal expansion coefficients like the zeolite layer. This gives an asymmetric membrane that has a coarse porous support for the required mechanical strength and a thin 1–20 μm zeolite layer for sufficiently

[5]Dalton (D or Da) is an alternative name for the unified atomic mass unit (u or amu). The SI accepts dalton as an alternative name for the unified atomic mass unit and specifies Da as its proper symbol. The unit honors the English chemist John Dalton (1766–1844), who proposed the atomic theory of matter in 1803.

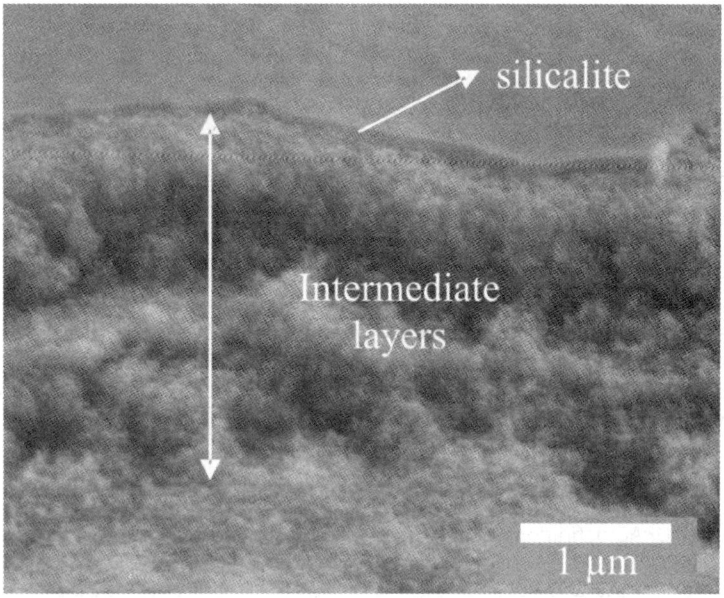

Figure 6 Cross section of a supported silicalite-1 membrane made by coating with nanoblocks [69]. The bar measures 1 μm.

high fluxes. Depending on the membrane type, various intermediate layers are employed to establish the necessary surface properties (surface smoothness, sufficiently small pores, matching of the thermal expansion coefficients between support and top layer) for successful coating of the membrane.

The most common supports for crystallization of zeolite layers are porous alumina and titania in planar and tubular geometry as plates, tubes, capillaries, multi-channel tubes. Corresponding test facilities have been developed (Fig. 7). It is the aim of the permeation tests to determine for the membranes under study the following crucial permeation parameters: from single component permeation experiments the fluxes N_i and the permselectivity (ideal selectivity) as a ratio of the single component fluxes (Table 2, Table 3). For binary mixtures, the mixture separation factor α and the components permeate fluxes from a mixed feed are determined (Table 2). The permeation measurements of single and mixed component feeds require different experimental setups (Table 3). Whereas for single component permeation studies a pressure recording is sufficient, in mixture permeation the change of the mixture composition of feed and permeate have to be analyzed (Table 3).

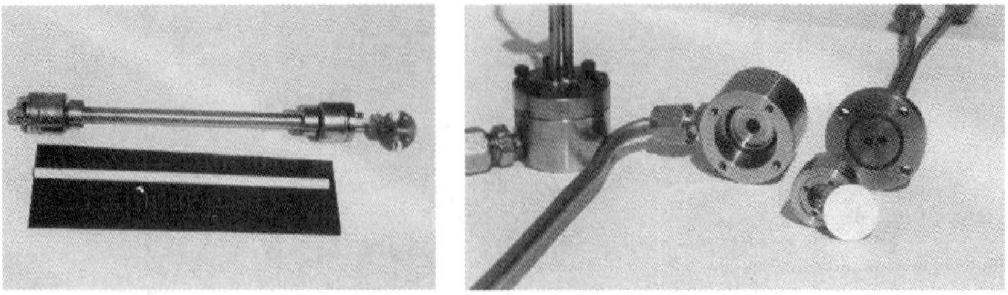

(a) (b)

Figure 7 Permeators for testing zeolite membranes: (a) stainless-steel housing for tubular membranes with water-cooled ends and (b) stainless-steel housing for disc membranes up to 180 °C. In both cases Kalrez or Viton-O-rings can be used for sealing the membranes to the housing.

Table 2 IUPAC definitions of flux, permeance, permeability, permselectivity, and separation factor [72]

Flux, N_i	mol m^{-2} h^{-1} or m^3(STP)m^{-2} h^{-1}
Permeance, Π_i = pressure normalized flux	mol m^{-2} h^{-1} Pa^{-1} or m^3(STP)m^{-2} $h^{-1}bar^{-1}$
Permeability[a,b] P_i = thickness normalized permeance (permeance multiplied by membrane thickness)	mol m m^{-2} h^{-1} Pa^{-1} or m^3(STP)mm^{-2} $h^{-1}bar^{-1}$
Permselectivity[c] (ideal selectivity) PS(i,j)	Calculated as ratio of the single component fluxes PS $(i,j) = \Pi_i/\Pi_j$
Mixture separation factor α (i,j)	Measured as α $(i,j) = (y_i{:}y_j)/(x_i{:}x_j)$ with y and x as mole fractions i and j in the permeate (y) and feed (x)

[a] The unit *Barrer* of gas permeability is the permeability represented by a flow rate of 10^{-10} cm^3 (STP) per second times 1 cm of membrane thickness, per square cm of area and cm of Hg difference in pressure which is 1 *Barrer* = 10^{-10} cm^3(STP)cmcm^{-2} s^{-1}cmHg^{-1} or 10^{-10} cm^2 s^{-1}cmHg^{-1} or in SI units 1 *Barrer* = 7.5×10^{-18} m^2 s^{-1} Pa^{-1}.

[b] ssuming a linear Henry-like absorption isotherm, for polymer membranes the permeability P_i (mol m m^{-2} h^{-1} Pa^{-1}) can be expressed as product of solubility S (mol Pa^{-1} m^{-3}) and diffusivity D (m^2 h^{-1}). For pore membranes, however, this simple relation is not valid since due to the limited pore volume the amount adsorbed does not increase linearly with the pressure (Henry-like behavior only for permanent gases at relative high temperatures) and the adsorption isotherm is usually curved (Langmuir-like).

[c] Instead of the ratio of fluxes, the permselectivity PS can be calculated as well as the permeance or permeability ratio of the components i and j, respectively.

Table 3 Measuring principles and definitions for membrane permeation

Flux, permeance, permeability, permselectivity	**Mixture separation factor α without pressure difference after Wicke and Kallenbach ($p_1 = p_2$)**	**Mixture separation factor α with pressure difference $p_1 > p_2$**
Pressure increase in an evacuated volume is determined	Permeate is transported by sweep gas into a GC or MS	Permeate streams at $p_2 < 1$ bar without sweep gas into GC, MS
Flux, permeance, and permeability are calculated		
Permselectivity as ratio of fluxes/permeances is calculated		

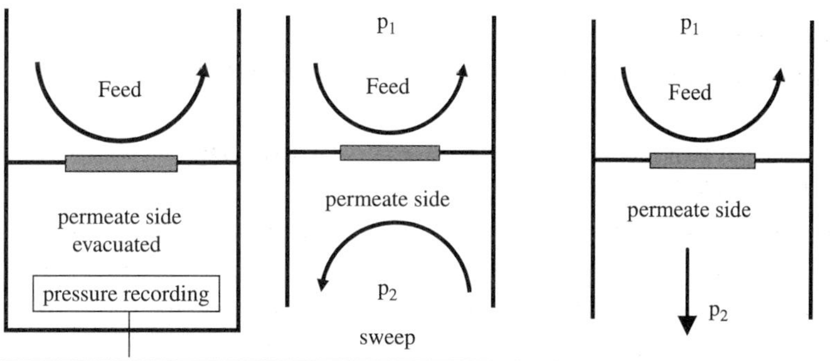

Often the zeolite membranes prepared are not perfect, and this results in an overlap of more or less selective mass transport contributions. After zeolite membrane crystallization, the following steps of testing a zeolite membrane are recommended:

1. The *as synthesized*-MFI membranes should be gas-tight for inert gases such as He or N_2 since the SDA tetrapropyl ammonium ions (TPA^+) used as template is located in the channel intersections of the MFI membrane thus blocking it for gas transport. In the case of the hydrophilic LTA and FAU membranes, no SDA is used and the organic additives are not incorporated into the zeolite structure. However, in these cases, water blocks the regular zeolite pores and makes the LTA

and FAU membranes gas-tight (if not calcined previously and working at room temperature).

2. After careful calcinations (e.g., 0.3 K min^{-1} to 450 °C in air for template removal), the membrane quality can be evaluated by permporosimetry as described in Section 3.2. Note that after mounting the membrane into the housing under ambient air, an *in situ* drying of the membrane in the cell at approximately 200 °C under vacuum is recommended.
3. When the single-gas permeation of probe gases of different kinetic diameters is measured, for perfect zeolite membranes, a clear molecular sieve effect with the expected cut-off should be found, which is to say that bulky molecules should be excluded from the membrane passage due to their size (Section 3.3). From a variation of pressure and temperature, the relative contributions of the different transport mechanisms can be evaluated.
4. If the mixture separation factor α is found to be different from the Knudsen separation factor, a high contribution of the transport though the regular zeolite pores can be expected. However, due to the interplay of mixture adsorption and mixture diffusion (which are p and T dependent), a quantitative evaluation is difficult (Section 3.4).

3.2 Characterization of zeolite membranes by permporosimetry

The basic concept of permporosimetry is that an inert non-condensable and less adsorbing gas (He, N_2) and a vapor that prefers to fill the regular micropores (*n*-hexane, water) are sent as co-feed through the membrane. The highly adsorbing vapor such as *n*-hexane for hydrophobic membranes such as silicalite-1 or water for hydrophilic membranes such as FAU or LTA is mixed to the inert gas with increasing p/p_s ratios of the strongly adsorbing component with p and p_s denoting the real and the maximum, that is, saturation vapor pressure at the given temperature, respectively. The vapor fills the regular micropore system of the membrane and blocks them for the passage of the less adsorbing He or N_2. A remaining He or N_2 flux indicates the presence of defect pores in the mesopore region.

Usually, for real zeolite membranes, a superimposition of two fluxes is observed: the intracrystalline shape-selective flux through the regular zeolite pores and an additional non-selective flux through defect mesopores that are larger than the zeolite pores. The flux through these defect pores in the mesopore range can spoil completely any shape selectivity and result in very low separation factors. For a quantitative evaluation of the flux through the defect pores, the permporosimetry can be used. Permporosimetry is similar to the terms permporometry [73–78], dynamic capillary condensation porometry [79,80], or dynamic flow-weighted pore-size distribution technique [81]. To the authors' knowledge, permporosimetry was originally developed for the characterization of pores where the Kelvin equation is valid ($r > 1.5$ nm). Later, permorosimetry was extended to microporous membranes [82,83], first applied to zeolite membranes by Deckman [84], and further developed by different groups [7,85,86]. According to the adsorption isotherm, at a certain p/p_s ratio the zeolite pores are filled and the remaining flux of the inert gas can be assigned exclusively to non-regular zeolite mesopores. One has to be aware of the effect that, when the p/p_s ratio of the strongly adsorbing component is continuously increased, also narrow mesopores of increasing diameter become filled according to the Kelvin equation and thereby blocked for the flux of the inert gas.

A typical permporosimetry experiment at constant temperature and pressure difference Δp across the membrane consists of the following steps:

1. Measure the flux of an inert non-condensable gas such as He or N_2 through an outgassed porous membrane and set this to 100%.
2. Select a suitable vapor species that is well adsorbed by the zeolite, for example, *n*-hexane for the hydrophobic silicalite-1 membrane and water for the hydrophilic LTA membrane.
3. Send a part of the non-condensable gas through a saturator filled with the well-adsorbing liquid, mix this gas stream with the pure non-adsorbing gas to fine-tune the p/p_s, and send the blended gas through the membrane.
4. Measure the relative decrease of the flux of the inert gas for increasing p/p_s of the well-adsorbing species.
5. Calculate the relative decrease of the inert gas flux.

As mentioned earlier, the well-adsorbing vapor is expected to fill completely the regular micropore structure of the zeolite membrane under study, so that the remaining flux can be attributed completely to defect pores. As an example, permporosimetry on two SiO_2 membranes will be compared: silicalite-1 and an SiO_2 sol-gel membrane (Fig. 8). It is found that *n*-hexane even at $p/p_s \approx 0.05$ completely fills the micropore volume of this rather perfect silicalite-1 membrane. However, there is a remaining nitrogen flux if water is used since water is not well adsorbed by the hydrophobic silicalite-1 membrane. On the contrary, in the case of the hydrophilic SiO_2 sol-gel membrane, water is better adsorbed than *n*-hexane

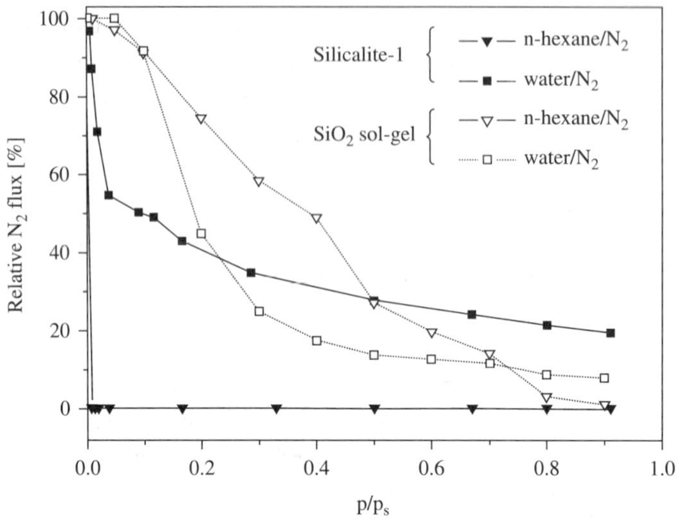

Figure 8 Comparison of permporosimetry measurements using water and *n*-hexane as micropore blocking probe molecules on hydrophilic SiO_2 sol-gel membranes (dashed lines) and on hydrophobic silicalite-1 membranes (straight lines) (after Ref. [90]).

and causes a steeper decrease of the nitrogen flux as compared to *n*-hexane. However, very recently it was found that *n*-hexane adsorption causes the size of the defects to decrease in MFI membranes [87,88]. Therefore, the use of *n*-hexane in permporosimetry is questionable and more inert molecules such as benzene should be used. The capillary condensation of dimethylbutane can be used to estimate relative sizes of non-zeolite pores [87,88].

After filling the regular micropores of a zeolite membrane, with increasing p/p_s also narrow mesopores can be filled. Assuming perfect wettability (i.e., contact angle $\approx 0°$), it can be estimated by the Kelvin equation, at which p/p_s a pore is "closed" by capillary condensation (Table 4). In calculating the pore size due to capillary condensation, it is important to consider the thickness of the already adsorbed molecules [89] (Fig. 9). It can be advantageous in permporosimetry to use molecules of different size. Small probe molecules such as water or *n*-hexane can completely fill at low p/p_s the regular micropores of a zeolite membrane and the remaining N_2 stream can be ascribed to the defect pores. In contrast, large probe molecules such as perfluoro tributyl amine ($\sigma = 1.03$ nm), triethyl amine ($\sigma = 0.74\,\text{nm}$), trimethyl benzene ($\sigma = 0.62\,\text{nm}$), triethyl benzene ($\sigma = 0.84\,\text{nm}$), tri-i-propyl benzene ($\sigma = 0.85$) can fill at high p/p_s the mesopores of the membrane by capillary condensation

Table 4 Pore diameters Ø for capillary condensation for various gases according to the Kelvin Equation, and under consideration of the pore narrowing of the pore radius r_P by the thickness of the so-called t-layer already adsorbed on the pore wall according to the Halsey Equation, that is, the multilayer thickness t, $r_{Pore} = r_{Kelvin} + 2t$ [91]

	0.01	0.05	0.1	0.2	0.3	0.4	0.5	0.6	0.7	0.8	0.9	0.95	p/p$_s$
H_2O	0.4	0.7	0.9	1.3	1.8	2.3	3.1	4.1	5.9	9.5	20.2	41.3	Pore
													$Ø_{Kelvin}$ (nm)
	0.9	1.3	1.6	2.1	2.6	3.2	4.1	5.3	7.2	11.0	22.1	43.7	$Ø_{Pore}+2t$ (nm)
n-hexane	0.8	1.3	1.7	2.4	3.2	4.3	5.6	7.6	10.9	17.5	37.1	75.9	Pore
													$Ø_{Kelvin}$ (nm)
	1.7	2.3	2.8	3.7	4.6	5.8	7.3	9.5	13.0	19.9	40.2	79.9	$Ø_{Pore}+2t$ (nm)
Perfluoro tributyl amine	2.0	3.1	4.1	5.7	7.7	10.0	13.3	18.1	25.9	41.4	87.9	180.0	Pore
													$Ø_{Kelvin}$ (nm)
	4.1	5.5	6.8	8.8	11.0	13.6	17.3	22.5	30.8	47.2	95.4	189.4	$Ø_{Pore}+2t$ (nm)

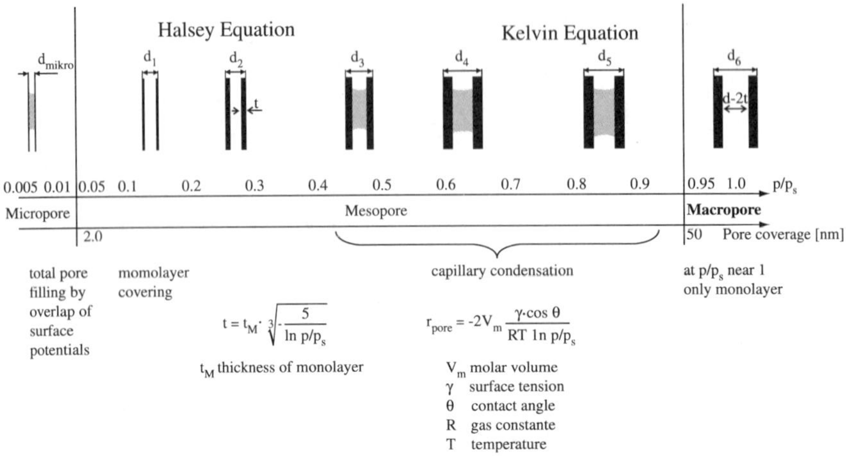

Figure 9 Schematic presentation of the pore filling process (after Ref. [90]).

(because of their bulkiness, these molecules can not enter micropores) and the remaining N_2 stream is due to the transport through the regular pore system of the zeolite membrane.

Permporosimetry measurements give a quick insight on the existence of defect pores and their pore size distribution and can forecast the separation behaviour of a membrane. As an example, the correlation of permporosimetry data and the mixture separation factor α for ZSM-5 membranes of different Al-content is shown in Fig. 10. It is found by permporosimetry that an increasing Al-incorporation into the MFI structure gives membranes with high concentrations of defects. Consequently, the membrane with the highest Al-content shows the highest residual nitrogen flux and has the lowest separation factor. On the contrary, the silicalite-1 membrane (Si/Al $= \infty$ has no measurable residual nitrogen flux at $p/p_s > 0.05$ and shows the highest separation factor (Fig. 10).

3.3 Permeation of single components

There are some detailed qualitative and quantitative descriptions of the permeation behavior of single gases in zeolite membranes [44,94–96], but a detailed theoretical treatment is difficult because of the lack of reliable experimental permeation data on perfect zeolite membranes. When probe molecules of different size are sent through a perfect zeolite membrane, molecular sieving can be expected, which is based on the principle that permeating molecules must be smaller (cut off) than the pores of the membrane [97,98] (Fig. 11). As an example, an MFI membrane with pores

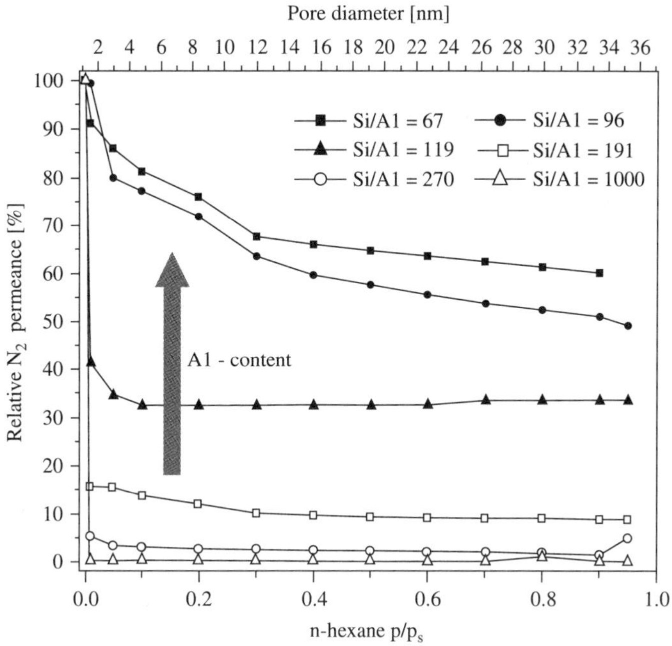

Figure 10 Permporosimetry characterization of MFI membranes of different Si/Al-ratio [92] showing that the residual nitrogen permeance as a measure of the defect concentration increases with the Al-content. The permporosimetry measurements correlate very well with the mixture separation factor α(n/i-pentane) on these ZSM-5 membranes (binary 50%/50% mixture, 110 °C): Si/Al $= 119 \rightarrow \alpha = 1.1$; Si/Al $= 270 \rightarrow \alpha = 15$, Si/Al $= \infty$: $\rightarrow \alpha = 120$ [93].

of 0.55 nm in the regular micropore zeolitic structure should allow the passage of small molecules with kinetic diameters <0.55 nm but should reject bulky probe molecule such as methyl-tertiary-butyl ether (MTBE) with a kinetic diameter of $\sigma = 0.63$ nm (Fig. 12). The remaining non-zero flux of MTBE is due to imperfections in the membrane or a result of an imperfect sealing of the membrane in the module. Note that due to the low vapor pressure at room temperature, in the case of high-boiling liquids, the use of a mass spectrometer for the detection of the permeate is recommended: triethyl amine ($\sigma = 0.74$ nm, $p = 91$ mbar), trimethyl benzene ($\sigma = 0.79$ nm, $p = 3$ mbar), tri-i-propyl benzene ($\sigma = 0.85$, $p < 1$ mbar).

The "kinetic diameter" (or "collision diameter"), σ, is the most commonly used measure of the size of probe molecules [98,100–102]. The kinetic diameter is the intermolecular distance of closest approach for two colliding molecules with zero initial energy. Numerical values for σ can be

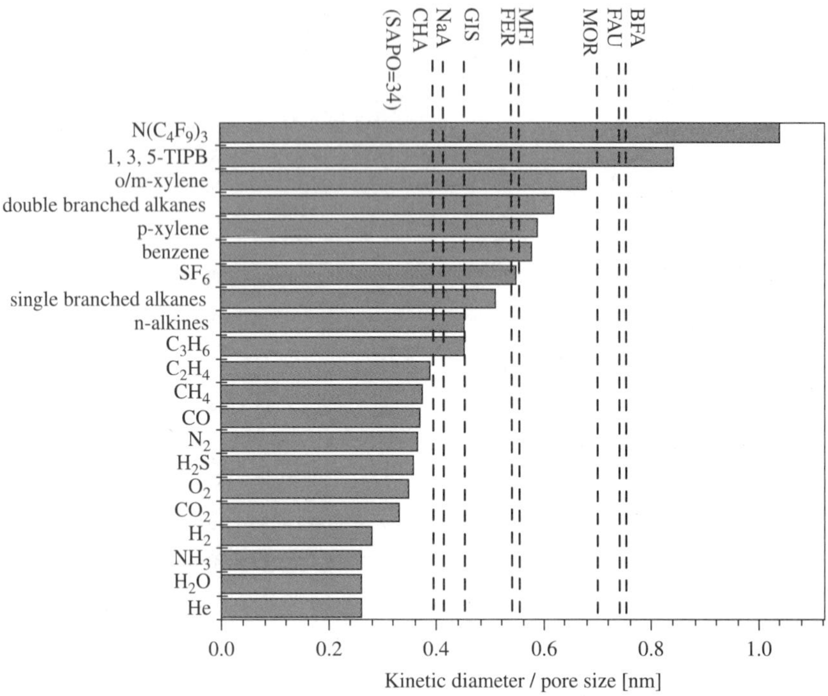

Figure 11 Comparison between the effective pore sizes of different zeolites and the kinetic diameters of gas molecules. Reproduced from Ref. [6].

calculated from viscosity or critical data or from the second virial coefficient. For spherical nonpolar molecules, the energy of interaction is usually described by the Lennard–Jones potential. For diatomic molecules, σ is calculated from van der Waals lengths. More accurately, the molecular geometry can be described by consideration of bond distances, bond angles, and van der Waals radii [30]. For more complex molecules such as *n*-paraffins, it is recommended that σ values are taken as the minimum cross-sectional diameters. Commonly used probe molecules for the characterization of porous materials in gas flow experiments are water ($\sigma = 0.26$ nm), methane ($\sigma = 0.38$ nm), *n*-hexane ($\sigma = 0.43$ nm), benzene ($\sigma = 0.585$ nm), and cyclohexane ($\sigma = 0.60$ nm).

In addition to the kinetic diameter σ, in the laboratory praxis in a permeation measurement the maximum vapor pressure p_s of the liquid to be studied at a given temperature turns out to be important (Table 5). The component with the lowest p_s determines the measuring conditions (maximum pressure differences Δp across the membrane, lowest permeation temperatures, heating of the permeation apparatus to avoid condensation, etc.).

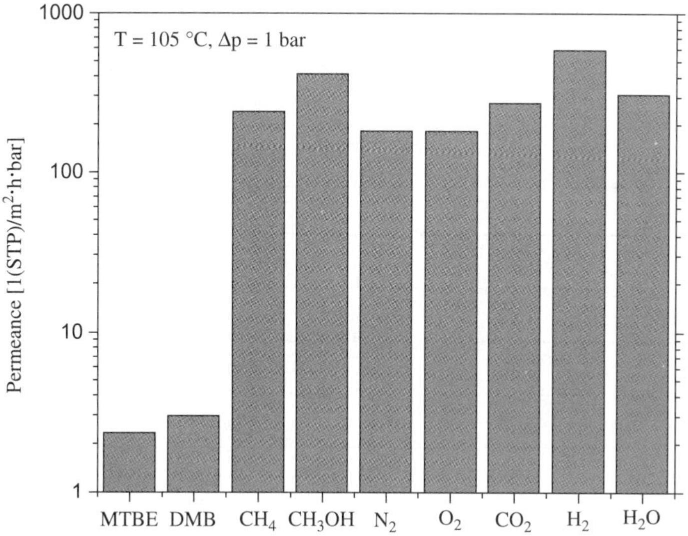

Figure 12 Single component permeances through a silicalite-1 membrane with 0.55 nm pore width as a function of the molecular size of probe molecules at 105 °C at $\Delta p = 1$ bar [5,99] (DMB, MTBE).

A simple test of a porous membrane is its molecular sieving ability in a permeation experiment using a binary or multi-component mixture. As an example, Table 6 shows the molecular sieving of binary mixtures of *n*-heptane and a second component of different size on an $AlPO_4$-5 molecular sieve membrane, which consists of oriented $AlPO_4$-5 single crystals in a nickel foil [5,103,104]. It can be seen that the integral fluxes of the mixtures *n*-heptane/toluene, and *n*-heptane/trimethyl benzene are much smaller than those for pure n-heptane. Furthermore, for these mixtures no separation is achieved since both components of the mixture can pass through the membrane because they are smaller than the pore diameter of $AlPO_4$-5 with 0.73 nm. Obviously, the bulky aromatic molecule blocks the more mobile *n*-heptane molecule, and since the more mobile *n*-heptane cannot "overtake" the less mobile aromatic molecule in the narrow one-dimensional (1D) channel, no separation can take place. However, if the aromatic mixture component (triethyl- or triisopropyl benzene, respectively) is bulkier than the channel diameter of $AlPO_4$-5, only *n*-heptane can pass through the pores of the membrane. This exclusion selectivity results in a reasonable separation and the fluxes of *n*-heptane are higher than those in the case if both mixture components can enter the membrane and block each other.

A high value of the separation factor α is not exclusively due to a molecular sieving effect and may as well result from the interplay of mixture

Table 5 Kinetic diameter and vapor pressure for different probe molecules

Molecule	Kinetic diameter [nm]	Saturation vapour pressure p_s (mbar)		
		25 °C	50 °C	100 °C
H_2	0.29	–	–	–
N_2	0.36	–	–	–
CH_4	0.38	–	–	–
H_2O	0.26	31.6	123	1010.8
n-hexane	0.43	200	538.9	1845
benzene	0.585	126.6	360.8	1796
SF_6	0.55	25,000	–	–
cyclohexane	0.60	129,8	361.5	1742.5

Note that above the critical temperature T_{crit} the compounds behave as permanent gases: $T_{crit}(H_2) = -239.9\,°C$, $T_{crit}(N_2) = -146.9\,°C$, $T_{crit}(CH_4) = -82.6\,°C$, $T_{crit}(SF_6) = -45.6\,°C$. The kinetic diameter σ is derived from 6-12-Lennard Jones potentials taken from Ref. [100].

Table 6 Fluxes *N* in % of the pure *n*-heptane flux (9×10^{-6} mole $s^{-1}cm^{-2}$) and separation factor α for binary mixtures *n*-heptane/toluene, *n*-heptane/trimethyl benzene, *n*-heptane/triethyl benzene, and *n*-heptane/triisopropyl benzene through an $AlPO_4$-5 molecular sieve membrane at 91 °C [5,103,104]

n-heptane	n-heptane/ toluene		n-heptane/ trimethyl benzene		n-heptane/ triethyl benzene		n-heptane/ triisopropyl benzene	
N/α	*N*	α	*N*	α	*N*	α	*N*	α
100%/-	22%	0.8	11%	1.7	47%	105	24%	1220

adsorption and mixture diffusion. A strong temperature dependence of flux and separation is then observed as with some microporous sol-gel-based metal oxide membranes [105–109]. However, as Weitkamp and Puppe have shown [98], the relationship between the molecular dimensions and the pore size is rather complicated. The transport through a microporous medium is determined by molecular shape (rather than kinetic diameter) in relation to the shape and size of the pore windows, channels, and/or intersections. Thus, although cyclohexane and 2,2-dimethyl butane (DMB) have the same kinetic diameter, due to its elliptical cross section cyclohexane shows much faster adsorption kinetics than DMB [110] into the elliptical zeolite pores of ZSM-5, ZSM-11, and ZSM-48.

Furthermore, one has to consider the thermal vibrations of both the adsorbent host (framework flexibility) and the guest molecules when comparing the results of different exclusion experiments at certain temperature and pressure conditions. Usually rigid zeolite frameworks are considered. It is generally believed that there is no influence of small molecules relative to the pore diameter of the zeolite [111,112]. Molecular dynamics simulations showed that the effects can be much larger indeed if the hydrocarbon fits tightly into the channels of the zeolite. For example, the diffusivity of aromatics in silicalite-1 changes by an order of magnitude if framework flexibility is taken into account [113]. A similar effect was found for *n*- and *i*-butane in silicalite-1 [114]. Also for adsorption it is believed that framework flexibility is only important if the guest molecules fit tightly into the zeolite pores. Examples are light hydrocarbons in DD3R [115], aromatics in silicalite-1 [116], and naphthalene in silicalite-1 [117]. In the latter case, naphthalene with a size of $0.72 \times 0.38\,\mathrm{nm}^2$ is adsorbed by silicalite-1 with $0.53 \times 0.56\,\mathrm{nm}^2$ and $0.51 \times 0.55\,\mathrm{nm}^2$ cross-sectional areas of the straight and sinusoidal pores, respectively.

The permeation of short-chain hydrocarbons reflects in an excellent way the interplay of molecular adsorption and diffusion controlling the permeation behavior. The fluxes of single components through MFI-type membranes as a function of temperature often exhibit maxima at certain temperatures [118–121], also weak minima are observed at higher temperatures [120] (Fig. 13a). Following Refs. [122–124], the limiting relations for the single component flux N_i of the component i are

Low loadings: $N_i = -q_i^{\text{sat}} \rho D_i K_i \frac{\Delta p_i}{\delta}$

High loadings: $N_i = -q_i^{\text{sat}} \rho D_i \frac{\Delta \ln p_i}{\delta}$

with N_i, D_i, q_i^{sat}, K_i, and p_i denoting the flux, diffusivity, saturation loading, adsorption equilibrium constant, and partial pressure of component i, respectively. δ and ρ are the membrane thickness and density. The maxima and minima of the flux as a function of temperature can be explained by the Arrhenius type temperature dependencies of K and D due to $K = K_0 \exp(-\Delta H/RT)$[6] and $D = D_0 \exp(-E_a/RT)$ with ΔH and E_a as enthalpy of

[6]Since adsorption is exothermic, the enthalpy of adsorption is negative, which results in a positive exponent. On the contrary, the activation energy of the diffusion is positive and results in a negative exponent.

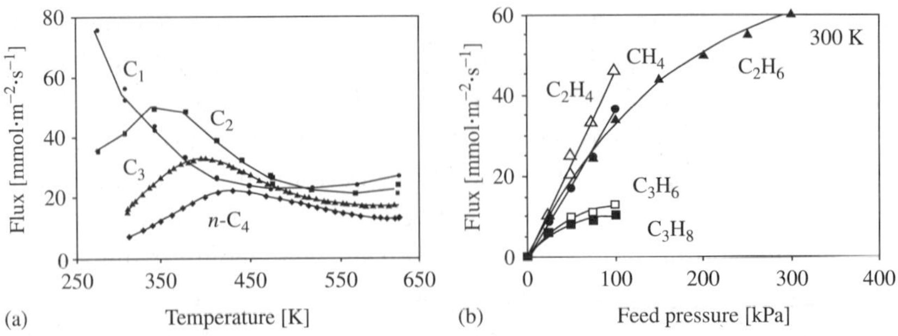

Figure 13 Single component fluxes of short-chain length hydrocarbons through a stainless-steel supported silicalite-1 membrane as function of the temperature (a) [125] and of the feed partial pressure (b) [122] using the Wicke-Kallenbach technique with He as sweep gas, both sides at 101 kPa, silicalite-1 layer facing the feed side.

adsorption and diffusivity activation energy. For permeation systems with Henry-like adsorption isotherms such as for CH_4 and C_2H_4 on silicalite-1, a linear relationship between flux and feed pressure can be expected whereas for curved Langmuir-like isotherms no linear relationship between pressure and flux exist (Fig. 13b).

As shown in Table 2, the permselectivity (ideal selectivity) is defined as the ratio of the single component fluxes. It is usual to characterize the quality of zeolite membranes by their permselectivities. A high value of the permselectivity of a highly permeable gas (H_2) and a bulky gas that can pass the zeolite membrane only with a low rate or is even unable to pass the membrane (SF_6, cyclohexane, DMB, etc.) is taken as a measure of membrane quality. Often the gas pairs N_2/SF_6 or H_2/SF_6 are studied. Whereas the permanent gases H_2 and N_2 are only weakly adsorbed on MFI zeolites and show, therefore, at room temperature almost linear adsorption isotherms (Henry-like), SF_6 adsorption on MFI is stronger and the adsorption isotherms are curved (Langmuir-like). As a consequence, the SF_6 permeances will depend decisively on the pressure range covered (Fig. 14). However, one should be very careful when comparing the permselectivities from different authors, even when the pressure difference Δp across the membrane is in every case the same [126]. Roughly speaking, the driving force of the flux (or permeability) is the concentration difference of the membrane between feed and permeate side which is determined by the adsorption isotherm. When we take the adsorption isotherms of SF_6 on silicalite-1 (Fig. 14) as an example, we can see that – working with the constant pressure difference of $\Delta p = 1$ bar – the driving force for permeation can be quite different. There is a much larger driving force for permeation

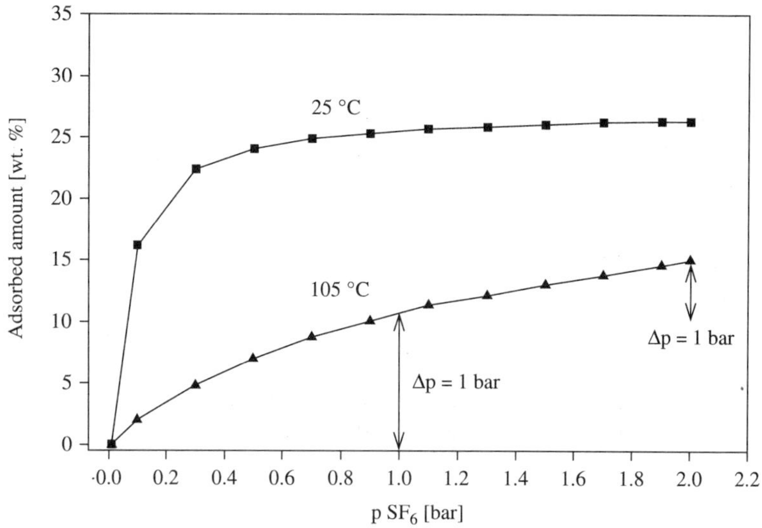

Figure 14 SF_6 adsorption isotherms on silicalite-1 [126]. Owing to the curved shape of the isotherm, a pressure difference of $\Delta p = 1$ bar across the membrane can cause quite different concentrations on the feed and permeate side of a membrane and, thus, a different driving force.

when working at pressures of 1 bar (feed) against vacuum (permeate) compared with 2 bar (feed) against 1 bar (permeate). Besides the permselectivities of the gas pair N_2/SF_6 (Fig. 15a), also the permselectivities of the gas pair *n*-/*i*-butane (Fig. 15b) is often used as a criterion for the quality of an MFI membrane.

3.4 Separation of binary mixtures

There are only a few examples for a real molecular sieving with zeolite membranes. As already shown in Section 3.3, Table 6, the separation of binary mixtures of *n*-heptane and different aromatic compound of various bulkiness on an 1D model membrane of oriented $AlPO_4$-5 single crystals in a nickel matrix follows the pattern of molecular sieving. If both, the *n*-heptane and a small aromatic compound can pass the 0.73 nm wide pores of the membrane, low separation factors and low fluxes are found for this case of "single file" diffusion. If the pore diameter is inbetween the size of the *n*-heptane and the bulky aromatic molecule, we have "molecular sieving" with high separation factors and high fluxes (Fig. 16). As another example, the separation behavior of two MFI-type membranes with 0.55 nm pore size will be compared: a silicalite-1 membrane that can be prepared in a rather good quality and a ZSM-5 membrane that theoretically should

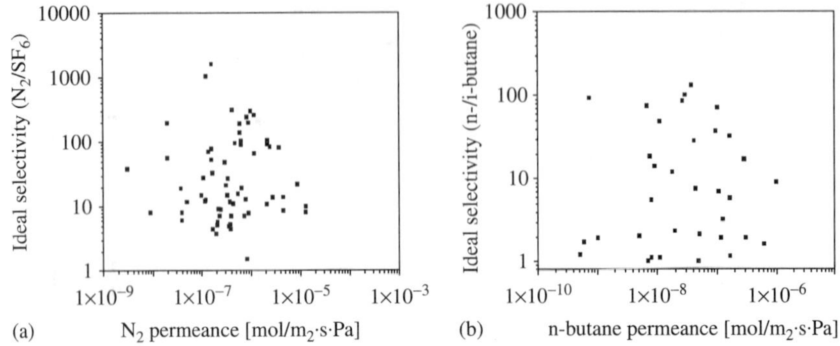

Figure 15 Collected permselectivities (ideal selectivities) from different authors (literature data) for the gas pairs N_2/SF_6 (a) [127] and n/i-butane (b) [128] on MFI membranes.

exhibit the same pore size of 0.55 nm, but it contains high concentrations of defect mesopores due to its Al-content (Table 7). From comparing the pore size of 0.55 nm with the gas kinetic diameters of methanol (0.39 nm); MTBE (0.62 nm); water (0.27 nm); and *i*-propanol (0.5 nm), it can be expected that reasonable separation of MeOH/MTBE mixtures due to molecular sieving takes place on MFI membranes with 0.55 nm pore diameter. Whereas the silicalite-1 membrane fulfils this forecast, the ZSM-5 membrane with more defects shows a lower separation factor for the MeOH/MTBE mixture. Now the same two MFI membranes are used for the separation of water/*i*-propanol mixtures. The molecular size of both water and *i*-propanol are lower than the pore size but, nevertheless, separation takes place due to the adsorptive interaction: the hydrophobic silicalite-1 membrane is *i*-propanol selective and the hydrophilic ZSM-5 membrane is water selective.

In another example, the combination of diffusion characteristics and the Ideal Adsorption Solution[7] (IAS) theory predict that *n*-butane would permeate at low temperatures much faster through a MFI-type silicalite-1 membrane than *i*-butane [high loading case in the Generalized Maxwell-Stefan[8] (GMS) model]. However, experimental data do not support this

[7]The IAS theory originated by Myers and Prausnitz (A.L. Myers, J.M. Prausnitz, *AIChE J.* 11 (1965) 121) derives mixed-gas adsorption equilibria exclusively from single-component isotherms. The IAS theory serves as a benchmark for the prediction of mixed gas–vapour adsorption equilibria (S. Sircar, *AIChE J.* 41 (1995) 1135).

[8]The GMS model employs the gradients of the chemical potentials as driving forces in agreement with the theory of irreversible thermodynamics (R. Krishna, *Chem. Eng. Sci.* 45 (1990) 1779).

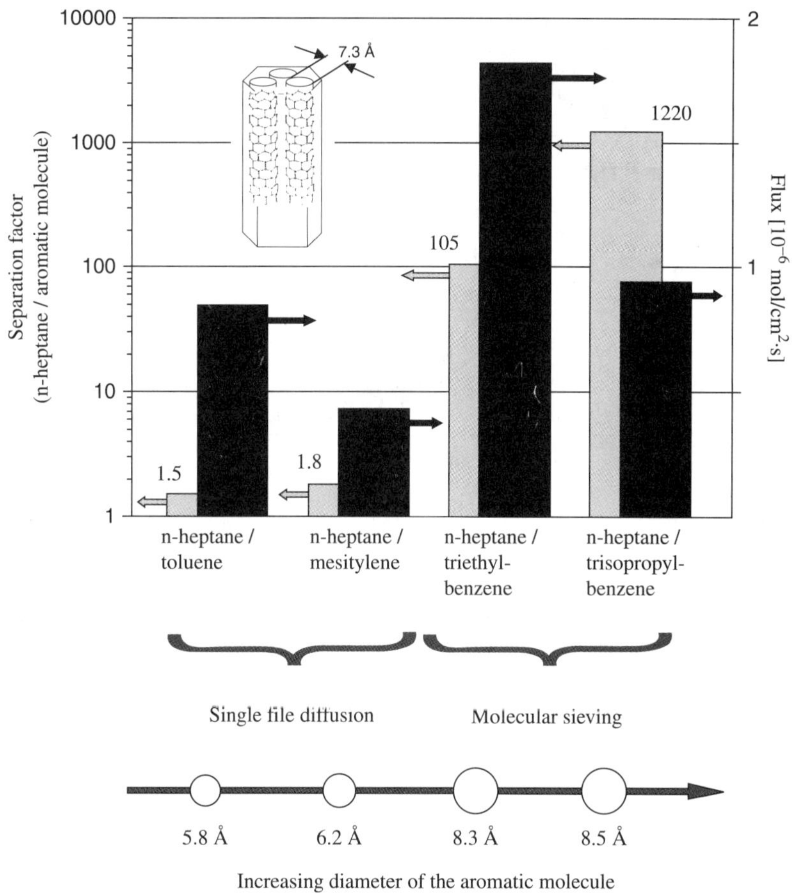

Figure 16 Permeation fluxes of binary mixtures (50%/50%) of *n*-heptane and various aromatic compounds of different bulkiness and the corresponding separation factors (*n*-heptane/aromatic compound) on a model membrane of vertically oriented large $AlPO_4$-5 single crystals in a nickel matrix (after Refs. [103,104]).

Table 7 Mixture separation behavior of two MFI-type membranes

MFI-type membrane	Si/Al	Separation factor α (MeOH/MTBE)[a]			Separation factor α (water/i-propanol)		
		5:95	50:50	95:5	5:95	50:50	95:5
Silicalite-1	> 1,000	250	160	55	5.5	0.6	0.04
ZSM-5	30	30	20	5	501	26	10

Shape-selective and hydrophobic separation for the silicalite-1 membrane and hydrophilic separation behaviour for the ZSM-5 membrane [129].

[a] Methanol/methyl-tertiary-butyl ether

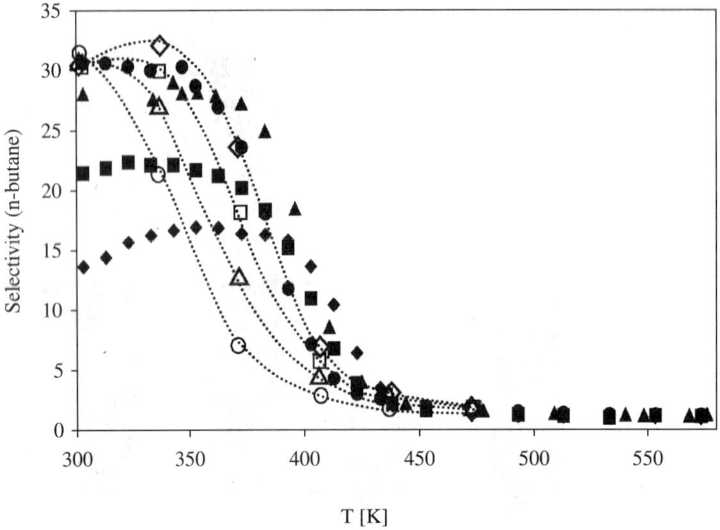

Figure 17 Selectivity toward *n*-butane for equimolar *n*/*i*-butane mixtures on a MFI-type silicalite-1 membrane as a function of temperature at different total hydrocarbon feed pressures: ●, 45 kPa; ▲, 101 kPa: ■, 150 kPa; ♦, 200 kPa. Dashed lines are adsorption selectivity calculations after the IAS theory for the corresponding feed pressures (open symbols) [122,130].

(Fig. 17) since with decreasing temperatures the *n*/*i*-butane selectivities even decrease. Assuming a loading-dependent diffusivity D_{ij}, that is to say a diffusion coefficient *D* of component *i* in the presence of *j*, the *n*/*i*-butane selectivities become equal to the adsorption selectivity according to the IAS theory, which predicts decreasing selectivities with decreasing temperature. This correlates very well with the experimental results (Fig. 17).

For a practical application, the thin zeolite membrane layer must be supported by a porous layer that provides the mechanical strength and allows for the handling of the membrane. This support can remarkably influence the binary separation behavior. In a modeling study, the permeation behavior of a methane/propane mixture through a stainless steel supported silicalite-1 membrane as a function of the thickness of the supporting layer was simulated [122]. It was found that the thicker the support, the lower the fluxes and the poorer the propane selectivity. For the membrane under discussion, the selectivity drops from initially 45 to 4 when the support thickness increases from initially 1 mm to 10 cm (Fig. 18). Note that for very thick supports even a reversal of the selectivity can occur since the support resistance dominates the membrane behavior and methane diffuses faster in the He used as sweep than propane.

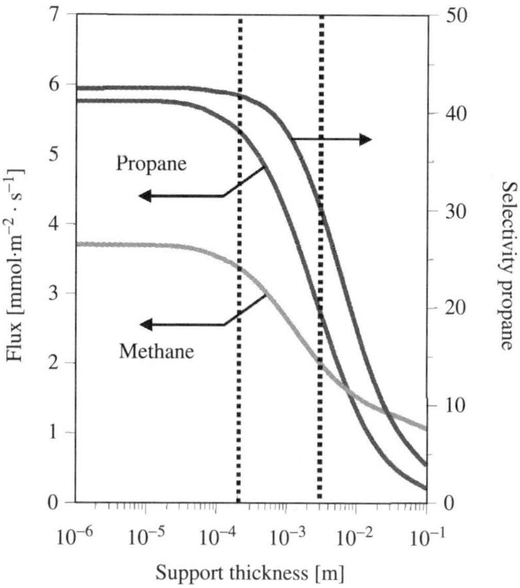

Figure 18 Simulation of the permeation behavior of a 95%/5% methane/propane mixture through a supported silicalite-1 membrane at 30 °C according to the full GMS model as a function of the thickness of the supporting layer (Trumen, sintered stainless steel with a titania intermediate layer): fluxes of both components and propane selectivity. Wicke-Kallenbach method with He as sweep gas, silicalite-1 membrane facing the feed mixture, both sides at 101 kPa. The dashed lines indicate the region of practical support thicknesses [122].

3.5 Case study: Hydrogen separation

Gas separation membranes are widely used for H_2 recovery [131]. Among the microporous membranes, the X-ray amorphous metal oxide membranes, mainly silica [109,132–134], and zeolite membranes, first of all of the MFI-type [5,6,135], are the most common ones. Carbon molecular sieve (CMS) membranes form the third group [136–142]. Since the kinetic diameter of H_2 is around 0.29 nm, the pore diameter of the membrane should be larger than this but smaller than the kinetic diameter of the molecules from which H_2 is to be separated. The microporous membranes based on amorphous metal oxides (SiO_2, TiO_2, ZrO_2) are most often prepared by a sol-gel technique using spin coating or dip coating. On the one hand, SiO_2 sol-gel layers show excellent separation factors and fluxes but have a very limited hydrothermal stability, which excludes their use for H_2 removal from atmospheres containing steam at high temperature. On the other hand, TiO_2 and ZrO_2 sol-gel layers are much more stable but it does not succeed to prepare highly selective narrow-pore membrane layers for gas separation.

In contrast to these X-ray amorphous metal oxide membranes, crystalline zeolite membranes offer much better thermal and hydrothermal stability. Gas permeation results for silicalite-1 membranes have been reported in many studies in literature [60,66,118,119,221,143–153]. The mixture separation factor α of H_2 from *i*-butane increases from approximately $\alpha \approx 1.5$ at room temperature to $\alpha \approx 70$ at 500 °C (Fig. 19). It must be considered here that both H_2 and *i*-butane can pass the 0.55 nm pores of the silicalite-1 membrane due to their smaller kinetic diameter (0.29 and 0.50 nm, respectively). At low temperature, mainly *i*-butane is adsorbed inside the silicalite-1 pores and the slowly moving *i*-butane blocks the diffusion paths of the rarely adsorbed highly mobile H_2. With increasing temperature, less *i*-butane is adsorbed and H_2 with its higher diffusivity can now move fast in the resulting free volume. It is also remarkable that no

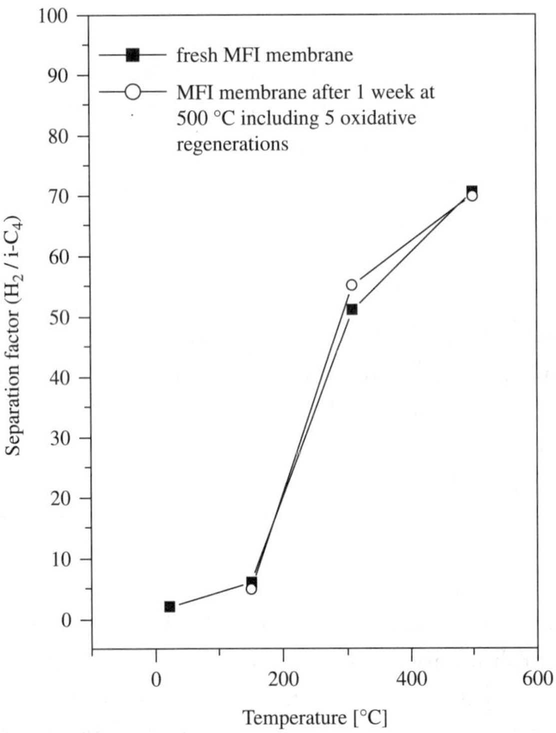

Figure 19 Mixture separation factor α for H_2/*i*-butane (feed composition 1:3, which is representative for the equilibrium composition of *i*-butane dehydrogenation at 500 °C) at different temperatures (after Ref. [154]).

degradation during 1 week of operation at 500 °C including five oxidative regenerations to burn off carbonaceous residues was observed [154]. The reasonable H_2 separation factor of $\alpha \approx 70$ and the H_2 permeance of $\Pi_{H2} \approx 1\,m^{-2}h^{-1}bar^{-1}$ at 500 °C suggest that this MFI membrane is a candidate for an extractor membrane reactor with selective H_2 removal.

CMS membranes are less frequently used because of the poor mechanical strength of carbon and its instability in O_2-containing atmospheres at elevated temperature. However, in the absence of O_2, CMS membranes offer excellent thermal and chemical stability as well as high separation factors [136]. Apart from CMS membranes, a second type of microporous carbon membranes exists, which is called *selective surface flow* (SSF) membranes since the separation mainly relies on the adsorption properties of the microporous carbon layer [155–157]. These membranes show preferred permeation of adsorbable over non-adsorbable components irrespective of the molecular dimensions and are recommended, therefore, for the removal of hydrocarbons or carbon dioxide from H_2 [156].

Dense metal membranes selectively absorbing and transporting atomic hydrogen are well-suited for high-temperature H_2 separation offering an infinite selectivity for H_2, Pd, and its alloys, for example, $Pd_{77}Ag_{23}$, $Pd_{60}Cu_{40}$, and $Pd_{94}Ru_6$ can be used for this purpose. In particular, the alloy with 23 wt.% Ag is the most common material for H_2-selective metal membranes. It has a higher H_2 permeability than pure Pd and an improved resistance against H_2 embrittlement at lower temperatures (pure Pd fails below 300 °C). The main benefit of an alloy with 40 wt.% Cu is the better resistance against H_2S, whereas the alloy with 6 wt.% Ru has an excellent high-temperature stability and strength. Metal membranes are produced in the form of self-supporting foils or thin tubes with a thickness of at least 10–50 μm to reach sufficiently low leak rates (typical thicknesses of foils are 25–100 μm), which limits the H_2 flux and causes high material costs. Group Vb metals (V, Nb, Ta) are known to exhibit an even higher H_2 permeability than palladium [158], but to facilitate entry and exit of H_2, their surface needs to be protected from oxide formation [159]. Therefore, composite membranes have been developed, consisting of a supporting foil made from a group Vb metal and a thin Pd coating (<1 μm) on both sides [160–164]. However, the temperature window for these membranes is narrow, that is, from approximately 30 °C to 350 °C [164]. At higher temperature, the H_2 permeability is gradually lost due to intermetallic diffusion between the base metal and the Pd coating, which is the reason why a metal diffusion barrier is placed in between [165–167]. At lower temperature, H_2 embrittlement leads to membrane failure [168].

Dense ceramic high-temperature inorganic mixed proton and electron conducting membranes consist of a ceramic solid oxide proton conductor and an electron conducting second phase, which could be a ceramic or a high-temperature resistant metal, for example, Pd, Ni, or an alloy. First studies by Iwahara et al. on the perovskite-type materials $SrCeO_3$ and $BaCeO_3$ doped with Y, Yb, or Gd were made in the early 1980s [169]. Since then, the mechanism of high-temperature proton conduction in solid oxides has been studied [170–173]. Supported films with thicknesses down to a few micrometer are reported [174,175]. A supported film of $SrCe_{0.95}Yb_{0.05}O_{3-\delta}$ with a thickness of 2 μm reached a H_2 flux of $8\,m^3\,m^{-2}\,h^{-1}$ at 677 °C [174]. To increase the electronic conductivity, which limits the H_2 flux, the composition of the perovskites was systematically varied [176–178]; $BaCe_{0.9}Gd_{0.1}O_{2.95}$ is one example for an optimized mixed proton and electron conducting material, others include $SrCe_{0.95}Tm_{0.05}O_{3-\delta}$, $SrCe_{1-x}Eu_xO_{3-\delta}$. Very recently, supported thin-film cermet[9] membranes based on Pd/Yttria-stabilized zirconia (YSZ) with thicknesses down to ~22 μm became known which reached a H_2 flux of $12\,m^3\,m^{-2}\,h^{-1}$ at 900 °C and 0.9 bar H_2 feed pressure [179]. The highest H_2 flux was obtained with so-called intermediate-temperature cermets consisting of a group Vb or another low cost H_2 permeable metal and a metal oxide [180]. Owing to the higher reactivity of these metals as compared to Pd, the formation of a cermet is rather difficult, and the temperature window for the resulting membranes is narrow (340–440 °C). However, supported thin-film membranes based on such materials were reported to show H_2 fluxes up to ~$254\,m^3\,m^{-2}\,h^{-1}$ at 400 °C and 33 bar H_2 feed pressure [181].

Summarizing, one can state that silicalite-1 as the only zeolite membrane with a neglecting intercrystalline defect flux developed so far can hardly compete with the other organic and inorganic membranes. However, zeolite membranes can become important tools for H_2 separation, if it succeeds to develop narrow pore all-silica zeolite membranes with pore sizes near 0.3 nm as a thin (μm thick) supported layer. These requirements are based on experiences obtained from a model membrane of aligned $AlPO_4$-5 crystals with a 1D pore system [103,104]. For the case of real molecular sieving, permeation of the component that is to be separated is not influenced by the presence of the other mixture components, and the flux of this component can be rather high. In the usual case for silicalite-1

[9]Cermet = Nano-composite material of a ceramic and a metal component.

membranes, all mixture components can enter due to their size, the pores of the zeolite membrane, and the observed separation effect is the result of a complicated interplay of mixture diffusion and mixture adsorption. In this transport mechanism, strongly adsorbed or bulky components can drastically reduce the permeation of more mobile components. On the contrary, the narrow pore size and the rather compactness (low density of pores per unit area) of the suitable zeolite structures for H_2 sieving require thin membrane layers for reasonable fluxes (Table 8).

3.6 Case study: Carbon dioxide separation

The development of proper separation technologies for the removal of CO_2 from exhaust gases and from natural gas is still a challenging problem. In applications for natural gas treatment, the feed gas usually stems directly from gas wells in a wide pressure range from 20 to 70 bar with 5–50% CO_2. The product gas must contain less than 2% CO_2. At the moment, it is not economic to produce from gas fields with CO_2 content higher 10%. Glassy polymer membranes are used for natural gas purification (removal of CO_2, H_2O, H_2S), but they suffer from swelling-induced plastification by incorporation of CO_2 and hydrocarbons [190] which reduces their selectivity. This kind of membrane failure would not happen with zeolite membranes since they are chemically stable toward organic solvents and plastification due to gas absorption.

Although polymer membranes with a high performance for CO_2/CH_4 separation exist, these membranes have only a rather low separation performance in the CO_2/N_2 separation because of low diffusivity and solubility selectivities due to the similar size of CO_2 and N_2 [191–193]. CO_2 (0.33 nm kinetic diameter), N_2 (0.364 nm kinetic diameter), and CH_4 (0.38 nm kinetic diameter) are relative small molecules, that is to say much smaller than the pores of large- and medium-pore zeolites. Therefore, the separation of CO_2 from N_2 or CH_4 using zeolite membranes will be based on competitive adsorption and the selectivities were found to be rather low. Nevertheless, most often the MFI-type membrane was studied [60,119,120, 152,194–203]. As an example, Lovallo et al. [60] obtained a selectivity of approximately 10 for a silicalite-1 membrane at 120 °C. CO_2 has a stronger electrostatic quadrupole moment than N_2 leading to a preferential adsorption of CO_2 from N_2/CO_2 mixtures [204]. Thus, it can be expected that surface diffusion of CO_2 contributes significantly to its permeation and simultaneously reduces the N_2 permeation flux. The best results for the separation of CO_2/N_2 mixtures on large-pore zeolite membranes were reported by Kusakabe et al. [205,206] using FAU-type membranes.

Table 8 Typical performance of different H_2-selective membranes [182]

Membrane type	T [°C]	Thickness of the separation layer (μm)	H_2-Flux at $\Delta p_{H_2} = 1$ bar (from 2 to 1 bar) (m^3(STP)m^{-2} h^{-1})	Separation factor (–)	Ref.
Organic polymer	<100	0.05–0.5	1–2	30–40 (H_2/CO)	[183]
Solid polymer electrolyte	<100–200	50–500	2–6	∞	
Self-supporting $Pd_{77}Ag_{23}$ foil (OMG)	300–450	20	6–11	∞	–
Pd/V/Pd composite foil	300	0.5/40/0.5	35.9	>50.000	[164]
Thin-film Pd alloy membrane (heraeus)	300–450	5	10–25	3.000	–
Molecular sieve silica	200	0.02–0.06	7–20 (He)	100–940 (He/N_2)	[134]
	200	0.03	4.1–16.1	4000–321 (H_2/CH_4)	[184]
	100–300	n.d.	0.8–5.6	54–132 (H_2/CH_4)	[185]
	450–550	~0.1	⩾20	30–75 (H_2/C_3H_8)	[133]
MFI zeolite (silicalite-1)	20–200	3	0.8–1.3	1.9–12.8 (H_2/CO_2)	[119]
	65–290	n.d.	0.2–24.2	0.1–3 (H_2/n-butane)	[186]
	30–210	~40	0.97–0.83	4.3–1.3 (H_2/n-butane)	[187]
	500	35	~1	70 (H_2/i-butane)	[154]

Molecular sieve carbon	80	~1	0.05–0.5	100–630 (H_2/CH_4)	[136]
	22	21.3	0.005	331 (H_2/N_2)	[138]
	25–150	n.d.	0.2–0.85	400–50 (H_2/CH_4)	[141]
	35	0.4	~4	290 (H_2/CH_4)	[142]
Single-phase ceramic mixed H^+/e^- conductor (SCYb)	677	2	0.7–5.4	∞	[174]
Dual-phase mixed H^+/e^- conducting cermets	950	160	0.64	∞	[188]
Ni-alloy/BCY (34 vol.% metal)	900	~22	5.3	close to ∞	[179]
Pd/YSZ (50 vol.%)	440	n.d.	~18.3	∞	[189]
Intermediate-temperature composite					

n.d.: not determined.

In contrast, small-pore zeolites such as zeolite T (0.41 nm pore size), DDR (0.36 nm × 0.44 nm), and SAPO-34 (0.38 nm) have pores that are similar in size to CH_4 but larger than CO_2. It can be expected, therefore, that these membranes show high CO_2/CH_4 selectivities due to a combination of differences in diffusion and adsorption. For T-type zeolite membranes, Cui et al. [207] found a mixture separation factor $\alpha = 400$ with a CO_2 permeance of $\Pi = 4.6 \times 10^{-8}\,\text{mol}\,\text{m}^{-2}\,\text{s}^{-1}\,\text{Pa}^{-1}$ at 35 °C. Tomita et al. [208] obtained a CO_2/CH_4 separation factor of $\alpha = 220$ with a CO_2 permeance of $\Pi = 7 \times 10^{-8}\,\text{mol}\,\text{m}^{-2}\,\text{s}^{-1}\,\text{Pa}^{-1}$ at 28 °C using a DDR membrane.

Very powerful SAPO-34 membranes were recently synthesized by *in situ* crystallization on a porous tubular stainless-steel support by Noble and Falconer [209]. For a SAPO-34 membrane synthesized from a Si/Al gel ratio of 0.1, a CO_2/CH_4 selectivity of $\alpha = 170$ with a CO_2 permeance of $\pi = 1.2 \times 10^{-7}\,\text{mol}\,\text{m}^{-2}\,\text{s}^{-1}\,\text{Pa}^{-1}$ was found at 22 °C. With decreasing temperature, the selectivity increases and at – 21 °C a CO_2/CH_4 separation factor $\alpha = 560$ was found. A SAPO-34 membrane prepared from a gel with (the higher) Si/Al of 0.15 had a slightly lower selectivity ($\alpha = 115$) but a higher CO_2 permeance ($\pi = 4 \times 10^{-7}\,\text{mol}\,\text{m}^{-2}\,\text{s}^{-1}\,\text{Pa}^{-1}$) at 35 °C. At 7 MPa, the SAPO-34 membrane showed a CO_2/CH_4 selectivity $\alpha = 100$ for a 50%/50% feed at room temperature over about a week [209]. In a previous paper, the same authors found that SAPO-34 membranes can separate CO_2 from CH_4 best at low temperatures with a selectivity of $\alpha = 270$ at −20 °C [210]. The SAPO-34 membranes effectively separate CO_2 from CH_4 for conditions at or near industrial requirements (Fig. 20). However, CO_2 flux and selectivity decrease in the presence of water since water has a strong affinity to the hydrophilic SAPO-34 membrane [211]. Therefore, hydrophobic small-pore zeolite membranes are more appropriate to separate CO_2 from humid gases. Consequently, DD3R membranes show high CO_2 flux and selectivity and a negligible water influence on the performance in the CO_2 separation from natural gas [212].

Studies of single and binary mixture permeation of CH_4 and CO_2 through silicalite-1 membranes have shown that the CO_2 selectivity in the permeation is due to the favorable CO_2 adsorption [194]. The GMS equations, in combination with the Ideal Adsorbed Solution (IAS) theory, were used to model their binary permeation. It was found that the use of accurate adsorption data is of utmost importance for extracting transport properties from the single-component permeation as well as for modelling multi-component permeation. In detail, both, the CH_4 and the CO_2 fluxes in the mixture increase with increasing total pressure at 30°C (Fig. 21a). For

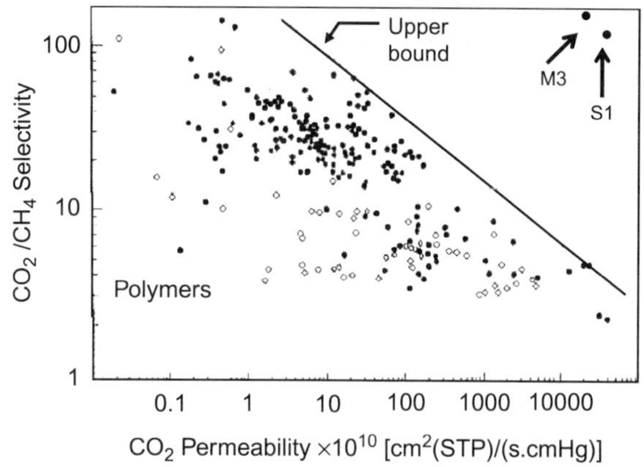

Figure 20 Comparison of the CO_2/CH_4 separation selectivity versus the CO_2 permeability for polymeric membranes and two SAPO-34 membranes (M3 and S1) at room temperature (feed and permeate pressures of 222 and 84 kPa, respectively) [209]. The unit of the abscissa is Barrer: 1 *Barrer* = 10^{-10} cm^3(STP)cmcm^{-2} s^{-1} cm Hg^{-1} or 10^{-10} cm^2 s^{-1} cm Hg^{-1} = 7.5 × 10^{-18} m^2 s^{-1} Pa^{-1}(cf. footnote 6).

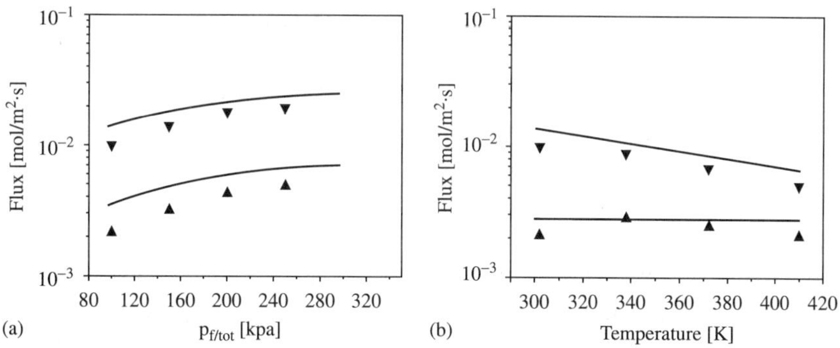

Figure 21 Component fluxes of the binary (50:50) mixture of CH_4 (▲) and CO_2 (▼) through a silicalite-1 membrane: (a) as a function of the total feed pressure at 30 °C; (b) as a function of temperature at a total feed pressure of 101.3 kPa. The solid lines are the full GMS model predictions [194].

a fixed total pressure of 101.3 kPa, the CO_2 flux in the binary permeate decreases monotonically with temperature whereas the CH_4 flux remains almost constant (Fig. 21b). Owing to these component fluxes in the binary mixture permeation, the mixture selectivity is almost constant around a value of 4 at 30°C with increasing gas pressure (Fig. 22a), but it decreases at 101.3 kPa with increasing temperature (Fig. 22b). Summarizing, the GMS

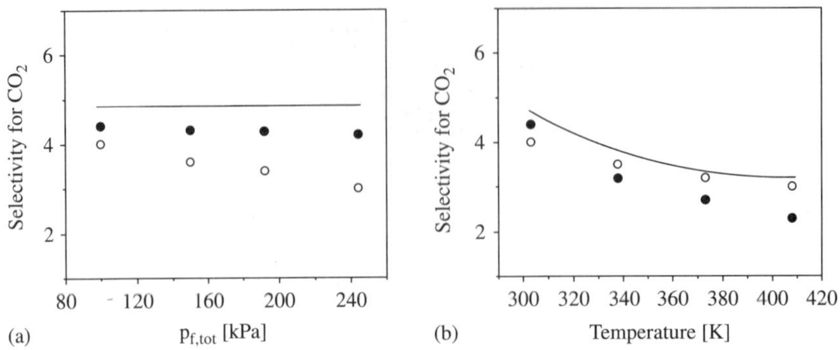

Figure 22 Mixture permeation selectivity for CO_2 based on the data of Fig. 21: (a) as a function of the total feed pressure at 30 °C; and (b) as a function of temperature at a total feed pressure of 101.3 kPa; open symbols (○) indicate the ideal selectivity, the filled ones (●) the real mixture selectivity, the lines are the GMS model predictions [194].

Equations in combination with the IAS theory enables one to predict the binary gas permeation through zeolite membranes.

Another separation problem with relevance to practical applications is the CO_2 removal from N_2 in exhaust gases for CO_2 sequestration. Because a CO_2 separation will take place at elevated pressures, the CO_2 permeation from pressurized feeds on a silicalite-1 membrane on different supports has been studied [213]. A maximum value of 12–13 for the mixture separation factor (CO_2/N_2) was found between 6 and 16 bar total retentate pressure (Fig. 23). The CO_2/N_2 selectivity was found to depend on (i) the kind of support and (ii) the modification of the MFI structure. Boron-ZSM-5 was found to have a higher selectivity toward CO_2 than Na-ZSM-5 indicating that the adsorption mechanism includes electrostatic components. Furthermore, MFI membranes prepared on stainless steel supports showed higher CO_2/N_2 selectivities than those deposited on alumina since aluminium is believed to leach from the support and to become incorporated into the MFI layer.

There are contradicting statements on the separability of CO_2 from gas mixtures by zeolite membranes. As an example, in [214] for equimolar mixtures CO_2/N_2 on FAU membranes the separation factor was determined to be α (CO_2/N_2) $\approx$ 2–5 at 30 °C whereas in [30], another selectivity with α (N_2/CO_2) $\approx$ 5–8 was reported. These different experimental findings can be explained by the role of moisture. ZSM-5-type zeolite membranes showed a permeance of approximately 3.6×10^{-8} mol $m^{-2} s^{-1} Pa^{-1}$ and a separation factor α (CO_2/N_2) $\approx$ 54.3 at 25 °C and α

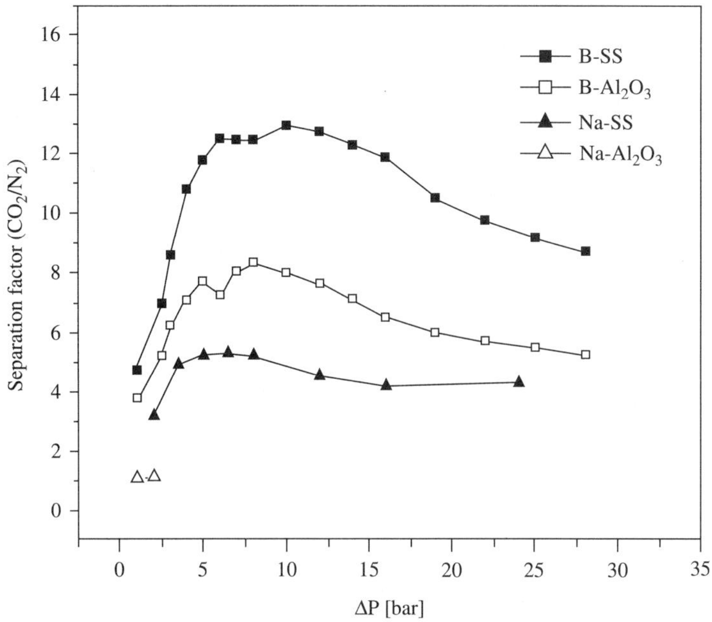

Figure 23 Separation factor for the separation of a CO_2/N_2 mixture (50%/50%) for variable retentate pressure at 25 °C [213]. B_SS and B_Al_2O_3: Boron containing MFI membrane in the H^+ form on stainless steel (SS) and alumina supports (Al_2O_3). Na_SS on stainless steel support and Na_Al_2O_3 on the alumina support are Al-containing MFI membranes in the Na^+ form.

(CO_2/N_2) ≈ 14.9 at 100 °C [215]. However, the separation factor α of the ZSM-5 membrane increases as the permeation time increases (Fig. 24). This experimental finding is explained by the mechanism that moisture occupies large pores through which mainly the N_2 flows and as a result, the separation factor α (CO_2/N_2) is higher for moisture-saturated feed gases than for dry feed gases. The same finding was observed independently by Gu et al. [216], namely that the presence of water vapor significantly enhances the CO_2 selectivity of a FAU membrane in the CO_2/N_2 mixture separation at 110–200 °C. Alternatively, a more simple explanation of this experimental finding would be that CO_2 is dissolved by a water film acting like a supported liquid film membrane. Another zeolite membrane for CO_2 separation is SAPO-34 as a silicon-substituted eight-membered ring aluminium phosphate. From CO_2/CH_4 mixtures, the smaller CO_2 preferentially permeates and high CO_2 selectivities under praxis-relevant test conditions of 30 bar at 50 °C were found [217].

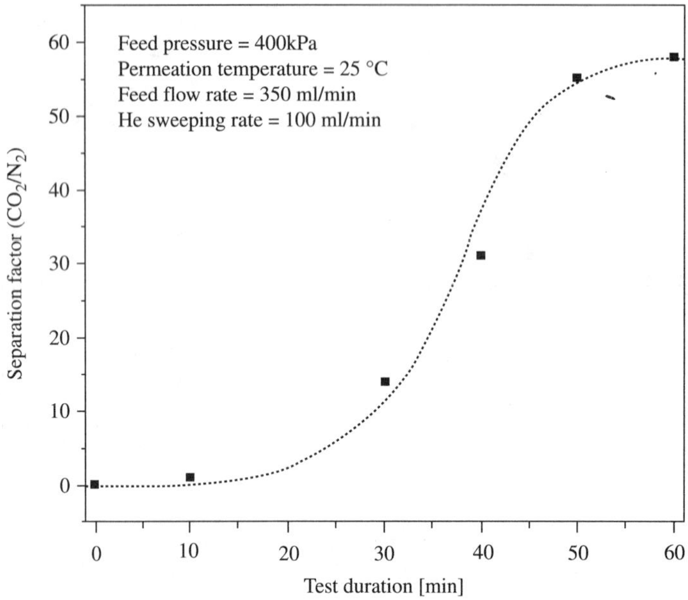

Figure 24 CO_2/N_2 separation factor versus gas permeation test duration for a ZSM-5-type zeolite membrane with moisture-saturated feed gases [215].

3.7 Membrane reactors on the laboratory scale

The classical concepts for the application of zeolite membranes in membrane reactors focus on the conversion enhancement by equilibrium displacement or by removing of inhibitors [218]. There are numerous examples for the application of zeolite membranes to enhance dehydrogenation, partial oxidation, isomerization, or esterification reactions. However, in only a few of these cases, the zeolite membrane acts as a real shape- and size-selecting molecular sieve membrane. Owing to their molecular sieve properties, zeolite membranes recommend themselves as a "membrane extractor reactor" removing under equilibrium controlled reaction conditions small product molecules such as hydrogen or water, thus increasing the conversion and the yield of a dehydrogenation or dehydration reaction, respectively.

In the first example, the use of H_2-selective membranes in dehydrogenations will be treated. Several studies showed that the X-ray amorphous H_2-selective sol-gel prepared silica membranes were not stable under the reaction conditions of a catalytic dehydrogenation. After several cycles of dehydrogenation at 535 °C followed by burning off the carbon deposits at 450 °C with 3% oxygen in nitrogen, the membranes showed some deterioration of the hydrogen permeance and the separation factor compared to the fresh state in the beginning [219]. Crystalline zeolite membranes are more stable under hydrothermal conditions than the amorphous silica membranes and have been tested, therefore, also for dehydrogenation membrane reactors. Experimental results were described, for example, in Refs. [220–223] silicalite-1 membranes were studied in *i*-butane dehydrogenation. MFI zeolite membranes can be used in catalytic dehydrogenations of, for example, *i*-butane although both H_2 and *i*-butane can pass the 0.55 nm pores due to their kinetic diameters (0.29 and 0.50 nm, respectively) since the interplay of mixture adsorption and mixture diffusion results in a H_2 selectivity at high temperatures (Fig. 19). In the conventional fixed-bed experiment, the thermodynamic equilibrium conversion was obtained (Fig. 25). As hydrogen was removed from the shell side of the membrane reactor through the sweep gas, the *i*-butane conversion increased by approximately 15% [154]. Removal of the hydrogen leads to hydrogen-depleted conditions as compared to the conventional fixed-bed. This has two positive effects: (i) the conversion of *i*-butane is increased,

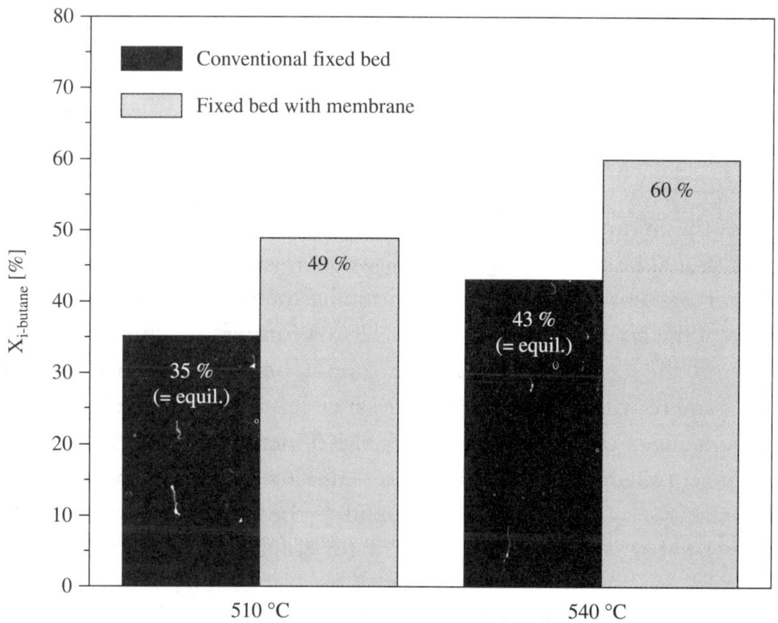

Figure 25 Increase of the *i*-butane conversion above the equilibrium limit if hydrogen is removed through a silicalite-1 membrane. Conditions: WHSV = 1 h^{-1}, Cr_2O_3/Al_2O_3 catalyst (Süd-Chemie), membrane area per unit mass of catalyst = 20 $cm^2\,g^{-1}$, data after 20 min time-on-stream (after Ref. [154]).

and (ii) the selectivity to *i*-butene is increased since hydrogenolysis is suppressed. As a result, at the beginning of the reaction the *i*-butene yield in the membrane reactor is higher by approximately 1/3 than in the conventional fixed-bed. However, because of the hydrogen removal, coking is promoted and after approximately 2 h time-on-stream the olefin yield of the membrane reactor drops below that of the classical packed-bed [223]. After an oxidative regeneration, however, the activity and selectivity of the membrane reactor (membrane and catalyst) are restored completely.

In a second example, the effect of water removal during an esterification reaction will be shown. There are different ways to increase the yield of an esterification. Most frequently, the cheapest reactant is present in a surplus concentration or the low boiling ester is removed by reactive distillation. Another concept is to keep the concentration of the product molecule water as low as possible by the use of adsorbents such as LTA zeolites or by the hydrolysis of aluminium tri-isopropylate. In the case of the low-temperature esterification of methanol or ethanol with short-chain monovalent hydrocarbon acids under equilibrium-controlled reaction conditions, hydrophilic organic polymer membranes can be used for the de-watering. However, to support esterifications at higher temperatures, hydrophilic inorganic membranes with high stability against strong acids have to be used. MFI-type zeolite membranes are suitable candidates to fulfil these demands. The benefits of a membrane-assisted esterification were shown for the reaction of *n*-propanol with propionic acid using a MFI-type ZSM-5 membrane with a molar ratio Si/Al = 96 (Fig. 26). The hatched area indicates the optimum working range of the membrane reactor. The water content should be reduced to values between 5% and 10%. The reduction of the water content to 5% by the hydrophilic membrane corresponds to an increase of the ester yield from 52% to 92%. A further reduction of the water content in the esterification mixture cannot be recommended since this would require very large membrane areas and would give only slight improvements of the ester yield. The ZSM-5 membrane with Si/Al = 96 is stable in acid media up to pH = 1 but – due to the low Al-contents – the hydrophilicity is low and, consequently, the resulting water flux of 72 $gm^{-2}h^{-1}bar^{-1}$ is still much too low for commercial applications.

3.8 Micromembrane reactor

The combination of the concepts of membrane reactor and process miniaturization provides new routes for chemical synthesis that promises to be more efficient, cleaner, and safer [224]. These smart integrated micro chemical systems are expected to bring into realization a distributed, on site,

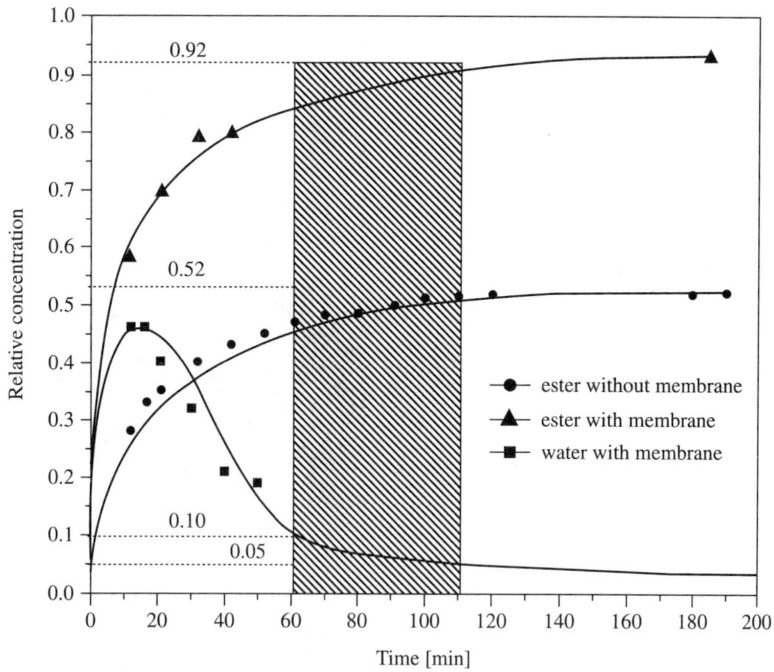

Figure 26 Conversion enhancement by water removal via a hydrophilic ZSM-5 membrane with Si/Al = 96 in a membrane reactor for the esterification of propionic acid with *i*-propanol to yield the corresponding ester and water at 70 °C [8].

and on demand production network for high value added products in the form of miniature factories and micropharmacies [225]. The incorporation of zeolites in microreactors as functional elements including catalysts [226–228] and membranes have been reported in previous works [229–231].

A recent example of fine chemical reaction carried out in a membrane microreactor is the Knoevenagel condensation reaction where the selective removal of the by-product water during the reaction led to a 25% improvement in the conversion [232]. The reaction between benzaldehyde and ethyl cyanoacetate to produce ethyl-2-cyano-3-phenylacrylate was catalyzed by a CsNaX zeolite catalyst deposited on the micro channel and the water was selectively pervaporated across a LTA membrane (Fig. 27a) [233]. All the water produced by the reaction was completely removed and the membrane was operating below its capacity [234]. This means that the performance of the membrane micro reactor is limited mainly by the kinetics, that is to say that both thermodynamic and mass transfer constraints were removed. A fourfold increase in reaction conversion was obtained

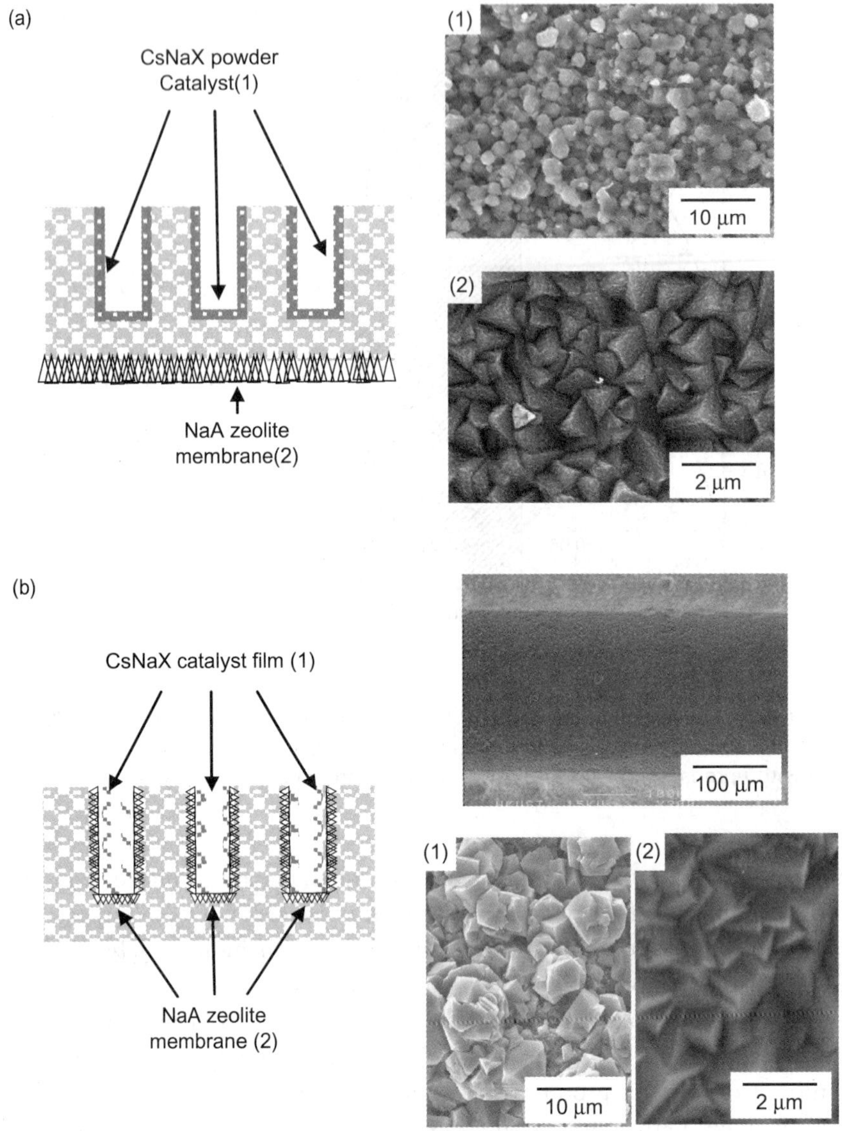

Figure 27 (a) Membrane microreactor design and SEM pictures of (1) the 3-μm thick CsNaX catalyst powder deposited on the micro channel wall and (2) the 6-μm thick NaA grown on the back of the stainless steel plate. (b) Membrane reactor design and SEM of the micro channel and the (1) 3-μm thick CsNaX film grown on top of the (2) 6.5-μm thick NaA membrane in the micro channel [232,233].

when the improved CsNaX-NH_2 catalyst was used instead of CsNaX [235]. Locating the separation membrane immediately next to the catalyst further improved the membrane micro reactor performance (Fig. 27b). From the selective removal of the by-product water in the membrane microreactors also benefited other Knoevenagel condensation reactions such as reactions between benzaldehyde and (i) ethyl acetoacetate (EAA) and (ii) diethyl malonate (DEM) [236].

A multichannel membrane microreactor for continuous selective oxidation of aniline by hydrogen peroxide on TS-1 nanoparticles was successfully demonstrated. The high surface area to volume ratio that can be attained in the microreactor (3000 m^2/m^3) facilitates the selective removal of water by-product, which reduces the effect of catalyst de-activation during the reaction. An improvement in the product yield and selectivity toward azobenzene was also observed. Azobenzene was obtained as by-product, and its formation was attributed to homogeneous reaction of nitrosobenzene with aniline. Increasing temperature was beneficial for both yield and selectivity, but beyond 67 °C, microreactor operation was ineffective due to bubble formation and hydrogen peroxide decomposition [237,238] .

Synthesis of advanced materials was also successfully carried out in zeolite membrane-enclosed microchannels [239]. The hollow silica nanospheres were successfully prepared within the zeolite-enclosed microchannels by a ship-in-a-bottle approach. The zeolite microchannels were fabricated by selective etching of the silicon below the zeolite membrane to create the membrane-enclosed microchannels. The nanospheres were then prepared *in situ* using ferrocene as catalyst and the silicon substrate as silica source.

4. INDUSTRIAL APPLICATIONS OF ZEOLITE MEMBRANES

4.1 De-watering of ethanol and propanol by hydrophilic zeolite membranes

There is a worldwide increase of the production of ethanol, mainly as bio-ethanol and to some extent by ethylene hydration. If used as a blend for gasoline, the water content must be reduced to 2000 ppm, for ethyl tertiary butyl ether (ETBE) production from i-butene and ethanol the water content must be below 500 ppm[10]. By distillation of an ethanol/water fermentation broth, an azeotrope with an ethanol content of approximately

[10]Owing to the Euronorm EN DIN 228, the bio-ethanol content in conventional fuel can be up to 5% and the ETBE content up to 15%. On the contrary, in Sweden E85 is offered which consists of 85% ethanol and 15% conventional fuel but it needs so-called Flexible Fuel Vehicles (FFV).

95.6 wt.% can be obtained (for economic reasons, a product with 92–93 wt.% ethanol is obtained). The conventional de-watering of alcohols by azeotropic, extractive or two-pressure distillations, is energy intensive and requires a complex process layout. Especially for ethanol/water and other water-containing azeotropes, two alternative processes are available: (i) pressure swing adsorption employing type 3A or type 4A LTA[11] molecular sieves, and (ii) steam permeation/pervaporation using hydrophilic organic or inorganic (4A molecular sieve) membranes. The application of membrane processes is especially beneficial for systems of low relative volatility [240].

A hydrophilic LTA zeolite layer is extremely selective in the separation of water from organic solutions by steam permeation and pervaporation and can be used, therefore, for the production of water-free ethanol. For the de-watering of the crude ethanol stream using membrane technology, Mitsui-BNRI (Bussan Nanotech Research Institute Inc., a 100% subsidiary of Mitsui & Co. Ltd., Japan) played a pioneering role in the cost reduction of the membrane separation by integrating distillation and membrane separation in the so-called Membrane Separation and Distillation (MDI process, see Fig. 28). In the MDI process, dehydrated ethanol containing 0.4 wt.% water is produced from a fermented liquid containing 8 wt.% ethanol starting from ligno cellulose. The aim is the production of 1 l dehydrated ethanol with less than 1000 kcal (4200 kJ) of energy. LTA membranes were developed and produced by BNRI. These hydrophilic LTA membranes have been applied in industrial plants for dehydration [8,18,17]. The water flux measured in pervaporation operation for 90 wt.% ethanol solution at 75 °C is approximately 7 kg m^{-2} h^{-1}. Ethanol scarcely leaks through the membrane resulting in a separation factor α (water/ethanol) $\approx$10,000. Two different tubular ZeoSepA membranes are produced: large-size elements with 16 mm outer diameter and 1 m length for the de-watering of bio-ethanol and small-size elements of 12 mm outer diameter and 0.8 m length for the recovery of *i*-propanol. The supports are in both cases porous α-alumina tubes. Recently, by combined SEM, TEM, FT-IR, focused ion beam, and XRD characterization, the molecular structure of the Mitsui-BNRI LTA membranes could be solved [241–244].

[11] LTA stands for zeolite Linde Type A which has in the as-synthesized Na^+-containing state the unit cell composition $Na_{12}[Si_{12}Al_{12}O_{48}]$ with a pore opening of about 0.41 nm (4A). By replacing the Na^+ by Ca^{++}, the pore openings are enlarged to about 0.45 nm (5A). On the contrary, K^+ exchange results in a pore narrowing to about 0.30 nm (3 A).

(a)

(b)

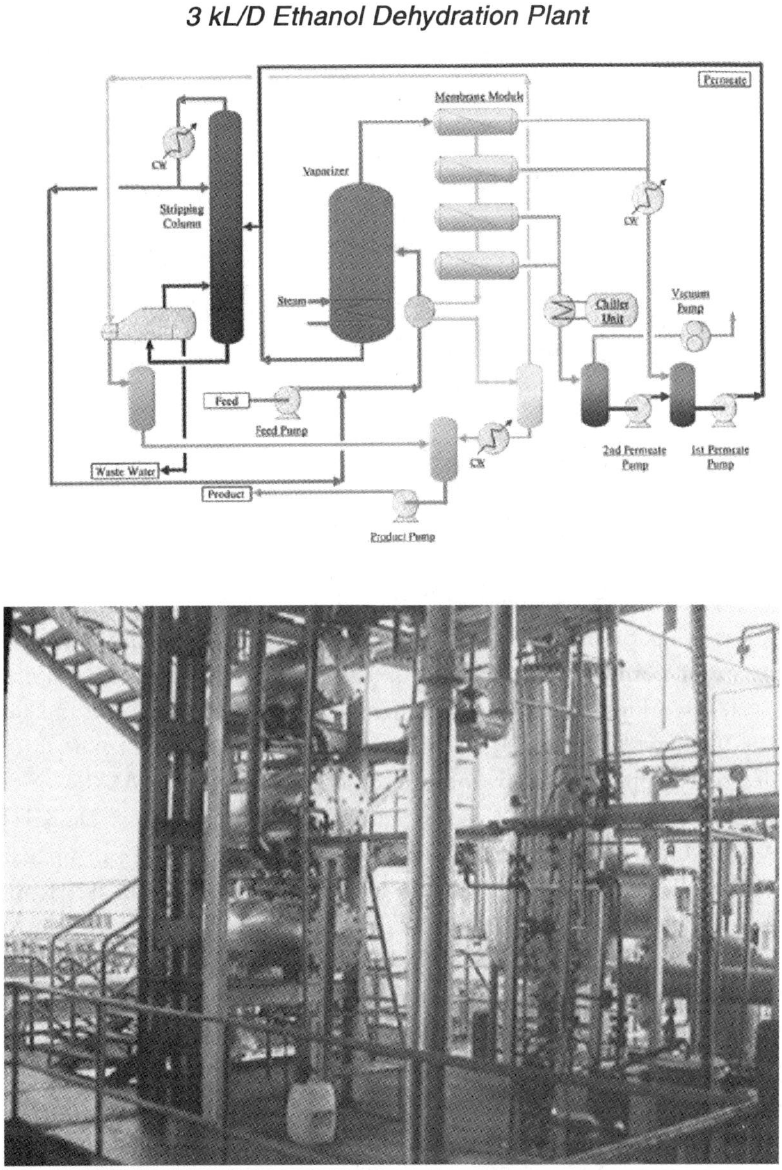

Figure 28 Pilot plant of the combined distillation/membrane process for fuel ethanol production from cellulose-based biomass. Feed: Fermentation liquid containing 8 wt.% ethanol, product: 99.6 wt.% dehydrated ethanol [248] (a) gives the flow sheet, (b) shows the pilot plant.

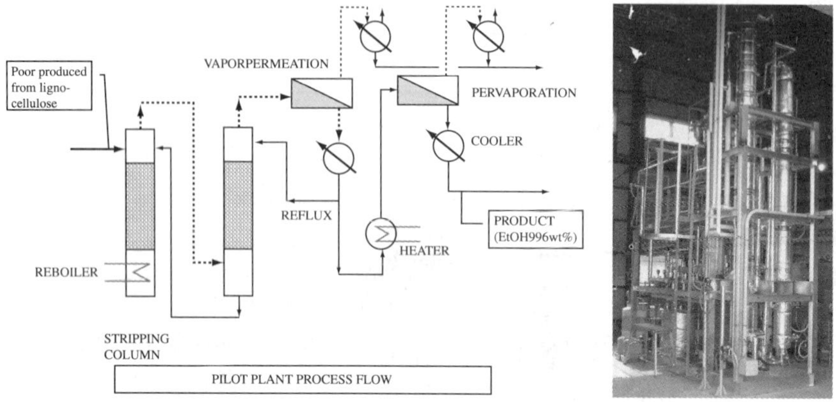

(a) (b)

Figure 29 Plant for the dehydration of bio-ethanol by steam permeation using LTA membranes at Daurala Sugar Works (Uttar Pradesh, India) with a capacity of 30,000 l/d [248]. (a) shows the flow sheet, (b) is a view of the plant.

From April to September 2003, Mitsui-BNRI tested successfully the de-watering of bio-ethanol in the pilot scale by using LTA membranes for vapor permeation in Piracicaba (Sao Paulo, Brazil). The capacity was 100 l/h working with a feed containing 93 wt.% ethanol and the product was 99.65 wt.% ethanol. The demonstration plant in Brazil is driven by electricity applying vapor compressor energy recycling. As the next step, Mitsui-BNRI installed a larger steam permeation capacity at Daurala Sugar Works (Uttar Pradesh, India). The capacity of 30,000 l/d can be achieved with a LTA membrane area of 30 m^2. Each of the membranes is inserted into a sheath tube (i.e., 19 mm). The feed is evaporated by heating with steam. The feed containing 93 wt.% bio-ethanol, the product purity is 99.8 wt.% ethanol suitable for blending with gasoline. The ethanol content in the permeate is below 0.1 wt.%. The operating pressure and temperature of the membrane are 600 kPa and 130 °C, respectively. Extremely high ethanol fluxes are reported: 11.9, 14.9, 17.6, and 22.4 $kg\,m^{-2}\,h^{-1}$ at 100, 110, 120, and 130 °C, respectively [245]. The plant is in permanent operation since January 2004 (Fig. 29).

The smaller ZeoSepA membranes are mainly used for the recovery of *i*-propanol (IPA process) in the Japanese electronic industry using vapor permeation (azeotrope: 87.9 wt.% *i*-propanol, 12.1 wt.% water). For a 90%/10% *i*-propanol/water mixture, the water flux at 75 °C is 3.5–3.7 $kg\,m^{-2}\,h^{-1}$ with a separation factor α (water/*i*-propanol) $\approx$10,000. Zeolites as crystalline materials are much more stable toward phase transformation and densification compared with X-ray amorphous metal oxide membranes from sol-gel techniques. Because of the high Al-content, LTA membranes should be operated at $6.5 < pH < 7.5$. Therefore, Mitsui-BNRI developed with the same supports used for the ZeoSepA element, a zeolite FAU membrane with lower Al-content (namely Si/Al between 1.5 and 1.6). This FAU membrane was tested successfully in vapor permeation for the de-watering of a spent IPA solution with a starting content of 12 wt.% water to 0.46 wt.%. The water flux of this membrane was evaluated at 75 °C in a pervaporation experiment with a model feed containing 90 wt.% ethanol and gave a water flux of 7–10 kg $m^{-2} h^{-1}$ and a separation factor α (water/ethanol) $\approx$300. For the LTA membrane tested under the same conditions, a flux of 13.5 kg $m^{-2} h^{-1}$ with a separation factor α (water/ethanol) $\approx$4,800 was found [246]. The described FAU membrane was also successfully tested in alcohol/ether separations. As an example, ethanol is separated from a 5/95 wt.% mixture of ethanol/ETBE with fluxes of 2.2 and 4.1 kg $m^{-2} h^{-1}$ and selectivities α (ethanol/ETBE) of 2800 and 1600 at 90 and 110 °C, respectively, by vapor permeation [232]. For a 10/90 wt.% mixture of methanol/MTBE at 100 °C in vapor permeation, methanol fluxes of 10 kg $m^{-2} h^{-1}$ with selectivities α (methanol/MTBE) of 3000 are found [247]. The stability of the hydrophilic FAU membranes could be increased having an Si/Al$\approx$2.2 by using USY[12] seed crystals [248]. A further increase of the stability of hydrophilic membranes was achieved by developing a weakly hydrophilic MFI-type membrane with Si/Al$\approx$120 [57,249]. In pervaporation tests of a 90/10 wt.% i-propanol/water model mixture, water fluxes of 3.1 kgm^{-2} h^{-1} with α (water/i-propanol) $\approx$690 were determined at 75 °C.

Despite these developments, in recent installations in Europe, the adsorptive drying of ethanol was implemented instead of the membrane technology. Here, ethanol is purified by distillation which is coupled with a zeolite adsorption section. By the so-called DELTA-T technology [250], ethanol is purified to less than 100 ppm water. The preference of the molecular sieve adsorption process with zeolites for the de-watering of ethanol may be because the azeotrope ethanol/water contains only a relative low amount of water (4.4 wt.%) so that the heat management of the cyclic adsorption processes can be managed. Since the azeotrope *i*-propanol/water contains more water (12.1 wt.%), steam permeation using

[12]USY stands for ultrastable Y. By various techniques the Al-content of Y zeolites can be reduced to make them ultrastable. USY is widely used as catalyst in FCC (fluid catalytic cracking).

hydrophilic organic or inorganic membranes has a higher chance for realization in comparison with the ethanol/water mixture.

In Europe the company inocermic GmbH, which is a 100% subsidiary of HITK, Hermsdorf, Germany, produces NaA membranes for de-watering processes, especially for bio-ethanol, by pervaporation and vapor permeation. The membrane layer is inside an α-Al_2O_3 4-channel support thus protected against mechanical damage. Organic solutions can be dried to water levels as low as 0.1 wt.% by pervaporation/steam permeation. For a feed with 90 wt.% ethanol, at 100–120 °C water fluxes of 7–12 $kg\,m^{-2}\,h^{-1}$ with separation factors H_2O/ethanol >1000 are found [251]. The de-watering behavior of these semi-industrially produced NaA membranes was tested by pervaporation with bio-ethanol feed stocks from real fermentation processes. The impurities in the bio-ethanol from grain fermentation or wine production lowered the specific permeate flux by only 10–15% as compared to synthetic ethanol/water mixtures [252]. All bio-ethanol samples were de-watered to >99.5 wt.% ethanol [251]. Much progress can be stated in the commercialization of bio-ethanol within the past few years by the German company inocermic GmbH producing NaA-zeolite membranes inside of a four-channel geometry in an industrial length of 1.2 m. The optimal implementation of inocermic NaA-zeolite membranes in the production of bio-ethanol results in the reduction of the overall steam consumption down to below 1 kg of steam per 1 l of dry ethanol. In cooperation with GFT Membrane Systems GmbH several industrial plants with 30,000 l/day, 60,000 l/day, 80,000 l/day, 100,000 l/day, and 200,000 l/day bio-ethanol production with inocermic NaA-zeolite membranes have been built since 2007.

4.2 Ethanol removal from fermentation batches by hydrophobic zeolite membranes

In Section 4.1 it was shown that several types of hydrophilic membranes such as LTA and FAU can be used for water extraction from aqueous ethanol or *i*-propanol mixtures to get concentrated alcohol. An opposite target can be the continuous removal of ethanol from the fermentation broth since the fermentation process stops at ethanol concentrations $\geqslant$ 15 wt.%. Hydrophobic membranes to solve this problem, such as the MFI-type, are under development (Figs. 30 and 31, Table 9). As Table 9 shows, the ethanol fluxes are between 0.2 and 1.4 $kg\,m^{-2}\,h^{-1}$ with separation factors between 30 and 70. A typical result is a flux of approximately 1 $kg\,m^{-2}\,h^{-1}$ of 85 wt.% ethanol from a feed with 8 wt.% ethanol, which corresponds to a separation factor of 57 [253]. The relative low ethanol

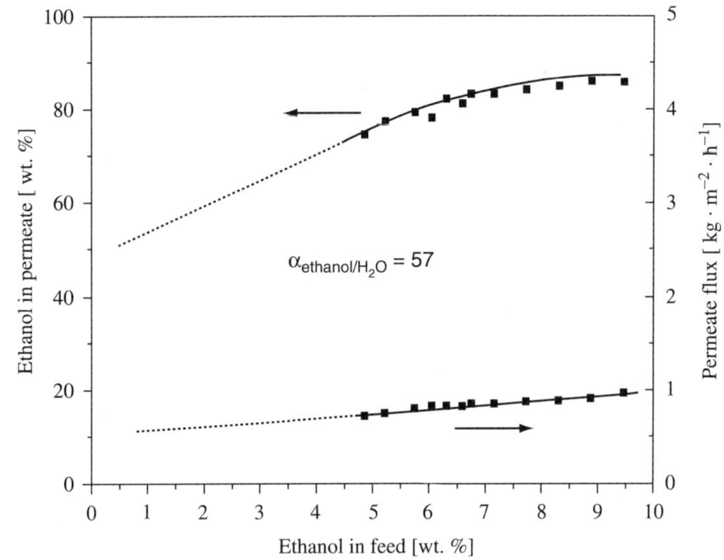

Figure 30 Permeation behavior of a ZSM-5 membrane (Si/Al $\approx$ 300) in the separation of ethanol from an aqueous fermentation broth by pervaporation at 40 °C [255].

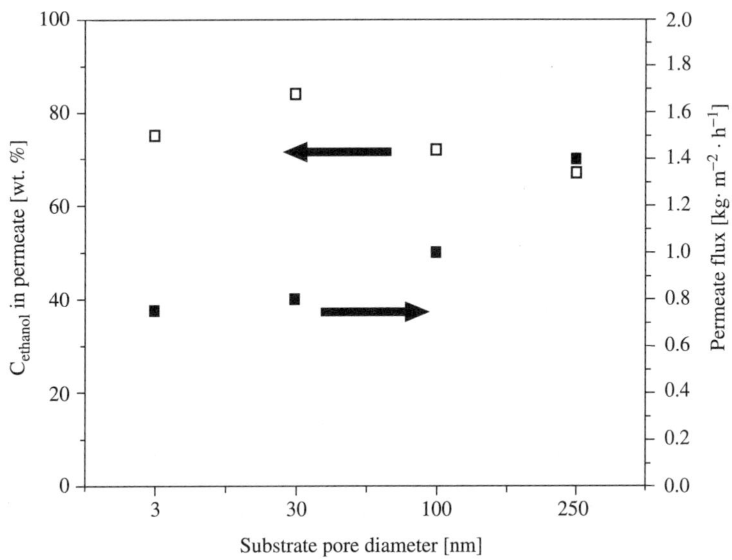

Figure 31 Influence of the pore size of the support on the ethanol permeance of a ZSM-5 membrane (Si/Al $\approx$ 300) in the pervaporation of an aqueous solution containing 5 wt.% ethanol at 40 °C [255].

Table 9 Comparison of the separation behavior of different membranes for synthetic mixtures of 5/95 wt.% ethanol/water [255]

Membrane	Support	T (°C)	Permeate pressure (mbar)	Flux (kg^{-2} h^{-1})	wt.% EtOH in permeate	α (EtOH/H_2O)	Ref.
Poly (trimethyl silyl propyne)	–	30	<1	0.33	55	19	[256]
ZSM-5 (Si/Ge ≈ 40)	Steel	30	–	0.22	71	47	[257]
Silicalite-1	Steel	30	–	0.68	63	32	[258]
ZSM-5 (Si/Al ≈ 300)	30 nm TiO_2	40	5	0.74	84	73	[255]
	250 nm TiO_2	40	7	1.39	71	48	

fluxes are due to the test conditions with real fermentation broths. By optimizing the support structure, reducing the membrane thickness and increasing the Si/Al-ratio of the MFI membrane, it should be possible to increase the ethanol fluxes. By changing the top layer of the support from a 5-nm γ-Al_2O_3 nanofiltration layer to a 250-nm α-Al_2O_3 microfiltration layer, the ethanol-enriched flux (between 70 and 80 wt.% ethanol) could be increased from 0.8 to 1.4 kg m^{-2} h^{-1} [254].

4.3 Further R&D on zeolite membrane-based separation processes

The Mitsui-BNRI LTA (Linde type A) with a pore size of 0.41 nm in the Na^+ form is the first zeolite membrane that has reached a commercial status in the de-watering of (bio-) ethanol and *i*-propanol by vapor permeation or pervaporation [8,17,18]. However, this successful application of LTA membranes in dehydration is based on the hydrophilic character of LTA resulting in a preferential adsorption of water from mixtures rather than on a real size-exclusion molecular sieving. When tested for hydrogen separation from gas mixtures, the LTA membranes show only Knudsen separation [259]. Recently, NGK announced the commercialization of another type of zeolite membrane, DDR (deca-dodecasil 3R) with narrow pores of 0.36×0.44 nm for CO_2 separation from CH_4 [260]. The gas separation characteristics of DDR membranes including hydrogen are reported in Ref. [261]. Also, ExxonMobil is active in the field of DDR membranes [262], and recently joint communications of ExxonMobil and NGK on DDR zeolite membranes have been issued [263]. Also promising is the development of H–SOD[13] (sodalite) membranes, with narrow pores of 0.28 nm. At $\Delta p = 20$ bar, they have shown a water flux of $N_{H_2O} = 4$ kg m^{-2} h^{-1} and a permselectivity PS (H_2O/other components) $\approx 10^6$ [264]. Such small-pore zeolite membranes are interesting candidates also for hydrogen separation. However, the most often used shape-selective zeolite membrane is MFI (silicalite-1) with a pore size of 0.55 nm. It is near to commercialization for isomer separation, for example, of xylenes [265] or *n*/*i*-hydrocarbons. The advanced state of development of silicalite-1 membranes was reached because its preparation is relatively easy, and this highly siliceous zeolite type provides excellent chemical stability and allows for oxidative regeneration [7].

[13]H-SOD is hydroxy sodalite with a Si/Al = 1.

4.4 Cost analysis: Need for cheaper supports

Although the manufacture of inorganic membranes is more expensive than the production of polymeric ones, the long-term cost implications due to their chemical and thermal stability make the use of inorganic membranes a viable option. Therefore, zeolite membrane costs are a major factor to be considered for industrial applications. An estimated price limit of membranes for petrochemical applications is €200m^{-2} [266,267]. In contrast, at present the prices for zeolite membranes are estimated to be close to €1000 m^{-2} (Table 10). There are several concepts to reduce the price for the formation of the zeolite layer. These concepts mainly aim at the saving of the chemicals needed for the membrane synthesis and comprise, for example, the template (SDA)-free synthesis and the continuous synthesis with a re-circulating synthesis solution. In another concept, chemicals for the zeolite synthesis can be saved if the autoclaves with the supports are only filled to approximately 1/3 with the synthesis solution and continuously moved (e.g., rotation, shaking). However, these attempts to lower the costs of the zeolite layer will not alone solve the problem since most of the costs do not stem from the manufacture of the zeolite layer but from the support (Table 10).

Therefore, to reduce the costs of supported zeolite membranes, the support costs must also be lowered. At present, mainly supports with asymmetric cross section for high fluxes and reduced pressure drop across the support are used (these supports are micro and ultra filters). A disadvantage of the sandwich-like layered support structures is that the individual layers have to be calcined after each layer deposition, which increases the production

Table 10 Estimated costs for the production of MFI membranes on 5 nm TiO2 supports in 19-channel geometry of 1.20 m length with gas-tight glass sealings at the two ends of the support [254][a]

Capacity (membranes per year)	Price of support (€)	Price of MFI membrane (€)	Sum of prices support + membrane (€)	Costs per MFI zeolite membrane area (€ m^{-2})
<5000	220	80	300	1190
>5000	180	20	200	800

[a] For comparison: The price of a typical polymer membrane in flat geometry is approximately €10 m^{-2} hollow fibers are approximately €5 m^{-2} (K.-V. Peinemann, GKSS Geesthacht, Germany, personal information).

costs. Regarding HITK e.V./inocermic GmbH (Germany) as one of the most prominent suppliers of high quality supports, the ceramic support is responsible for at least 70% of the zeolite membrane price (cf. Table 10). There is ongoing R&D, therefore, for a cheaper and automatic one-step production of relative simple porous supports, which can be coated subsequently by a micrometer-thick top layer of, for example, a zeolite, palladium, perovskite, or carbon membrane layers (Fig. 32). Recently, in the development of hydrophobic zeolite membranes, HITK e.V./inocermic GmbH substituted their ceramic multi-layer by a one-layer support; the asymmetric multi-layer support was replaced by a thick 3-μm-coarse one-layer support (Fig. 33). Owing to a reduced pressure drop in the support structure, the EtOH-enriched permeate fluxes increased to 1.4 kgm^{-2}h^{-1} bar^{-1} at a constant separation factor α (EtOH/H_2O) = 73 [255].

Producing tubular ceramic supports by conventional methods such as extrusion or isostatic pressing followed by sintering are acceptable manufacturing techniques, but unroundness, insufficient microstructural homogeneity and considerable surface roughness may impose problems for the crystallization of a thin zeolitic top layer. Centrifugal casting represents a

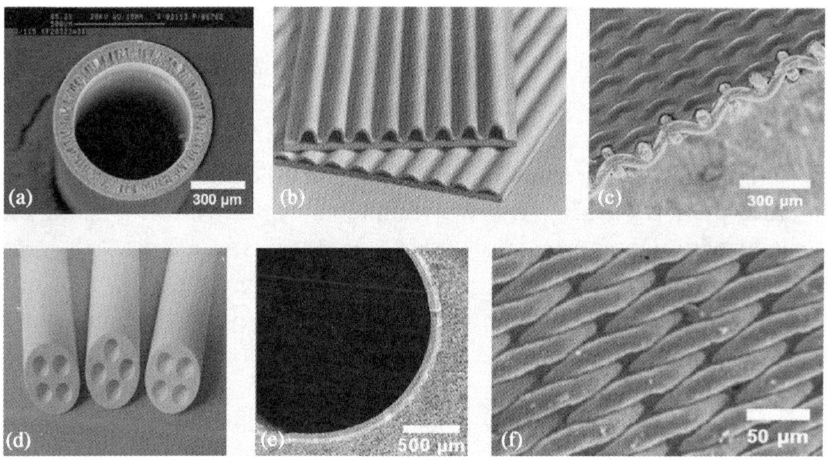

Figure 32 Recent developments of full-material supports by a low-step production: Al_2O_3 hollow fiber prepared by a wet spinning process with a wall thickness of approximately 120 μm, 46% porosity, mean pore size 0.45 μm, bending strength 105 MPa (a) [268]; SiC multi-channel element that can be continuously produced and calcined by co-firing (b) [269]; flexible ceramic foil with a stainless steel web as mechanical support for the ceramic particles (c) [270,271], 4-channel alumina monoliths (d) [272]; ceramic capillary coated inside with a ZSM-5 membrane layer (e) [273] and stainless steel grid with silicalite-1 coating (f) [218].

Figure 33 Silicalite-1 membrane: substitution of an asymmetric multi-layer support with a 5 nm TiO_2 top layer (a) by a homogeneous 3 μm-coarse α-Al_2O_3 support (b). The surface roughness is compensated in the latter case by a thick layer of 0.7 μm sized silicalite-1 seed crystals [255].

novel concept for the cost effective one-step formation of asymmetric supports. The production steps comprise preparation of a colloidal polydisperse suspension of ceramic particles with a stabilizer in water, centrifugal casting with approximately 17,000 rpm, drying, calcination, and sintering steps [274–276]. By this technique, high-quality tubes can be obtained with a homogeneous packing of particles and a smooth inner surface, ideally suited for the deposition of a thin top-layer (Fig. 34).

One should consider, however, that changing the support can dramatically influence and even reverse the separation behavior (Fig. 18). The support thickness, porosity, tortuosity, and pore size affect the resistance that the support causes, which is in series with the resistance of the selective zeolite layer. Owing to the support resistance, the local concentration at the zeolite-support interface can differ from that in the permeate stream. This affects the adsorption coverage and changes the driving force for permeation over the zeolite layer, thus altering the fluxes and selectivities (Section 3.3) [122,146,277,278].

Another problem to be solved is the module design: An advantage of inorganic membranes compared with polymeric ones is the high temperature stability, which allows high-temperature separations and applications as

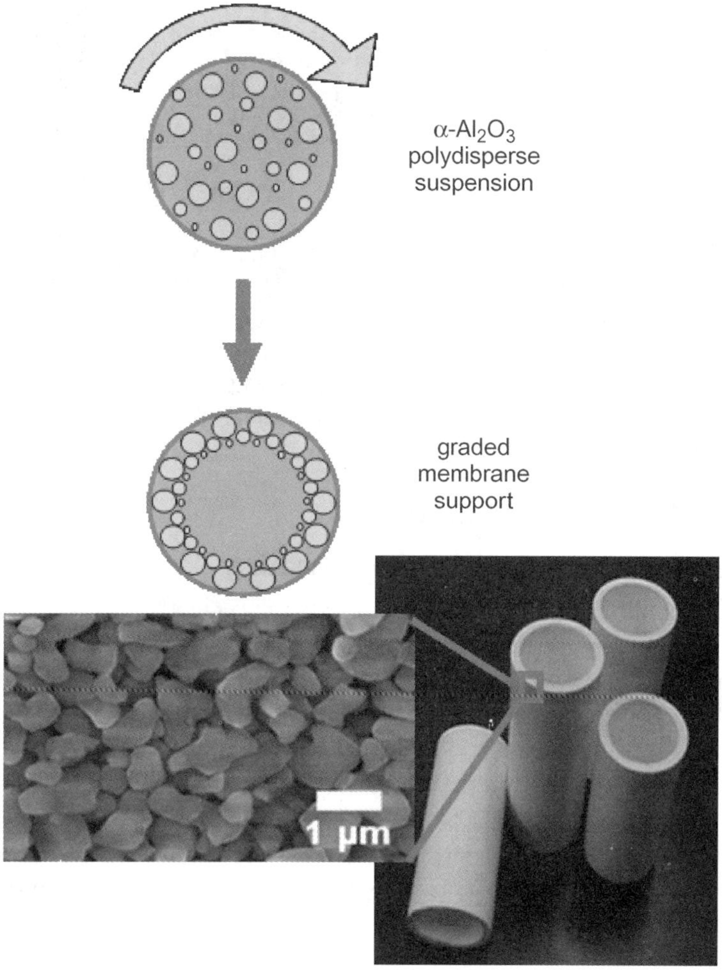

Figure 34 Scheme of centrifugal casting producing an asymmetric support by using a polydisperse suspension of ceramic particles [279].

chemical membrane reactors including an oxidative *in situ* regeneration. To keep this advantage, inorganic membranes should be sealed into modules by avoiding organic polymers. A new solution is the full ceramic module (Fig. 35a), where tubular membranes are gas-tight embedded into a ceramic housing by a ceramic binder. In the case of zeolite membranes, this technology requires the seeding and zeolite layer growth after completion of the housing. Another development is the arrangement of bunches of capillaries/tubes fixed by a ceramic plate with distance holes (Fig. 35b).

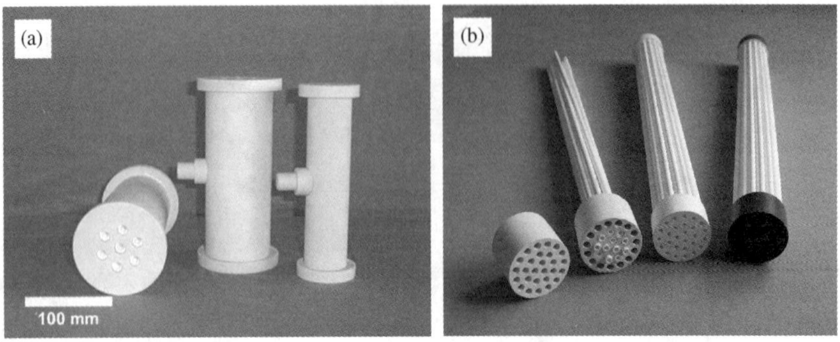

Figure 35 (a) Full ceramic module and (b) bunches of capillaries of the Hermsdorf Institute of Technical Ceramics HITK e.V./inocermic GmbH [272].

5. NOVEL SYNTHESIS CONCEPTS

5.1 Crystallization by microwave heating

In the beginning of the 1990s, good experience in the crystallization of large and phase-pure zeolite crystals was made by using microwaves for heating the autoclaves. Pioneering papers reported the successful synthesis of MFI (ZSM-5) and FAU(Y) [280,281], LTA and FAU(X) [282], AFI (CoAPO-5[283] and $AlPO_4$-5 [104,284]). Surprisingly, large single crystals were obtained in relative short synthesis times, and there was some mystery about the molecular understanding of microwave heating.

- *Homogeneous heating:* There can be a very high energy input by microwave absorption compared with the classical heating of autoclaves in air conditioned ovens. Therefore, microwave absorption leads – at least for a short time scale – to immense temperature gradients. From the measurements of the dielectric loss factor ε – as a measure of energy absorption – for zeolite NaA and NaY synthesis mixtures, it can be calculated that the penetration depth (following the Lambert Beer absorption rule, after this distance, the initial microwave energy irradiated onto the sample is attenuated to 1/e of its initial energy by absorption) at 2.45 GHz and 20 °C is approximately 6 mm for NaA and 10 mm for NaY [42].
- *Freely rotating water molecules:* The energy of one OH-bridging bond in water is approximately 20 kJ/mol [285], that is, 11.3×10^{-20} J per one water molecule. Since statistically every water molecule has approximately 3.4 OH-bridges, the energy of one OH-bridge is close to 3.3×10^{-20} J which is much larger than as the energy of one microwave quantum (1.6×10^{-24} J). Therefore, the existence of freely rotating water molecules as super-solvent seems not to be realistic [129].

- *Microwave effect:* Probably there is no intrinsic microwave effect leading to a reduction of the crystallization time compared to conventional heating if the latter can be done quick enough, for example, by using induction heating which is based on the mobility of ions in an alternating field and results in a really homogeneous heating [42].

However, there are other microwave effects that originate from the quick energy input and the resulting fast heating rate, which brings the zeolite batch quickly to the crystallization temperature and suppresses kinetically the formation of nuclei. Hence, microwave heating can shorten the nucleation period. Furthermore, because of the accelerated heating of the synthesis mixture, the silicate species, due to a kinetic effect, are not in their thermal equilibrium. It is assumed, therefore, that the transport and the reactivity of alumina and silica species to precursor building units are influenced by microwave heating. Increasingly, microwave heating is understood and can be applied in zeolite membrane synthesis [30].

Recent progress was achieved by Julbe and co-workers in the microwave assisted hydrothermal synthesis of silicalite-1 seeds for membrane preparation [286] and the silicalite-1 membrane crystallization itself [287]. Silicalite-1 membranes with a controllable thickness and high crystallinity can be derived within a few hours when seeded supports are microwave heated. By different synthesis temperatures and different methods for seeding the support, oriented silicalite-1 layers with a (101) channel orientation are obtained [288].

A remarkable progress in the utilization of microwave heating was achieved by Yang et al. in the past few years [25,289,290] who developed the "*in situ* aging – microwave synthesis" method (AM method) [291,292]. In a "first-stage synthesis," the polished, ultrasonicated, and calcined support is contacted with a clear solution synthesis mixture. The gel layer formed is aged *in situ* in an air conditioned oven. Then the membranes are microwave treated for crystallization. This process is repeated in a so-called second-stage synthesis (Fig. 36a). This way, LTA membranes could be synthesized with high reproducibility. The "*in situ* aging" step was found to be necessary for the subsequent successful microwave synthesis. The LTA membrane consists of spherical grains without well-developed crystal faces. This procedure takes into account that the support does not absorb microwaves and remains unheated but the microwaves selectively couple with the gel layer because of its higher dielectric loss factor. The gel layer first formed on the support after *in situ* aging, contains plenty of pre-nuclei. During the following microwave heating, these pre-nuclei rapidly and simultaneously develop into crystal nuclei. Then, crystal growth proceeds

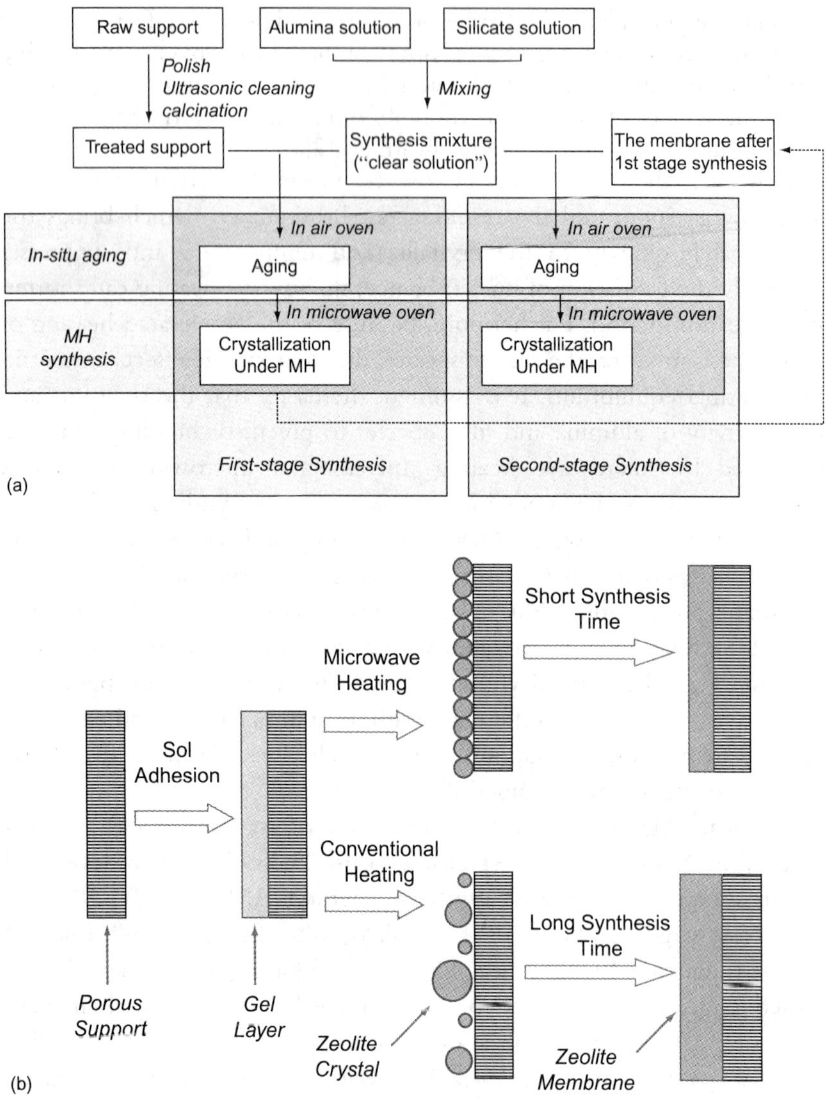

Figure 36 (a) Illustration of the "*in situ* aging – microwave synthesis method" (after Refs. [291,292]). MH stands for *M*icrowave *H*eating. (b) Comparative synthesis model of zeolite membrane preparation by microwave and conventional heating [389].

by propagation through the amorphous primary particles (size of approximately 50 nm) and, finally these particles transform into LTA crystals of about the same size. In this way, compact LTA zeolite membranes of spherical grains with undefined crystal facets are formed.

A recent review reflects the state of the art of microwave synthesis of zeolite membranes [389]. Whereas most zeolite membrane preparations deal with the MFI (silicalite-1, ZSM-5) structure, the majority of the microwave heating membrane preparations are focussed on LTA membranes. The general concept is to shorten the zeolite crystallization time so as to reduce the membrane thickness and to improve the flux (Fig. 36b). Pre-seeding the support with nano-LTA was needed to overcome the nucleation-related bottle-neck [390,391]. Compared to conventional heating, the synthesis time was shortened by 8–12 times by using microwaves and the Permeance was increased by 4 times while keeping comparable permeselectivity for H_2/n-C_4H_{10} [392,393]. To further improve the permeance of LTA membranes, the macroporous alumina support was covered with a thin mesoporous top-layer to prevent the penetration of the reagent into the support [394]. However, despite the remarkable progress in the LTA membrane synthesis, the permselectivities of the membranes are so far only slightly superior over the Knudsen separation factor.

5.2 Use of intergrowth supporting substances

The International Zeolite Association (IZA) data base contains more than 152 different zeolite structures [22]. It is estimated that for approximately 15 structures (Table 1) the preparation of a zeolite membrane has been tried. It was found experimentally that only the high-silica types show a real shape-selective separation behavior, especially silicalite-1 (as the Al-free MFI structure), and the DDR type. Most progress in the development of molecular sieve membranes was achieved, therefore, for silicalite-1 membranes since their preparation is relatively easy, and these highly siliceous zeolite membranes provide chemical stability and allow oxidative regeneration. On the contrary, when the high Al-containing zeolite membranes such as LTA and FAU are tested in shape-selective gas or steam permeation, usually Knudsen separation pattern is found, which indicates a high contribution of defect meso- and macropores to the mass transport.[14] For a ZSM-5 membrane series with systematically increasing Al-content, it was found that the intercrystalline defect transport is enhanced (both mixture separation factors and permselectivities decrease with increasing

[14]The successful application of LTA membranes in the dehydration of alcohols is primarily based on differences in the mixture adsorption behavior rather than a molecular-sieving effect. In a recent paper (P.J. Feibelman, *Langmuir*, 20 (2004) 1239), the experimental finding is reported that water becomes extremely immobilized in narrow pores of oxide materials because of strong hydrophilic interactions with the oxide surface. This phase of "frozen" water could effectively block mesopores in LTA membranes thus increasing their selectivity. Water-loaded LTA membranes can tolerate, therefore, a certain concentration of defects.

Al-content) and high residual nitrogen permeances are found in permporosimetry. It seems to be a general problem, therefore, to crystallize thin defect-free Al-containing zeolite membrane layers as shown by recent papers [93,126].

Searching for the reasons for this behavior, an increase of the negative surface charge (zeta potential) of zeolite crystals with enhanced Al-content was found, which is independent on the structure type. Since the zeolite precursors in the synthesis solution are negatively charged like the growing zeolite layer, it is assumed that a hindered diffusive transport and attachment of the precursors into narrow slits between growing crystals is hindered. In the case of narrow distances between the crystals, the negative surface charges overlap and block the diffusive transport of the negatively charged silicate species. This mechanism seems to cause the poor intergrowth of the Al-containing crystals to a continuous tight membrane layer. By use of ISS, the crystal surface can be re-charged and the crystal intergrowth is improved. The strong negative surface charge can be indeed compensated by adsorption of an ISS (Fig. 37). Suitable ISS are small positively charged molecules, stable under the alkaline conditions during the membrane

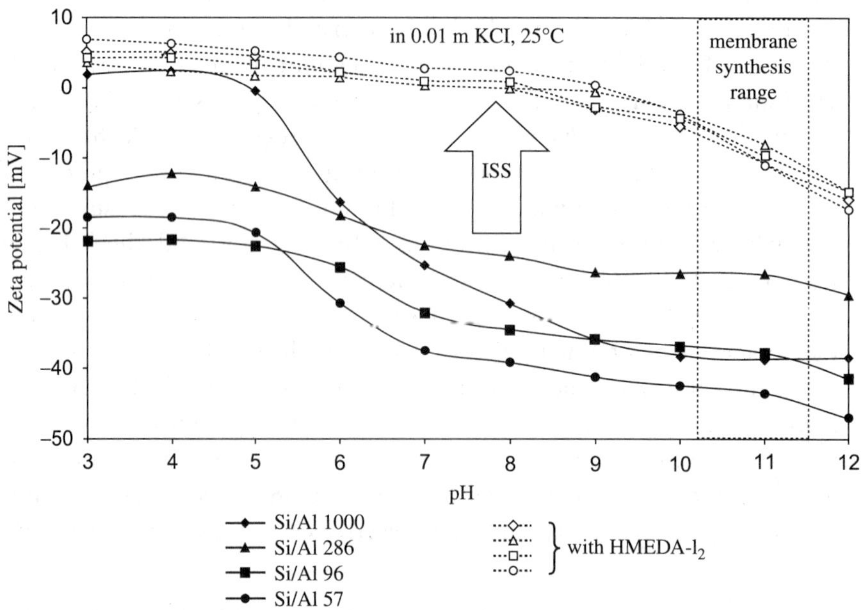

Figure 37 Zeta potentials of suspended MFI crystals of different Si/Al-ratios at room temperature. After addition of hexamethyl ethylene diammonium di-jodide (HMEDA-J_2) (0.01 m in the electrolyte) as an ISS, the zeta potentials become less negative [92].

Table 11 Chemical structures and abbreviations of possible ISS types [92]

Cationic molecule structure	Name	Acronym
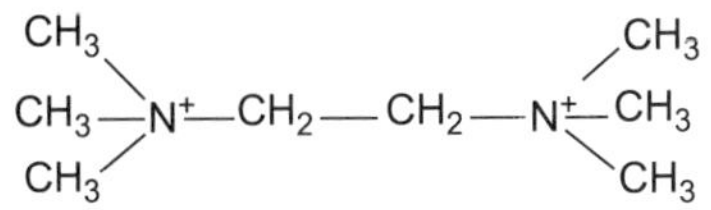	N,N,N,N′,N′,N′-hexamethyl ethylen diammonium diiodid	HMEDA-I_2
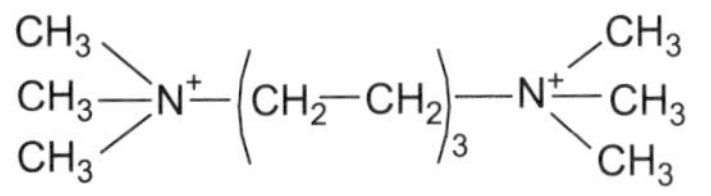	N,N,N, N′,N′,N′-hexamethyl hexylen diammonium diiodid	HMHDA-I_2
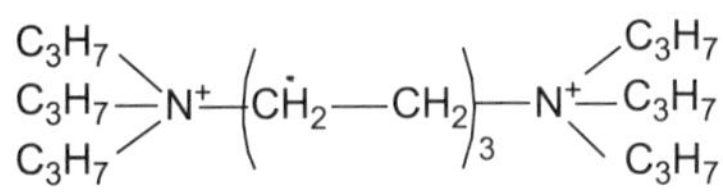	N,N,N,N′,N′,N′-hexapropyl hexylen diammonium diiodid	HPHDA-I_2
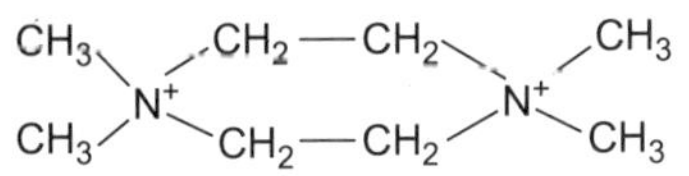	N,N,N′,N′-tetramethyl diethylen diammonium diiodid	TMDEDA-I_2
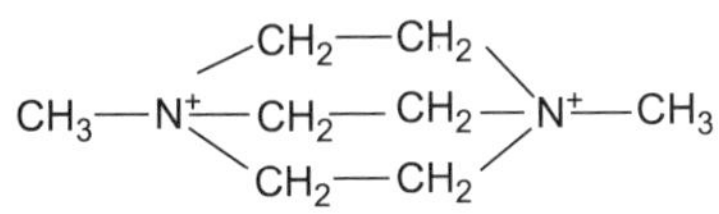	N,N′-Dimethyl triethylen diammonium diiodid	DMTEDA-I_2

synthesis (e.g., at 180 °C in the case of MFI membrane crystallization) and can be decomposed by calcination. Several ISS have been evaluated in MFI membrane preparation (Table 11).

After evaluating different ISS (Table 11) and determining their optimum concentration range [92], the effect of using ISS on the membrane quality was studied for HMEDA-I_2 in the preparation of MFI membranes of different Si/Al-ratio (Table 12).

The concept of enhanced crystal intergrowth by using an ISS is an effective tool for decreasing the intercrystalline defect transport thus increasing the selectivity of Al-containing MFI membranes. The effect of the ISS molecules to improve crystal intergrowth decreases in the order $TMDEDA^{2+} > HMEDA^{2+} > DMTEDA^{2+} > HMHDA^{2+} \gg HPHDA^{2+}$.

Table 12 Increase of the permselectivities (PS) derived for different gas mixtures from the corresponding single gas permeances at 105 °C for two MFI membranes when HMEDA-I_2 is used as ISS [92]

Si/Al	Permselectivity PS					
	H_2/n-butane		H_2/i-butane		H_2/SF_6	
	without ISS	with ISS	without ISS	with ISS	without ISS	with ISS
57	2.1	2.3	49	98.5	6.4	12.1
96	2.9	562	2.6	68.2	5.0	128.6

By the use of 0.1 m ISS solution, the zeta potential is changed to values close to the isoelectric point (IEP) but at this high ISS concentration, ISS molecules become incorporated during the membrane synthesis and create mesopores upon thermal decomposition of the ISS during calcination. The optimal ISS concentration is found to be 0.01 m for $HMEDA^{2+}$. So the improvement of the permeation properties when using an ISS varies and depends on the Al-content.

For Si/Al > 200:	nearly no ISS effect, already good permeation properties without ISS
200 > Si/Al > 96:	strong ISS effect, improved selectivities
96 > Si/Al > 57:	nearly no ISS effect

This ISS concept was first developed for Al-containing MFI membranes (ZSM-5) and later successfully transformed to the synthesis of LTA and FAU membranes. LTA and FAU membranes can separate water/organic mixtures in an excellent way but they fail in shape-selective gas separations. Therefore, many attempts were made to improve the separation properties of LTA and FAU membranes for gases. Zeta potential measurements on the Al-rich crystals of zeolites LTA and FAU also show a strong negative surface charge like it was found for Al-rich MFI crystals (ZSM-5). By adsorption of an ISS this negative zeta potential can be shifted close to the IEP, which improves the intergrowth of the seed crystals on the support to a continuous membrane layer. This improvement of the LTA and FAU membrane quality can be concluded from permporosimetry measurements (Fig. 38). By using an ISS, an improvement of the permeation selectivity of LTA and FAU membranes was found (Table 13). Nevertheless, the LTA

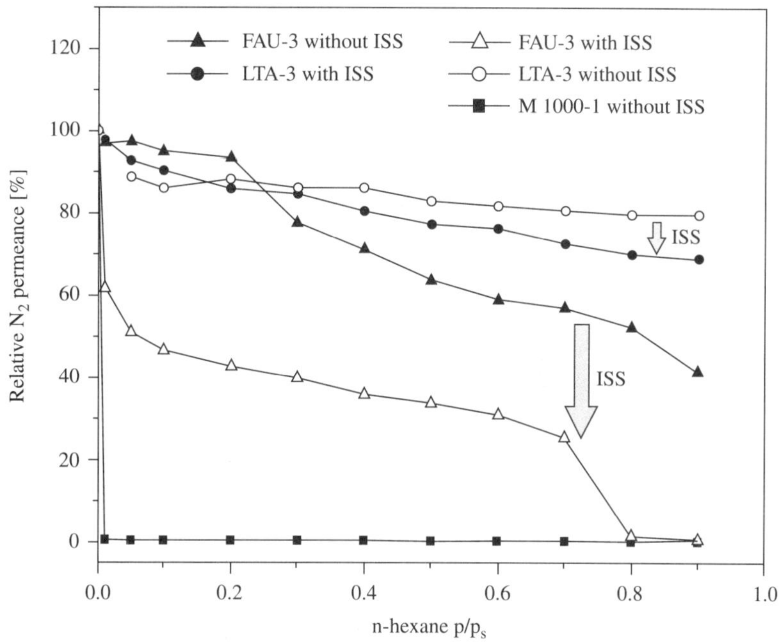

Figure 38 Improvement of the LTA and FAU membrane quality by using an ISS as measured by permporosimetry [91]. The arrows indicate the shift of the residual N_2 flux as a measure for reduced defect formation when an ISS is used. Meaning of the abbreviations: FAU-3 and LTA-3 denote three-layer FAU- and LTA-type membranes obtained by repeating three times the membrane synthesis. M 1000-1 denotes a one-layer MFI-type membrane with Si/Al-ratio of approximately 1000.

and FAU membranes prepared with ISS are still far from being defect-free and their permselectivities are in the range of the Knudsen Factor.

Improvement of the LTA and FAU membrane quality by using an ISS in the membrane synthesis is shown in Fig. 38 by permporosimetry studies [91]. The arrows indicate the shift of the residual N_2 flux as a measure for reduced defect formation when an ISS is used. Meaning of the abbreviations: FAU-3 and LTA-3 denote three-layer FAU- and LTA-type membranes obtained by repeating three times the membrane synthesis. M 1000-1 denotes a one-layer MFI-type membrane with Si/Al-ratio of approximately 1000.

5.3 Growth of oriented zeolite layers on supports

In Sections 2.2 and 2.3 it was shown that both *in situ* growth as well as secondary growth can give different orientations of the zeolite layer. By seeding of the support, MFI membrane layers of different crystallographic

Table 13 Comparison of the permselectivities (PS) derived from single gas permeances at 105 °C for FAU and LTA type membranes synthesized with and without HMEDA-I_2 as ISS [91]

Type	Permselectivity (PS)							
	H_2/CH_4		H_2/i n-butane		H_2/i-butane		H_2/SF_6	
	Without ISS	With ISS	Without ISS	With ISS	Without ISS	With ISS	Without ISS	With ISS
FAU	2.0	2.0	2.3	3.7	2.4	3.7	4.6	5.7
LTA	1.8	2.2	2.3	3.3	2.3	3.3	3.7	5.5

orientation can in principle be obtained, but most often the MFI-type zeolite membranes show a crystallographic orientation of the *c*-axis of the zeolite layer perpendicular to the plane of the support surface [54,68], as described in Section 2.3. The *c*-orientation can be explained by the competitive growth model [68]. Usually, the nanocrystallites used as seeds do not show developed crystal faces, and, therefore, these crystallites are randomly oriented. If crystal growth is anisotropic, those crystallites with their fastest growth direction pointing away from the seeded surface will grow faster than crystallites in other orientations. Finally, the crystals with the fastest growth direction perpendicular to the plane of the membrane will dominate. For MFI crystals, usually the *c*-axis is the longest dimension, and, consequently, the *c*-axis is the fastest growth direction. Therefore, most MFI-membranes are *c*-oriented with a columnar structure as shown in Fig. 5. Under certain growth conditions, other crystallographic orientations were observed like *a*-orientation [293,294], *b*-orientation [295,296], or intermediate orientations [152,297].

From studies on the diffusion anisotropy of the MFI structure, it can be expected that permeation through *c*-oriented MFI membranes perpendicular to the support is less favorable [40]. A *b*-oriented MFI layer is expected to exhibit higher fluxes. Recently, Tsapatsis et al. [54] have prepared a *b*-oriented MFI silicalite-1 membrane. They used relative large seeds ($0.5 \times 0.2 \times 0.1\ \mu m^3$) with developed crystal faces and attached the seeds as a *b*-oriented mono-seed layer to the support surface. By using di- and trimers of tetrapropyl ammonium hydroxide (TPAOH), the growth of the *b*-oriented seeds in *b*-direction could be enhanced. The resulting polycrystalline silicalite-1 films are approximately 1 μm thin and consist of large *b*-oriented single crystals with straight channels in the direction of the thickness of the membrane. This very careful membrane preparation results in a superior separation performance, which was demonstrated in the separation of xylene isomers (Fig. 39). The development of high-flux and high-selectivity MFI membranes for xylene separation by the Tsapatsis group [54] shows the importance of channel orientation and the significant influence of a seeded growth of an oriented particle monolayer. Whereas a *c*-orientation results in a separation factor $\alpha \approx 1$ (that is to say, there is no separation at all), the *b*-orientation gives an *o*/*p*-xylene separation factor $\alpha \approx 500$ at 200 °C. It is interesting to note that in the latter case the separation factor increases with increasing temperature. This experimental finding is characteristic for the interplay of adsorption and diffusion effects. At low temperatures, the zeolite pores are filled to a certain degree and *single file*-like behavior is observed. That is to say that the more mobile *p*-xylene

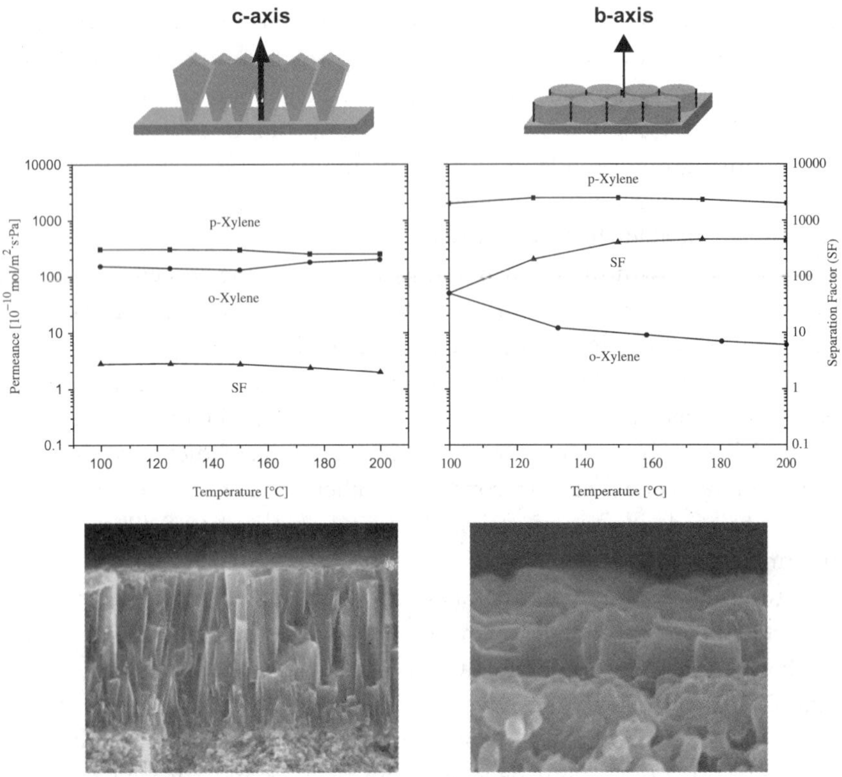

Figure 39 Supported silicalite-1 membrane in the separation of *p*/*o*-xylene mixtures: Influence of the preparation mode and channel orientation on flux and selectivity [54].

isomer cannot move faster through the pore network than the less mobile *o*-xylene. This situation changes dramatically at lower pore filling, which is found at higher temperatures and/or lower partial pressures. Now, the mobile *p*-xylene can move more or less independently from the presence of *o*-xylene. The permeation experiments were carried out at very low loadings corresponding to a low total pressure of xylene ($p/p_s \approx 0.007$) due to a high content of inert gas in the feed stream and the high temperature.

The concept of Tsapatsis [298] was further developed and highly *b*-oriented and intergrown MFI films could be produced by carrying out secondary growth of *b*-oriented seed layers under hydrothermal conditions using trimeric tetrapropyl ammonium iodide as SDA (Fig. 40). To deposit the seeds in *b*-orientation, the stainless steel support had to be smoothed by using an intermediate layer of mesoporous silica. The MFI seed monolayer was covalently attached to the intermediate silica layer by using

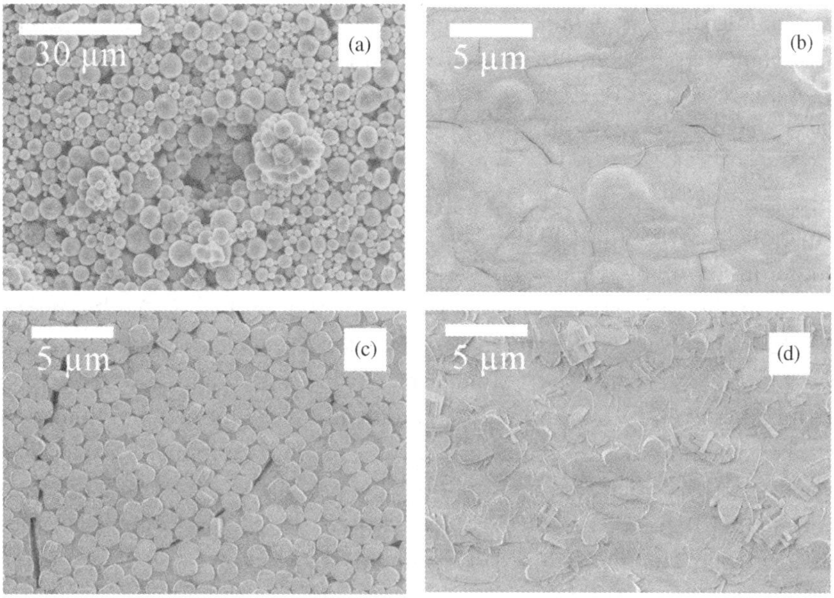

Figure 40 SEM of (a) surface of the stainless-steel support, (b) the support coated with mesoporous silica, (c) a MFI seed layer, and (d) the obtained *b*-oriented MFI film [298].

3-chloropropyl trimethoxy silane. XRD measurements showed the strong *b*-orientation of the seeds on the silica smoothed stainless-steel support. This *b*-orientation is preserved during secondary growth using trimeric tetrapropyl ammonium iodide as the SDA.

5.4 Bi-layer membranes

Different aims are followed when synthesizing multi-layer zeolite membranes:

1. Improved separation selectivity by repeated crystallization of one and the same zeolite type.
2. Novel properties by combination of layers of different zeolite types.
3. New fields of application by combination of zeolite layers with other inorganic membrane layers.

First, to improve the quality of MFI membranes, Vroon [48] proposed to repeat the crystallization step. Whereas a two-step growth was found to be beneficial for the quality of MFI membranes, further repetitions of the crystallization step did not improve the membrane quality since, as a result of the oxidative template removal, crack formation was observed for increased membrane thickness (cf. Section 2.2).

Second, different zeolite structure types have been investigated in zeolite membrane applications. In particular, MFI-, LTA-, and FAU-, but also BEA-, MOR-, FER-, OFF-, ANA-, CHA-, and ERI-type membranes have been studied [299]. On the contrary, chemical modifications have been made primarily for MFI-type membranes, for example, isomorphous substitution to give Al-, Fe-, B-, and Ge-ZSM-5 membranes [300,301], variation in Si/Al-ratios [93,302,303], and ion exchange [147,304]. These numerous possibilities allow a fine-tuning of the membrane characteristics to tackle many different liquid and gas separation problems. Multi-layered zeolite membranes with gradients of chemical composition or structure in the zeolite layers have the potential to expand the applications of zeolite membranes even further. Such membranes allow the intimate combination of different functions or characteristics in a single membrane, for example, catalytic activity/inertness, hydrophobic/hydrophilic character, and different pore sizes. Membranes that combine a catalytically active zeolite layer with an inert one are interesting for membrane reactors because they possess reactive and inert environments adjacent to each other, which is a prerequisite for staged reaction/non-reactive separation and/or for the passivation of non-shape-selective catalytic sites at the external surface [305–307]. Lai and Corcoran [306] patented the fabrication of multilayered zeolite membranes and demonstrated both seeded and epitaxial growth of ZSM-5 on silicalite-1 layers supported on porous alumina and stainless steel supports. They reported that the growth of ZSM-5 layers on calcined silicalite-1 layers led to partial erosion of the underlying silicalite-1 layer and that this erosion could be prevented when the silicalite-1 layer was not calcined prior to the synthesis of the second layer. Gora et al. [308] reported the seeded synthesis of silicalite-1 layers on top of zeolite LTA layers on porous Trumen supports by identifying conditions that allowed for the growth of the second layer without dissolution of the first one. The preparation of the opposite layer sequence was not successful. The crystallization of LTA and FAU layers on a silicalite-1 layer turned out to be more complicated because the high alkalinity of the LTA and FAU synthesis batches causes the dissolution of the silicalite-1 layer already formed [308]. Bi-layered ZSM-5/silicalite-1 films were also prepared on non-porous quartz and silicon substrates by Li et al. [309]. Recently, bi-layered silicalite-1/ZSM-5 membranes were synthesized and their permeation and separation properties were examined [310]. If the first layer on the support is silicalite-1 (shape-selective, hydrophobic), a second ZSM-5 layer (Bronsted acid sites, hydrophilic) can be crystallized onto the first one. Further synthesis work seems to be necessary for successful preparation of bi-layered membranes.

Third, there are ambitious attempts to combine zeolite layers with other inorganic membrane layers. As an example, for shape-selective oxidations, a thin silicalite-1 layer was crystallized on an oxygen-transporting perovskite membrane [311]. Assuming that a mixture of the xylene isomers would be in contact with this bi-layer membrane facing the Ti-modified silicalite-1 layer, mainly the *p*-xylene isomer would enter the silicalite-1 layer and could be oxidized to terephthalic acid with the oxygen released from the perovskite membrane.

5.5 Metal organic frameworks as molecular sieve membranes

Metal organic frameworks (MOFs) represent an interface between organic and inorganic compounds since they consist of metal ions linked together by organic molecules (ligands). MOFs comprise ionic inorganic–organic hybrid materials [312], especially coordination polymers based on bi- to tetravalent carboxylic acids [313–316]. This novel approach in preparing porous materials exhibits a rational and even more flexible design of the network compared to the already known inorganic materials. The first reports of MOFs in potential industrial processes have already been published [317]. Possible applications for MOFs are catalysis [318], gas purification, and gas storage, which needs information on the molecular transport in MOFs [319]. Despite the considerable attention devoted to these materials, only a handful MOFs with permanent porosity have been reported so far, in part because framework stability after template removal has emerged as a serious problem. Reports on functional aspects, especially the application as membranes or coatings are still rare [320–328]. Won et al. [320] embedded a Cu(II) complex in a polymer matrix yielding membranes with high H_2-selectivities and remarkable permeabilities. In another paper by Car et al. [321], the permselectivities and permeances could be slightly improved by incorporating different MOFs into the rubbery polydimethyl siloxane (PDMS) and the glassy polysulfone (PSf). In Ref. [328], it has been shown that the synthesis of supported MOF composite membranes based on manganese(II) formate ($Mn(HCO_2)_2$) is possible (Fig. 41). It was found that the amount of crystals grown on the supports and the orientation of the 1D channel system relative to the surface strongly depends on the selected support as well as on the synthesis route. Although membrane-type coatings could not yet be prepared, this latter study has identified some factors that seem to be important for producing continuous layered $Mn(HCO_2)_2$ membranes.

By testing various supports for the growth of a $Mn(HCO_2)_2$ membrane layer, it was found that best manganese (II) formate layers can be obtained

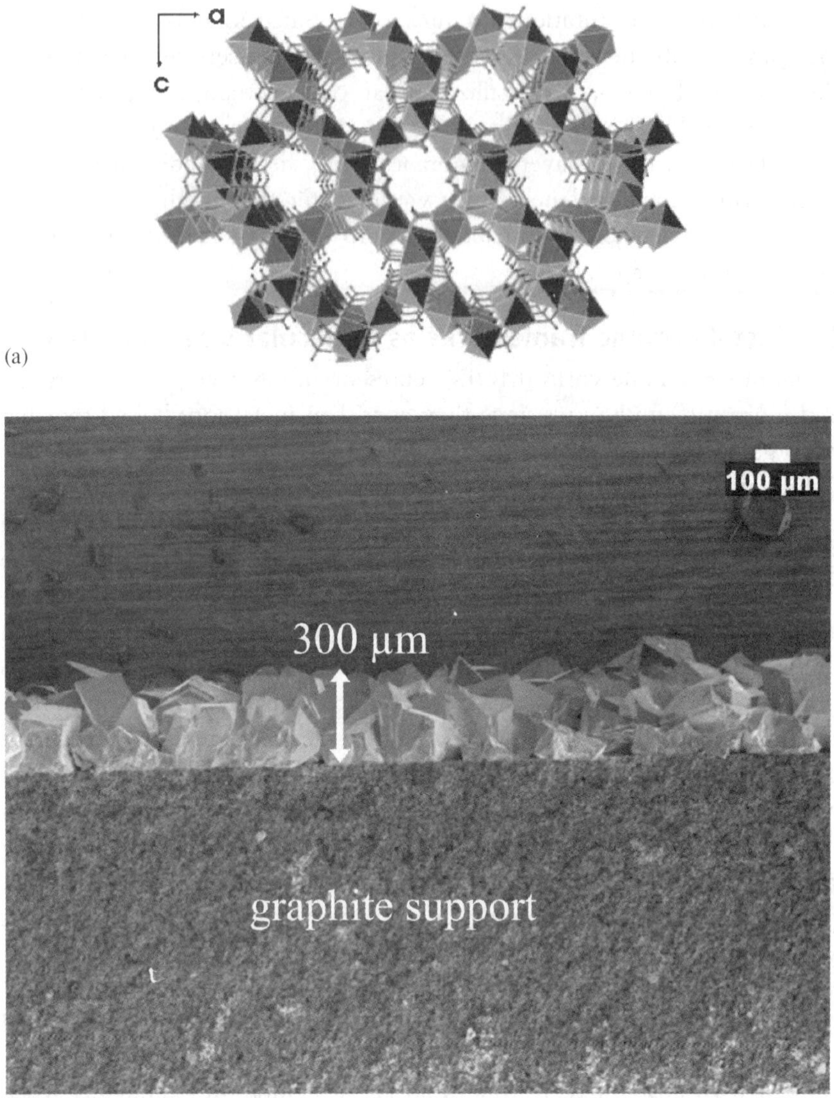

Figure 41 (a) Schematic drawing of the guest-free framework of the manganese (II) formate $Mn(HCO_2)_2$ along the *b*-axis (direction of the 1D pore system). Shown are corner- and edge-shared MnO_6 octahedra as well as the formate ligands, according to Ref. [329]. (b) Profile of $Mn(HCO_2)_2$ crystals grown on a graphite support disc after two *in situ* crystallizations according to Ref. [328,329].

on oxidized carbon supports using a seed technique route. A reasonable tilt angle of 34° of the 1D channel system of the manganese (II) formate membrane layer perpendicular to the support surface was found. Future work should focus on a detailed examination of surface charges during synthesis, and the obtained results can be applied in a direct modification of *in situ* synthesis and/or surface treatment to increase the amount of crystals on the supports. Finally, *in situ* crystallization of other promising MOFs for new molecular sieving membranes seems to be reasonable [320,330–332]. For instance, Pan et al. [333] published the synthesis of a porous lanthanum-containing MOF with a thermal stability up to 450 °C and with high selectivities and fast rates for H_2 adsorption. Although there have been a number of attempts to synthesize polycrystalline MOF layers on porous supports [328,334,335], only very few reports come up with a dense coating [336–339]. For membrane synthesis, not only the problems with growing a dense polycrystalline layer on porous ceramic or metal supports but also the thermal and chemical stability of a MOF have to be considered. Among the zeolitic imidazolate frameworks (ZIFs), a new subclass of MOFs, there are a number of members that exhibit exceptional thermal and chemical stability [340–344]. A prominent representative is ZIF-8 of formula $Zn(mim)_2$ (mim = 2-methylimidazolate), which crystallizes with a sodalite-related structure and is thermally stable up to 360°C [340–344]. Owing to a narrow size of the six-membered ring pores of the sodalite cage (~3.4 Å), it can be anticipated that a ZIF-8 membrane is capable to separate H2 (kinetic diameter ~2.9 Å) from larger gas molecules.

Figure 42 shows a crack-free, dense polycrystalline layer of ZIF-8 on a porous titania support [345]. The cross-section of the membrane shows a continuous and well intergrown layer of ZIF-8 crystals on top of the

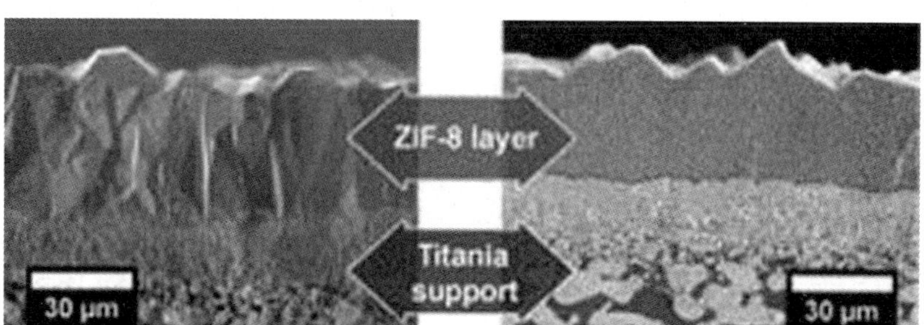

Figure 42 SEM image of the cross-section of the ZIF-8 membrane simply broken (left); EDXS-mapping of the sawn and polished ZIF-8 membrane (right), color code: orange, Zn; cyan, Ti [345].

support. Energy-dispersive X-ray spectroscopy (EDXS) reveals that there is a sharp transition between the ZIF-8 layer (Zn signal) and the titania support (Ti signal). A comparison of the X-ray diffraction (XRD) patterns of the ZIF-8 layer and the corresponding crystal powder sedimented in the course of membrane synthesis indicates that the membrane layer consists of randomly oriented crystallites.

The volumetric flow rates of the single gases H_2, CO_2, O_2, N_2, and CH_4 as well as a 1:1 mixture of H_2 and CH_4 through the membrane were measured using the Wicke-Kallenbach technique (Fig. 43). It can be seen that the permeances clearly depend on the molecular size of the gases. In addition, although the pore size of ZIF-8 is estimated from crystallographic data to be 0.34 nm, even larger molecules such as CH_4 (kinetic diameter ~0.38 nm) can - although only slowly - pass through the pore network, and consequently, there exists no sharp cut-off at 0.34 nm. This indicates that the framework structure of ZIF-8 is in fact more flexible rather than static in its nature, in accordance with recent findings by inelastic neutron scattering [346]. Comparison of the H_2 single gas permeance with its mixed gas ones reveals that there is only a small difference, meaning that the larger CH_4 molecules only slightly influence the permeation of the mobile H_2 molecules. This experimental finding is different to mixture diffusion in zeolites where an immobile component usually reduces the mobility of a co-adsorbed more mobile component. As an example, the presence of i-butane reduces the self-diffusivity of n-butane in MFI zeolites by orders of

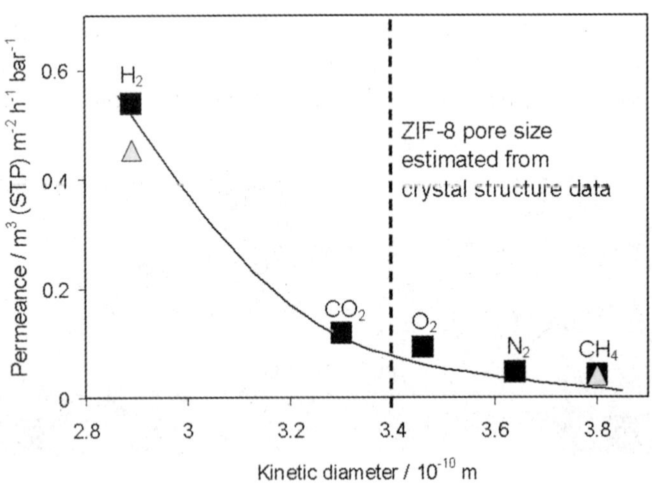

Figure 43 Single (squares) and mixed (triangle) gas permeances for a ZIF-8 membrane versus kinetic diameters of permeating probe molecules [345]

magnitude [347]. This observation can be understood when considering that the pore size of ZIF-8 is indeed narrow but the cages are rather large (~1.14 nm in diameter) and free of cations. Thus, although a CH_4 molecule may block the pore entrance for an H_2 molecule, as soon as it has entered the cage it does not restrict the H_2 diffusion any more. By the Wicke-Kallenbach permeation studies with gas chromatographic control of the 1:1 H_2/CH_4 mixture a mixture separation factor of $\alpha = 11.2$ (at 298 K and 1 bar) was determined. This value not only considerably exceeds the Knudsen separation factor for H_2/CH_4 (~2.8), it is as yet by far the highest H_2 separation factor reported for a MOF membrane. Only recently, a H_2/CH_4 separation factor of ~6 was reported by Guo et al. [339] for a supported $Cu_3(btc)_2$ membrane (btc = benzene-1,3,5-tricarboxylate).

5.6 Functional zeolite films

In addition to its use as separation membrane and catalytic membrane reactor, zeolite layers can act as functional film in chemical sensors, as electrode, as opto-electronic device or low dielectric constant material, as protection or insulation layer, as corrosion-resistant coatings [348], hydrophilic antimicrobial coatings [349], or sulfonated zeolite BEA for proton exchange membranes [350].

As an alternative method for crystallizing a zeolite layer in the liquid phase, by the so-called dry gel conversion, a dry (alumino) silicate gel can be converted into a zeolite layer in the presence of vapors [351]. The vapor phase can be only steam or a mixture of steam and a SDA such as TPAOH. In contrast to the conventional steam-assisted crystallization [53] in which the substrate is coated with all the nutrients and then steam-treated, by a novel steam-assisted method, the oxidized surface layer of a silicon wafer can be transformed into a silicalite-1 zeolite film [352,353]. A silicon wafer is coated with 1 M TPAOH and then zeolite films were crystallized under steam by adding some water to the autoclave. In this method, the silicon wafer serves both as support of the grown zeolite film and as the Si source for the formation of the silicalite-1 film. Pure silcalite-1 films with preferential *a*- and *b*-out-of-plane orientation were obtained in a temperature window from 100 to 200 °C (Fig. 44). This method seems to be very useful for zeolite film applications as chemical sensor or for optoelectronics.

Porous pure-silica materials are attractive as insulator material in on-chip interconnects due to their high porosity, hydrophobicity, acceptable heat conductivity, and low dielectric constant. A remarkable improvement could be achieved by the UV treatment of spin-on silicalite-1 films to

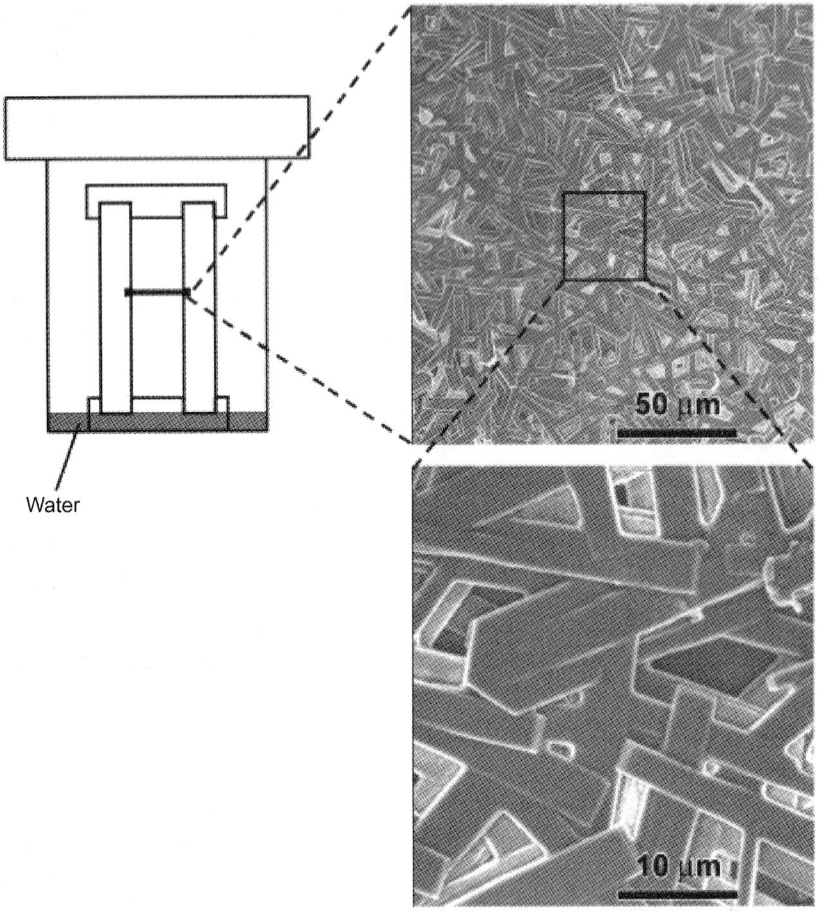

Figure 44 FE-SEM image of an *a*-, *b*-oriented silicalite-1 film obtained by steaming a SDA-oxidized silicon wafer [352].

induce hydrophobization [354]. In this post-treatment method during the removal of the organic template in combination with a thermal treatment, UV radiation decreases drastically the quantity of silanols. Parallel, methylation of the silica surface is obtained by decomposition and reaction of the TPA ions as SDA. By this method, the formation of cracks during the removal of the organic template is minimized. Pure-silica zeolites have a remarkably higher mechanical strength and hydrophobicity than amorphous porous silicas due to their crystalline structure making them a likely dielectric material for enabling smaller feature sizes in future generation of microprocessors [355].

Zeolite nanocrystals can be coated on substrates to form a transparent film with approximately 20 nm surface roughness. The resulting coating showed a broadband anti-reflection effect with less than 1% average reflection over the visible range. With proper control of the film thickness, one can shift the reflection minimum to achieve a neutral color [356]. This broadband neutral color anti-reflection coating was achieved in a single step sol-gel process and can find application in the display industry.

Last but not least, high-silica zeolite coating on metals and metal alloys can be a promising technology for corrosion protection of metals [357]. The as-synthesized SDA-containing MFI films are non-porous and allow the coating of complex shapes and in confined spaces by *in situ* crystallization [358].

5.7 Mixed matrix membranes

Soon after the first preparation of synthetic zeolites, the idea was born to incorporate zeolite crystals as modifier into a polymer matrix thus using the easy processing of polymers [359–361]. Mixed matrix membranes are interesting systems to enhance the properties of the host matrix taking advantage of the peculiar properties of specific inorganic fillers [362,363]. Today a renaissance of this concept can be observed [364–370]. Current polymeric membranes seem to have reached a limit in the trade-off between permeability and selectivity. Therefore, research efforts have been focused on mixed matrix membranes that contain porous (nano) particles [371] in a polymeric matrix and that can be processed by the usual spinning technology [372–374]. By surface treatment of the inorganic modifiers [375] and by polymer chain rigidification [376], mixed matrix membranes with improved separation patterns could be obtained.

Various especially small-pore zeolites have been used for their application in mixed matrix membranes. Although early studies were focused on LTA zeolite, this hydrophilic zeolite turned out to be less attractive for gas separation from humid feeds because of the pore blocking by water [377]. Recent studies have focused, therefore, on more hydrophobic zeolites with a high molar SiO_2/Al_2O_3 ratio. As an example, zeolite CHA with an SiO_2/Al_2O_3 ratio >30 with a particle size $<1\ \mu m$ can be processed to mixed matrix membrane layers of approximately 1 μm thickness [378].

However, only low selectivities were found for zeolite/polymer mixed matrix membranes, mainly because of the poor wetting [379,380]. This interfacial problem does not occur for mesoporous MCM-41 in PSf [381] and mesoporous ZSM-5 in Matrimid [382], which suggests that the polymer chains can penetrate into the mesopores. As a result of this penetration, both the glass temperature T_g and the Young's modulus of the mixed-matrix are higher than those of the pure polymers. In comparison with the pure polymers, the mixed-matrix membranes show substantially increased permeability and permselectivity.

5.7.1 Suppression of stress due to irregular expansion/shrinking during membrane activation

Whereas the hydrophilic membranes of type LTA and FAU do an excellent job in water separation, they fail in shape-selective separations. From *in situ* XRD studies there were indications from literature that during the drying of zeolite powders of the types LTA [383], FAU [384], or MOR [385,386], there occur extreme irregular expansions/shrinkages of the unit cell. Recently, the change of the unit cell (u.c.) of LTA and FAU crystals was studied by heating wet zeolite samples and measuring the change of the u.c. by *in situ* XRD [387]. For LTA an extreme shrinking of the u.c. and for FAU an extreme expansion of the u.c. as a result of the de-watering was observed for a temperature range between room temperature and 100 °C. So hydrated LTA shrinks from 2.460 nm at 50 °C to 2445 nm at 100 °C, whereas hydrated FAU expands due to the de-watering when heated in air from 2.478 nm at 50 °C to 2.490 nm at 100 °C. In contrast to the drastic changes due to the de-watering, only slight changes of the u.c. were found when the dried LTA and FAU crystals were heated and cooled, respectively. This extreme shrinkage/expansion behavior was only found for the hydrophilic zeolites like LTA and FAU. More hydrophobic ones such as MFI or MOR did not show this behavior. As Fig. 45 shows, the amount of expansion and shrinkage, respectively, can be controlled by the heating rate.

6. OUTLOOK

There is an impressive progress in the development of zeolite membranes during the past decade. The technologies are now available to prepare zeolite membranes of sufficient quality and reliability. In the near future, in a hard competition with other separation techniques, the exploitation of hydrophilic zeolite membranes for the de-watering will go on, and there will be first attempts to use small pore membranes such as zeolites for the separation of small molecules such as hydrogen. The main field of application for zeolite membranes is believed to be the shape-selective separation of C_4 to C_8 hydrocarbon isomers because no other separation technique for this task is available. On a medium time scale, therefore, zeolite membranes will be developed, which can do the unique job that no

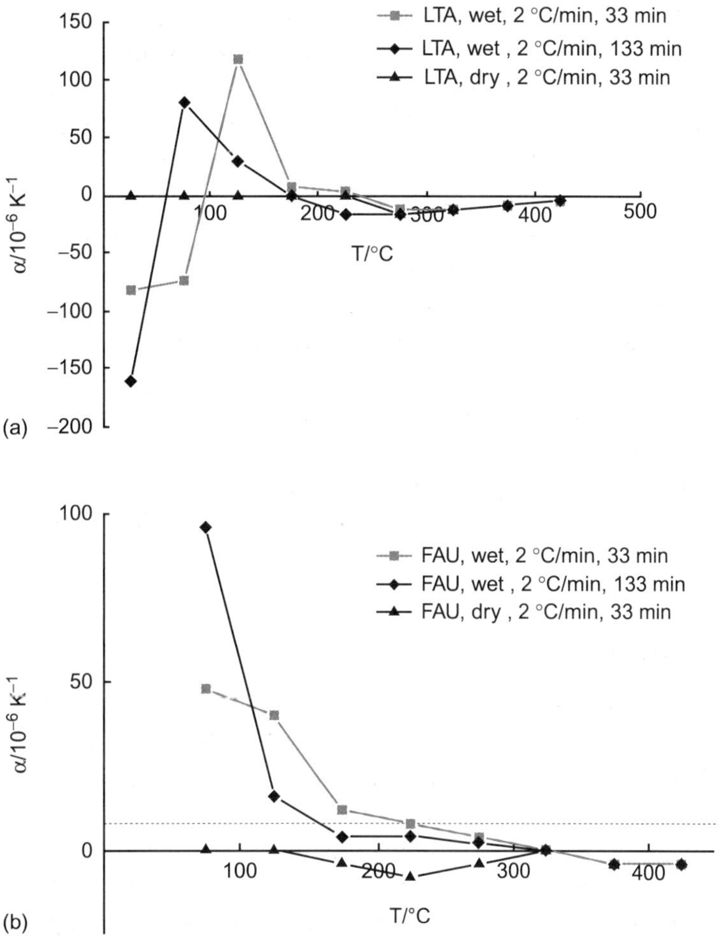

Figure 45 Change of the expansion coefficient during drying wet zeolites LTA (a) and FAU (b) in an *in situ* XR diffractometer [387]. Notation: The powder samples were heated with a rate of 2 °C/min and at every 50 °C the XRD were recorded. The time necessary for taking the XRD was 133 min (complete data set for Rietveld refinement at every 50 °C step or 33 min (reduced data set for Rietveld)). The dashed line indicates the temperature-independent thermal expansion coefficient of α-Al_2O_3 ($\approx 8 \times 10^{-6} K^{-1}$).

other membrane can do: molecular sieving of molecules of almost identical or similar mass but different size and shape.

Gas separation is highly competitive both within the membrane field itself and with other gas separation technologies. The key properties of a membrane process are flux, selectivity, processability, stability, and costs. The current rise in energy costs makes membrane separations – which can be generally low in costs – more attractive. However, at present there is no large-scale gas-separation based on inorganic membranes in industrial operation, but there are promising ongoing developments using Pd alloy-based membranes for hydrogen separation, perovskite-like mixed ionic and electronic conducting ceramics for oxygen separation from air and last but not least zeolite membranes for shape-selective separations. Porous sol-gel-derived X-ray amorphous metal oxides and carbon membranes are also promising candidates for an industrial use on a time scale of 5 and 10 years ahead. A huge impact on inorganic membrane R&D would be a successful application of inorganic membranes in such important environmental and energy-related large processes as the cost-effective purification of hydrogen and methane [388].

After the successful realization of an industrial separation process using zeolite membranes, the development of a (catalytic) membrane reactor becomes possible. Most probably, here again the shape-selective separation behavior of zeolite membranes will be exploited, which recommends the application of an extractor-type membrane reactor.

ACKNOWLEDGMENT

J.C. thanks Deutsche Forschungsgemeinschaft and the European Union for financing the project Ca 147/10-1 and the Network of Excellence InsidePores, respectively. M.N. thanks the Federal Ministry of Education, Science, Research and Technology of Germany, the Senate of Berlin, Department of Science, Research and Culture and the European Union, EFRE 2000 2005 1/0 for financial support of project no. 03C3014.

We thank our wives Marion and Uta for their patience to give us the time to write this contribution.

REFERENCES

[1] E.G. Derouane, CATTECH 5 (2001) 214.
[2] R. Singh, CHEMTECH (1998) 33.
[3] A. Tavolaro, E. Drioli, Adv. Mater. 11 (1999) 975.
[4] T. Bein, Chem. Mater. 8 (1996) 1636.
[5] J. Caro, M. Noack, P. Kölsch, R. Schäfer, Micropor. Mesopor. Mater. 38 (2000) 3.
[6] E.E. McLeary, J.C. Jansen, F. Kapteijn, Micropor. Mesopor. Mater. 90 (2006) 198.
[7] M. Noack, J. Caro, in: F. Schüth, K.S.W. Sing, J. Weitkamp (eds.), Handbook of Porous Solids, Wiley-VCH, Weinheim, 2002, p. 2478.
[8] J. Caro, M. Noack, P. Kölsch, Adsorption 11 (2005) 215.
[9] J.D.F. Ramsay, S. Kallus, in: N.K. Kanellopoulos (ed.), Recent Advances in Gas Separation by Microporous Ceramic Membranes, Membr. Sci. and Techn. Ser. 6,, Elsevier, Amsterdam, 2000, p. 373.

[10] A. Julbe, Zeolite Membranes – A Short Overview, in: J. Cejka, H. van Bekkum (eds.), Zeolites and Ordered Mesoporous Materials: Progress and Prospects, Stud. Surf. Sci. Catal., vol. 157 (2005) 135 pp.
[11] R.M. Barrer, J. Chem. Soc. Lond. Faraday Trans. 86 (1990) 1123.
[12] R.M. Barrer, in: G. Öhlmann, H. Pfeifer, R. Fricke (eds.), Stud. Surf. Sci. Catal. 65 (1990), 257.
[13] R.M. Barrer, Langmuir 3 (1987) 309.
[14] R.M. Barrer, J. Chem. Soc. Lond. Faraday Trans. 88 (1992) 1463.
[15] R.M. Barrer, Activated diffusion through silica glass, J. Chem. Soc. (1934) 378.
[16] R.M. Barrer, R. Ash, A. Vernon, J. Edge, T. Foley, Thermo-osmosis of sorbable gases in porous media, J. Membr. Sci. 125 (1997) 41.
[17] Y. Morigami, M. Kondo, J. Abe, H. Kita, K. Okamoto, Sep. Purif. Technol. 25 (2001) 251.
[18] J. Caro, Stud. Surf. Sci. Catal. 154 (2004) 154.
[19] G. Sumelius, Attempted translation of the original old-Swedish paper by Cronsted Kongl. Svenska Vetenskaps Academiens Handlingar 17 (1756) 120, in: M.L. Ocelli, H. Robson (eds.), Synthesis of Microporous Materials, Vol. I Molecular Sieves, Van Nostrand Reinhold, New York, 1992, p. 1.
[20] F. Kapteijn, 9th International Conference on Inorganic Membranes, Proceedings of the Workshop on Zeolite Membranes, Lillehammer, Norway, 2006.
[21] H. Robson (ed.) Verified Syntheses of Zeolitic Materials, Micropor. Mesopor. Mater. 22 (1988).
[22] Homepage of the Int. Zeolite Assoc., http://www.iza-online.org
[23] H. Robson, IZA-commission, verified syntheses of zeolitic materials, Elsevier Sciences, 20010-444-50703-5.
[24] X. Xu, Y. Bao, C. Song, W. Yang, J. Liu, L. Lin, J. Membr. Sci. 249 (2005) 51.
[25] X. Chen, W. Yang, J. Liu, L. Lin, J. Membr. Sci. 255 (2005) 201.
[26] M. Lassinanti, J. Hedlund, J. Sterte, Micropor. Mesopor. Mater. 38 (2000) 25.
[27] H. Kita, M. Kondo, N. Miyake, Y. Matsua, US Patent 5,554,286, 1996.
[28] H. Kita, K. Fuchida, T. Horida, H. Asamura, K. Okamoto, Sep. Purific. Technol. 25 (2001) 261.
[29] B. Jeong, Y. Hasegawa, K. Sotowa, K. Kusakabe, S. Morooka, J. Chem. Eng. Jap. 35(2) (2002) 167.
[30] K. Weh, M. Noack, I. Sieber, J. Caro, Microp. Mesop. Mater. 54 (2002) 27.
[31] V. Nikolakis, G. Xomeritakis, A. Abibi, M. Dickson, M. Tsapatsis, D. Vlachos, J. Membr. Sci. 184 (2001) 209.
[32] J. Kumakiri, T. Yamaguchi, S. Nakao, Ind. Eng. Chem. Res. 38 (1999) 4282.
[33] S. Mintova, N.H. Olson, V. Valtchev, Th. Bein, Science 283 (1999) 958.
[34] S. Mintova, N.H. Olson, Th. Bein, Angew. Chem. 111(21) (1999) 3405.
[35] J.C. Jansen, W. Nugroho, H. van Bekkum, in: H.B. Higgins, R. von Ballmoos, M.M.J. Treacy (eds.), Proceedings of the 9th International Conference on Zeolite Conference, Butterworth-Heinemann, Montreal, 1993, p. 247.
[36] J.H. Koegler, A. Arafat, H. van Bekkum, J.C. Jansen, in: H. Chong, S.-K. Ihm, Y.S. Uh (eds.), Stud. Surf. Sci. Catal. 105 (1997) 2163.
[37] E.R. Geus, H. van Bekkum, W.J.W. Bakker, J.A. Moulijn, Micropor. Mater. 1 (1993) 131.
[38] J.C. Jansen, G.M. Rosmalen, J. Cryst. Growth 128 (1993) 1150.
[39] F. Kapteijn, W.J.W. Bakker, J. van de Graaf, G. Zheng, J. Poppe, J.A. Moulijn, Catal. Today 25 (1995) 213.
[40] J. Caro, M. Noack, F. Marlow, D. Petersohn, M. Griepentrog, J. Kornatowski, J. Phys. Chem. 97 (1993) 13685.
[41] J.C. Jansen, D. Kashchiev, A. Erdem-Senetalar, in: J.C. Jansen, M. Stöcker, J. Weitkamp, H.G. Karge (eds.), Stud. Surf. Sci. Catal. 85 (1994) 215.

[42] J.C. Jansen, 9th International Conference on Inorganic Membranes, Proceedings of the Workshop on Zeolite Membranes, Lillehammer, Norway, 2006.
[43] E. Piera, A. Giroir-Fendler, J.A. Dalmon, H. Moueddeb, J. Coronas, M. Mendenez, J. Santamaria, J. Membr. Sci. 142 (1998) 97.
[44] J. Coronas, J. Santamaria, Separ. Purif. Methods 28 (1999) 127.
[45] D. Uzio, J. Peureux, A. Giroir-Fendler, J.A. Dalmon, J.D.F. Ramsay, Stud. Surf. Sci. Catal. 87 (1994) 411.
[46] J. Coronas, J.L. Falconer, R.D. Noble, Ind. Eng. Chem. Res. 37 (1998) 166.
[47] Y. Yan, M.E. Davis, G.R. Gavalas, J. Membr. Sci. 126 (1997) 53.
[48] Z.A.E.P. Vroon, K. Keizer, M.J. Gilde, H. Verweij, A.J. Burggraaf, J. Membr. Sci. (1996) 127.
[49] M. Tsapatsis, G.R. Gavalas, in: V.N. Burganos (ed.), Synthesis of Porous Inorganic Membranes, MRS Bulletin, 24 (1999) 30.
[50] M. Matsukata, M. Ogura, T. Osaki, P. Raja, H.P. Rao, M. Nomura, E. Kikuchi, Topics Catal. 9 (1999) 77.
[51] J. Dong, T. Dow, X. Zhao, L. Gao, J. Chem. Soc., Chem. Commun. 2 (1992) 1065.
[52] T. Matsufuji, N. Nishiyama, K. Ueyama, M. Matsukata, Catal. Today 56 (2000) 265.
[53] S. Alfaro, M. Arruebo, J. Coronas, M. Menendez, J. Santamaria, Micropor. Mesopor. Mater. 50 (2001) 195.
[54] Z. Lai, G. Bonilla, I. Diaz, J.G. Nery, K. Sujaoti, M.A. Amat, E. Kokkoli, O. Terasaki, R.W. Thompson, M. Tsapatsis, D.G. Vlachos, Science 100 (2003) 456.
[55] J. Hedlund, F. Jareman, A.J. Bons, M. Anthonis, J. Membr. Sci. 222 (2003) 163.
[56] J. Hedlund, J. Sterte, M. Anthonis, A.J. Bons, B. Carstensen, N. Corcoran, D. Cox, H. Deckman, W.D. Gijnst, P.P. de Moor, F. Lai, J. McHenry, W. Mortier, J. Reinoso, J. Peters, Micropor. Mesopor. Mater. 52 (2002) 179.
[57] G. Li, E. Kikuchi, M. Matsukata, Micropor. Mesopor. Mater. 62 (2003) 211.
[58] W.F. Lai, H.W. Deckman, J.A. McHenry, J.P. Verduijn, US Patent 5,871,650, filed on July 8, (1994).
[59] K. Horii, K. Tanaky, K. Kita, K. Okamoto, Proceedings of the 26th Autumn Meeting of Soc. Chem. Eng., Japan, 1993, 99.
[60] M.C. Lovallo, A. Gouzinis, M. Tsapatsis, AIChE J. 44 (1998) 1903.
[61] A. Gouzinis, M. Tsapatsis, Chem. Mater. 10 (1998) 2497.
[62] G. Xomeritakis, A. Gouzinis, S. Nair, T. Okubo, M. He, R.M. Overney, M. Tsapatsis, Chem. Eng. Sci. 54 (1999) 3521.
[63] J. Hedlund, B.J. Schoeman, J. Sterte, Stud. Surf. Sci. Catal. 105 (1997) 2203.
[64] S. Mintova, J. Hedlund, V. Valtchev, B. Schoeman, J. Sterte, Chem. Commun. 2 (1997) 15.
[65] S. Mintova, J. Hedlund, V. Valtchev, B. Schoeman, J. Sterte, J. Mater. Chem. 10 (1998) 2217.
[66] J. Hedlund, M. Noack, P. Kölsch, D. Creaser, J. Caro, J. Sterte, J. Membr. Sci. 159 (1999) 263.
[67] M. Noack, P. Kölsch, R. Schäfer, P. Toussaint, J. Caro, Micropor. Mesopor. Mater. 49 (2001) 25.
[68] W.C. Wong, L.T.Y. Au, C.T. Ariso, K.L. Yeung, J. Membr. Sci. 193 (2001) 141.
[69] A. Buekenhoudt, F. Servaes, K. Wyns, A. Aerts, J. Martens, P. Jacobs, Proceedings of the 9th International Conference on Inorganic Membranes, Lillehammer, Norway, June 25–29, 2006, p. 286.
[70] C. Kirschock, V. Buschmann, S. Kremer, R. Ravishankar, C. Houssin, B. Mojet, R. Van Santen, P. Grobet, P. Jacobs, J. Martens, Angew. Chemie Int. Ed. 40 (2001) 2637.
[71] T. Van Gestel, C. Vandecasteele, A. Buekenhoudt, C. Dotremont, J. Luyten, R. Leysen, B. Van der Bruggen, G. Maes, J. Membr. Sci. 207 (2002) 73.

[72] W.J. Koros, Y.H. Ma, T. Shimidzu., Terminology for membranes and membrane processes, IUPAC recommendations from 1996, http://www.che.utexas.edu/nams/IUPAC/iupac.html
[73] R. Ash, R.M. Barrer, C.G. Pope, Proc. Royal Soc. Lond. A. 271 (1963) 19.
[74] R. Ash, R.M. Barrer, R.T. Lowton, J. Chem. Soc. Farad. Trans. 1 (1973) 2166.
[75] M.G. Katz, G. Baruch, Desalination 58 (1986) 199.
[76] A. May-Marom, M.G. Katz, J. Membr. Sci. 27 (1986) 119.
[77] F.P. Cuperus, D. Bargeman, C.A. Smolders, J. Membr. Sci. 71 (1992) 57.
[78] G.Z. Cao, J. Meijerink, H.W. Brinkman, A.J. Burggraaf, J. Membr. Sci. 83 (1993) 221.
[79] G.R. Gallaher, P.K.T. Liu, J. Mater. Sci. 92 (1994) 29.
[80] C.L. Lin, D.L. Flowers, P.K.T. Liu, J. Membr. Sci. 92 (1994) 45.
[81] D.E. Fain, Proceedings of the 1st International Conference on Inorganic Membranes. Montpellier, France, p. 199 (1989).
[82] A. Julbe, J.D.F. Ramsay, in: A.J. Burggraaf, L. Cot (eds.), Fundamentals of Inorganic Membrane Science and Technology, Elsevier, 2000.
[83] T.A. Steriotis, K.L. Stefanopoulos, A.C. Mitropoulos, N.K. Kanellopoulos, in: N.K. Kanellopoulos (ed.), Recent Advances in Gas Separation by Microporous Ceramic Membranes, Elsevier, 2000.
[84] H.W. Deckman, et al., Int. Workshop Zeolitic and Microporous Membranes, Purmerend, The Netherlands, p. 9 (2001).
[85] J. Caro, in: F. Schüth, K.S.W. Sing, J. Weitkamp (eds.), Handbook of Porous Solids, Wiley-VCH, 2002, p. 363.
[86] K.G. Beltsios, Th.A. Steriotis, K.L. Stefanopoulos, N.K. Kanellopoulos, in: F. Schüth, K.S.W. Weitkamp, J. Weitkamp (eds.), Handbook of Porous Solids, Wiley-VCH, 2002, p. 2382.
[87] M. Yu, J.L. Falconer, T.J. Amundsen, M. Hong, R.D. Noble, Adv. Mater. 19 (2007) 3032.
[88] M. Yu, T.J. Amundsen, M. Hong, J.L. Falconer, R.D. Noble, J. Membr. Sci. 298 (2007) 182.
[89] K.S.W. Sing, Pure Appl. Chem. 54 (1982) 2201.
[90] M. Noack, J. Caro, 9th International Conference on Inorganic Membranes, Proceedings of the Workshop on Zeolite Membranes, Lillehammer, Norway, 2006.
[91] M. Noack, P. Kölsch, A. Dittmar, M. Stöhr, G. Georgi, M. Schneider, U. Dingerdissen, A. Feldhoff, J. Caro, Micropor. Mesopor. Mater. 102 (2007) 1.
[92] M. Noack, P. Kölsch, A. Dittmar, M. Stöhr, G. Georgi, R. Eckelt, J. Caro, Micropor. Mesopor. Mater. 97 (2006) 88.
[93] M. Noack, P. Kölsch, V. Seefeld, P. Toussaint, G. Georgi, J. Caro, Micropor. Mesopor. Mater. 79 (2005) 329.
[94] A.J. Burggaaf, in: A.J. Burggraaf, L. Cot (eds.), Fundamentals of Inorganic Membrane Science and Technology, Membrane Science and Technology Series 3, Elsevier, Amsterdam, 1996, p. 331.
[95] F. Kapteijn, J.M. van de Graaf, J.A. Moulijn, AIChE J. 46 (2000) 1096.
[96] J.M. van de Graaf, F. Kapteijn, J.A. Moulijn, AIChE J. 45 (1999) 497.
[97] H.P. Hsieh, Inorganic Membranes for Separation and Reaction, Membrane Science and Technology Series 3, Elsevier, Amsterdam, Lausanne, New York, Oxford, Shannon, Tokyo, 1996.
[98] J. Weitkamp, S. Ernst, L. Puppe, in: J. Weitkamp, L. Puppe (eds.), Catalysis and Zeolites – Fundamentals and Applications, Springer-Verlag, Berlin, Heidelberg, New York, 1999, p. 327.
[99] M. Noack, P. Kölsch, M. Schneider, P. Toussaint, I. Sieber, J. Caro, Microp. Mesop. Mater. 35/36 (2000) 253.

[100] D.W. Breck, Zeolite Molecular Sieves – Structure, Chemistry and Use, 2nd edn., R.E. Krieger, Malabar, Florida, 1984.
[101] J. Koresh, A. Soffer, J. Chem. Soc. Faraday Trans. 76 (1980) 2473.
[102] A. Soffer, J.E. Koresh, S. Saggy, US Patent 4,685,940, 1987.
[103] M. Noack, P. Kölsch, D. Venzke, P. Toissant, J. Caro, Microp. Mater. 3 (1994) 201.
[104] I. Girnus, M.-M. Pohl, J. Richter-Mendau, M. Schneider, M. Noack, J. Caro, Adv. Mat. 7 (1995) 711.
[105] R.M. de Vos, H. Verweij, Science 79 (1998) 1710.
[106] R.S.A. de Lange, J.H.A. Hekkink, K. Keizer, A.J. Burggraaf, J. Membr. Sci. 99 (1995) 57.
[107] S. Kita, M. Asaeda, J. Chem. Eng. Jpn. 23 (1990) 367.
[108] N.K. Raman, C.J. Brinker, J. Membr. Sci. 105 (1995) 273.
[109] R.M. de Vos, H. Verweij, J. Membr. Sci. 143 (1998) 3781.
[110] E.L. Wu, G.R. Landolt, A.W. Chester, in: Y. Murakami, A. Iijima, J.W. Ward, A. Kodansha (eds.), New Developments in Zeolite Science and Technology, Tokyo, Elsevier, Amsterdam, 1986.
[111] S. Fritzche, M. Wolfsberg, R. Haberlandt, P. Demontis, G.B. Suffritti, A. Tiloca, Chem. Phys. Lett. 296 (1999) 253.
[112] D.I. Kopelevich, H.C. Chang, J. Chem. Phys. 114 (2001) 3776.
[113] T.R. Forester, W. Smith, J. Chem. Soc. Faraday Trans. 93 (1997) 3249.
[114] A. Bouyermaouen, A. Bellemans, J. Chem. Phys. 108 (1998) 2170.
[115] W. Zhu, F. Kapteijn, J.A. Moulijn, M.C. den Exter, J.C. Jansen, Langmuir 16 (2000) 3322.
[116] L.A. Clark, R.Q. Snurr, Chem. Phys. Lett. 308 (1999) 155.
[117] K.R. Kloetstra, H.W. Zandbergen, J.C. Jansen, H. Van Bekkum, Micropor. Mater. 6 (1996) 287.
[118] J. Coronas, J.L. Falconer, R.D. Noble, AIChE J. 43 (1997) 1797.
[119] A.J. Burggraaf, Z.A.E.P. Vroon, K. Keizer, H. Verweij, J. Membr. Sci. 144 (1998) 77.
[120] W.J.W. Bakker, L.J.P. van de Broeke, F. Kapteijn, J.A. Moulijn, AIChE J. 43 (1997) 2203.
[121] J. van de Graaf, F. Kapteijn, J.A. Moulijn, J. Membr. Sci. 144 (1998) 87.
[122] F. Kapteijn, W. Zhu, J.A. Moulijn, T.Q. Gardner, Zeolite membranes: modeling and application, in: A. Cybulski, J.A. Moulijn (eds.), Structured Catalysts and Reactors, M. Dekker, New York, 1998, p. 701 ff., ISBN 0-824-79921-6.
[123] F. Kapteijn, W.J.W. Bakker, G. Zheng, J.A. Moulijn, Micropor. Mater. 3 (1994) 227.
[124] F. Kapteijn, W.J.W. Bakker, G. Zheng, J. Poppe, J.A. Moulijn, Chem. Eng. J. 57 (1995) 145.
[125] F. Kapteijn, J. Van de Graaf, J.A. Moulijn, AIChE J. 46 (2000) 1096.
[126] M. Noack, G.T.P. Mabande, J. Caro, G. Georgi, W. Schwieger, P. Kölsch, A. Avhale, Micropor. Mesopor. Mater. 82 (2005) 147.
[127] A. Culfaz, Middle East Technical University, Ankara, Turkey, personal communication.
[128] P.Z. Culfaz, A. Culfaz, H. Kalipcilar, Desalination 199 (2006) 357.
[129] J. Caro, M. Noack, 9th International Conference on Inorganic Membranes, Proceedings of the Workshop on Zeolite Membranes, Lillehammer, Norway, 2006.
[130] N. Nishyama, L. Gora, V.V. Teplyakov, F. Kapteijn, J.A. Moulijn, Sep. Purif. Technol. 22/23 (2001) 295.
[131] R. Qi, M.A. Henson, Sep. Purif. Technol. 13 (1998) 209.
[132] Y.-S. Kim, K. Kusakabe, S. Morooka, S.-M. Yang, Kor. J. Chem. Eng. 18 (2001) 106.
[133] R. Schäfer, M. Noack, P. Kölsch, S. Thomas, A. Seidel-Morgenstern, Sep. Purif. Technol. 25 (2001) 3.

[134] T.A. Peters, J. Fontalvo, M.A.G. Vorstman, N.E. Benes, R.A. van Dam, Z.A.E. Vroon, E.L.J.T.A. van Soest-Vercammen, J.T.F. Keurentjes, J. Membr. Sci. 248 (2005) 73.
[135] W.J.W. Bakker, F. Kapteijn, J. Poppe, J.A. Moulijn, J. Membr. Sci. 117 (1996) 57.
[136] Y. Kusuki, H. Shimazaki, N. Tanihara, S. Nakanishi, T. Yoshinaga, J. Membr. Sci. 134 (1997) 245.
[137] T.A. Centeno, A.B. Fuertes, J. Membr. Sci. 160 (1999) 201.
[138] M.B. Shiftlett, H.C. Foley, Science 285 (1999) 1902.
[139] M.B. Shiftlett, H.C. Foley, J. Membr. Sci. 179 (2000) 275.
[140] T.A. Centeno, A.B. Fuertes, Carbon 38 (2000) 1067.
[141] H. Wang, L. Zhang, G.R. Gavalas, J. Membr. Sci. 177 (2000) 25.
[142] H. Kita, K. Nanbu, M. Yoshino, K. Okamoto, M. Funaoka, Polym. Mat. Sci. Eng. 85 (2001) 296.
[143] C. Bai, M. Jia, J.L. Falconer, R.D. Noble, J. Membr. Sci. 105 (1995) 79.
[144] A. Giroir-Fendler, J. Peureux, H. Mozzanega, J.-A. Dalmon, in: C.H. Bartholomew, G.A. Fuentes (eds.), Stud. Surf. Sci. Catal. 111 (1996) 127.
[145] K. Kusakabe, S. Yoneshige, S. Murata, S. Morooka, J. Membr. Sci. 116 (1996) 39.
[146] J.M. van de Graaf, E. van der Bijl, A. Stol, F. Kapteijn, J.A. Moulijn, Ind. Eng. Chem. Res. 37 (1998) 4071.
[147] K. Aoki, V.A. Tuan, J.L. Falconer, R.D. Noble, Micropor. Mesopor. Mater. 39 (2000) 485.
[148] T. Matsufuji, N. Nishiyama, M. Matsukata, K. Ueyama, J. Membr. Sci. 178 (2000) 25.
[149] P. Ciavarella, H. Moueddeb, S. Miachon, K. Fiaty, J.-A. Dalmon, Catal. Today 56 (200) 253.
[150] M.P. Bernal, J. Coronas, M. Menéndez, J. Santamaria, J. Membr. Sci. 195 (2002) 125.
[151] M.P. Bernal, J. Coronas, M. Menéndez, J. Santamaria, Ind. Eng. Chem. Res. 41 (2002) 5071.
[152] Y. Takata, T. Tsuru, T. Yoshioka, M. Asaeda, Micropor. Mesopor. Mater. 54 (2002) 257.
[153] M. Hanebuth, R. Dittmeyer, G. Mabande, W. Schwieger, Chem. Ing. Tech. 75 (2003) 221.
[154] U. Illgen, R. Schäfer, M. Noack, P. Kölsch, A. Kühnle, J. Caro, Catal. Commun. 2 (2001) 339.
[155] S. Sircar, W.E. Waldron, M.B. Rao, M. Anand, Sep. Purif. Technol. 17 (1999) 11.
[156] S. Sircar, M.B. Rao, C.M.A. Thaeron, Sep. Sci. Technol. 34 (1999) 2081.
[157] A.B. Fuertes, Adsorption 7 (2001) 117.
[158] A.C. Makrides, M.A. Wright, D.N. Jewett, US Patent 3,350,846, 1967.
[159] N. Boes, H. Züchner, Surf. Tech. 7 (1978) 401.
[160] D.L. Edlund, W.A. Pledger, J. Membr. Sci. 77 (1993) 255.
[161] D.L. Edlund, D.T. Friesen, US Patent 5,217,506, 1993.
[162] D.L. Edlund, W.A. Pledger, Gas. Sep. Purif. 8 (1994) 131.
[163] N.M. Peachey, R.C. Snow, R.C. Dye, J. Membr. Sci. 111 (1996) 123.
[164] T.S. Moss, N.M. Peachey, R.C. Snow, R.C. Dye., Int. J. Hydrogen Energy 23 (1998) 99.
[165] D.J. Edlund, US Patent 5,393,325, 1995.
[166] D.J. Edlund, J. McCarthy, J. Membr. Sci. 107 (1995) 147.
[167] R.C. Dye, R.C. Snow, US Patent 6,214,090, 2001.
[168] R.E. Buxbaum, A.B. Kinney, Ind. Eng. Chem. Res. 35 (1996) 530.
[169] H. Iwahara, T. Esaka, H. Uchida, N. Maeda, Solid State Ionics 3–4 (1981) 359.
[170] S. Hamakawa, T. Hobino, H. Iwahara, J. Electrochem. Soc. 141 (1994) 1720.
[171] J. Guan, S.E. Dorris, U. Balachandran, M. Liu, Solid State Ionics 110 (1998) 303.
[172] X. Qi, Y.S. Lin, Solid State Ionics 130 (2000) 149.

[173] E.D. Wachsmann, N. Jiang, US Patent 6,296,687, 2001.
[174] S. Hamakawa, L. Li, A. Li, E. Iglesia, Solid State Ionics 48 (2002) 71.
[175] S. Cheng, V.K. Gupta, J.S. Lin, Solid State Ionics 35–36 (2005) 2653.
[176] J.H. White, M. Schwartz, A.F. Sammells, US Patent 5,821,185, 1998.
[177] J.H. White, M. Schwartz, A.F. Sammells, US Patent 6,037,514, 2000.
[178] J.H. White, M. Schwartz, A.F. Sammells, US Patent 6,281,403, 2001.
[179] U. Balachandran, T.H. Lee, L. Chen, S.J. Song, J.J. Picciolo, S.E. Dorris, Fuel 85 (2006) 150.
[180] M.V. Mundschau, US Patent 6,899,744 B2, 2005.
[181] A.F. Sammels, Ionically conducting membranes for hydrogen production..., DOE hydrogen Separations Workshop, Arlington (VA), USA, September 8, 2004.
[182] R. Dittmeyer, J. Caro, chapter 10.7: Catalytic Membrane Reactors, in: G. Ertl, H. Knözinger, J. Weitkamp (eds.), Handbook of Heterogeneous Catalysis, in press.
[183] Th. Melin, R. Rautenbach, Membranverfahren. Grundlagen der Modul- und Anlagenauslegung, 2. Aufl., Springer, Berlin, 2004, p. 451.
[184] R.M. de Vos, H. Verweij, Science 279 (1998) 1710.
[185] Y.-S. Kim, K. Kusakabe, S. Morooka, S.-M. Yang, Kor. J. Chem. Eng. 18 (2001) 106.
[186] J. Coronas, J.L. Falconer, R.D. Noble, AIChE J. 43 (1997) 1797.
[187] A. Giroir-Fendler, J. Peureux, H. Mozzanega, J.-A. Dalmon, Stud. Surf. Sci. Catal. 111 (1996) 127.
[188] S. Roark, R. Mackay, A.F. Sammels, Hydrogen Separation Membranes for Vision 21 Energy Plants, 28th International Technical Conference on Coal Utilization & Fuel Systems, Clearwater/FL, USA, March 10–13, 2003.
[189] A.F. Sammels, Ionically Conducting Membranes for Hydrogen Production and Separation, DOE Hydrogen Separations Workshop, Arlington/VA, USA, September 8, 2004.
[190] J.D. Wind, D.R. Paul, W.J. Koros, J. Membr. Sci. 228 (2004) 227.
[191] V. Sebastian, I. Kumakiri, R. Bredesen, M. Menendez, Desalination 199 (2006) 464.
[192] K. Tanaka, Y. Osada, H. Kita, K. Okamoto, J. Polym. Sci. 33 (1995) 1907.
[193] H. Suzuki, K. Tanaka, H. Kita, K. Okamoto, H. Hoshino, T. Yoshinaga, Y. Kusuki, J. Membr. Sci. 146 (1998) 31.
[194] W. Zhu, P. Hrabanek, L. Gora, F. Kapteijn, J.A. Moulijn, Ind. Eng. Chem. Res. 45 (2006) 767.
[195] Y. Yan, M.E. Davis, G.R. Gavalas, Ind. Eng. Chem. Res. 34 (1995) 1652.
[196] K. Kusakable, S. Yoneshige, A. Murata, S. Morooka, J. Membr. Sci. 116 (1996) 39.
[197] J.C. Poshusta, R.D. Noble, J.L. Falconer, J. Membr. Sci. 160 (1999) 115.
[198] J. Hedlund, M. Noack, P. Kölsch, D. Creaser, J. Caro, J. Sterte, J. Membr. Sci. 159 (1999) 263.
[199] L.J.P. van den Broeke, W.J.W. Bakker, F. Kapteijn, J.A. Moulijn, Chem. Eng. Sci. 54 (1999) 245.
[200] L.J.P. van den Broeke, F. Kapteijn, J.A. Moulijn, Chem. Eng. Sci. 54 (1999) 259.
[201] T.Q. Gardner, A.I. Flores, R.D. Noble, J.L. Falconer, AIChE J. 48 (2002) 1155.
[202] C. Algieri, P. Bernardo, G. Golemme, G. Barbieri, E. Drioli, J. Membr. Sci. 222 (2003) 181.
[203] F. Bonhomme, M.E. Welk, T.M. Nenoff, Micropor. Mesopor. Mater. 66 (2003) 181.
[204] T.J.H. Vlugt, Ph.D. Thesis, University of Amsterdam, 2000.
[205] K. Kusakabe, T. Kuroda, A. Murata, S. Mooroa, J. Membr. Sci. 148 (1998) 13.
[206] K. Kusakabe, T. Kuroda, A. Murata, S. Mooroka, Ind. Eng. Chem. Res. 36 (1997) 649.
[207] Y. Cui, H. Kita, K.-I. Okamoto, J. Mater. Chem. 14 (2004) 924.
[208] T. Tomita, K. Nakayama, H. Sakai, Micropor. Mesopor. Mater. 68 (2004) 71.
[209] S. Li, J.L. Falconer, R.D. Noble, Adv. Mater. 18 (2006) 2601.

[210] S.G. Li, J.G. Falconer, R.D. Noble, T.Q. Gardner, Ind. Eng. Chem. Res. 44 (2005) 3220.
[211] J.C. Poshusta, R.D. Noble, J.L. Falconer, J. Membr. Sci. 186 (2000) 779.
[212] T. Tominaga, K. Nakayama, H. Sakai, Micropor. Mesopor. Mater. 68 (2004) 71.
[213] V. Sebastian, I. Kumakiri, R. Bredesen, M. Menendez, Proceedings of the 9th International Conference on Inorganic Membranes, Lillehammer, Norway, June 25–29, 2006, p. 290.
[214] G. Clet, L. Gora, N. Nishiyama, J.C. Jansen, H. van Bekkum, T. Maschmeyer, Chem. Commun. 2 (2001) 41.
[215] D.W. Shin, S.H. Hyun, C.H. Cho, M.H. Han, Micropor. Mesopor. Mater. 85 (2005) 313.
[216] X. Gu, J. Dong, T.M. Nenoff, Ind. Eng. Chem. Res. 44 (2005) 937.
[217] R.D. Noble, J.L. Falconer, Ind. Eng. Chem. Res. 44 (2005) 3220.
[218] J. Coronas, J. Santamaria, Topics in Catal. 29 (2004) 29.
[219] R. Schäfer, M. Noack, P. Kölsch, S. Thomas, A. Seidel-Morgenstern, J. Caro, Sep. Purif. Technol. 25 (2001) 3.
[220] D. Casanave, A. Giroir-Fendler, J. Sanchez, R. Loutaty, J.-A. Dalmon, Catal. Today 25 (1995) 309.
[221] D. Casanave, P. Ciavarella, K. Fiaty, J.-A. Dalmon, Chem. Eng. Sci. 54 (1999) 2807.
[222] P. Ciavarella, D. Casanave, H. Moueddeb, S. Miachon, K. Fiaty, J.-A. Dalmon, Catal. Today 67 (2001) 177.
[223] R. Schäfer, M. Noack, P. Kölsch, M. Stöhr, J. Caro, Catal. Today 82 (2003) 15.
[224] M.S.M. Kwan, A.Y.L. Leung, K.L. Yeung, Proc. IZMM July 2007, Zaragoza, Spain, p. 15.
[225] K.F. Jensen, Chem. Eng. Sci. 56 (2001) 293.
[226] Y.S.S. Wan, J.L.H. Chau, K.L. Yeung, A. Gavriilidis, Micropor. Mesopor. Mater. 42 (2001) 157.
[227] J.L.H. Chau, Y.S.S. Wan, A. Gavriilidis, K.L. Yeung, Chem. Eng. J. 88 (2002) 187.
[228] J.L.H. Chau, K.L. Yeung, Chem. Commun. 9 (2002) 960.
[229] J.L.H. Chau, A.Y.L. Leung, K.L. Yeung, Lab-on-a-Chip 3 (2003) 53.
[230] A.Y.L. Leung, K.L. Yeung, Chem. Eng. Sci. 59 (2004) 4809.
[231] A.Y.L. Leung, K.L. Yeung, Stud. Surf. Sci. Catal. 154 (2004) 671.
[232] S.M. Lai, R. Martin-Aranda, K.L. Yeung, Chem. Commun. 2 (2003) 218.
[233] S.M. Lai, C.P. Ng, R. Martin-Aranda, K.L. Yeung, Micropor. Mesopor. Mater. 66 (2003) 239.
[234] X.F. Zhang, S.M. Lai, R. Martin-Aranda, K.L. Yeung, Appl. Catal. A 261 (2004) 109.
[235] L. Yeung, X.F. Zhang, W.N. Lau, R. Martin-Aranda, Catal. Today 110 (2005) 26–37.
[236] W.N. Lau, K.L. Yeung, X.F. Zhang, R. Martin-Aranda, in: R. Xu, Z. Gao, J. Chen, W. Yan (eds.), Stud. Sur. Sci. Catal. 170 (2007) 1460.
[237] Y.S.S. Wan, K.L. Yeung, A. Gavriilidis, in: E. van Steen, L.H. Calanan, M. Claeys (eds.), Stud. Surf. Sci. Catal. 154 (2004) 285–293.
[238] Y.S.S. Wan, K.L. Yeung, A. Gavriilidis, Appl. Catal. A 281 (2005) 285.
[239] W.Y. Ho, K.L. Yeung, in: R. Xu, Z. Gao, J. Chen, W. Yan (eds.), Stud. Surf. Sci. Catal. 170B (2007) 1558.
[240] T. Brinkmann, J. Puhst, H. Pingel, Desalination 200 (2006) 262.
[241] T. Kyotani, K. Sato, T. Mizino, S. Kakui, M. Aizawa, J. Saito, S. Ikeda, S. Ichikawa, T. Nakane, Anal. Sci. 21 (2006) 321.
[242] T. Kyotani, S. Inoue, S. Kakui, J. Saito, Anal. Sci. 22 (2006) 317.
[243] T. Kyotani, N. Shimotsuma, S. Kakui, Anal. Sci. 22 (2006) 325.
[244] T. Kyotani, S. Kajui, T. Mizuno, N. Shimotsuma, S. Inoue, J. Saito, Anal. Sci. 22 (2006) 1031.

[245] S. Inoue, T. Mizuno, J. Saito, S. Ikeda, M. Matsukata, Proceedings of the 9th International Conference on Inorganic Membranes, Lillehammer, Norway, June 25–29, 2006, p. 416.
[246] K. Sato, M. Take, K. Sugimoto, T. Nakane, Proceedings of the 9th International Conference on Inorganic Membranes, Lillehammer, Norway, June 25–29, 2006, p. 670.
[247] K. Sato, K. Sugimoto, T. Nakane, Proceedings of the 9th International Conference on Inorganic Membranes, Lillehammer, Norway, June 25–29, 2006, p. 480.
[248] S. Ikeda, Bussan Nanotech Research Institute, personal information.
[249] K. Sato, K. Sugimoto, T. Nakane, M. Matsukata, Proceedings of the 9th International Conference on Inorganic Membranes, Lillehammer, Norway, June 25–29, 2006, p. 333.
[250] www.deltatcorp.com and http://www.sasol.com (accessed on July 2009).
[251] H. Richter, I. Voigt, J.-T. Kühnert, Proceedings of the 9th International Conference on Inorganic Membranes, Lillehammer, Norway, June 25–29, 2006, p. 552.
[252] H. Richter, I. Voigt, J.-T. Kühnert, Desalination 199 (2006) 92.
[253] M. Weyd, H. Richter, I. Voigt, C. Hamel, A. Seidel-Morgenstern, Desalination 199 (2006) 308.
[254] I. Voigt, H. Richter, Hermsdorf Institute for Technical Ceramics, HITK, personal communication.
[255] H. Richter, M. Weyd, G. Fischer, P. Puhlfürß, I. Voigt, Proceedings of the 9th International Conference on Inorganic Membranes, Lillehammer, Norway, June 25–29, 2006, p. 329.
[256] A.G. Fadeev, S.S. Kelley, J.D. McMillan, Y.S. Selinskaja, V.S. Khotinsky, V.V. Volkov, J. Membr. Sci. 214 (2003) 229.
[257] T.C. Bowen, S. Li, V.A. Tuan, J.L. Falconer, R.D. Noble, Desalination 147 (2002) 327.
[258] T. Ikegami, H. Yanagishita, D. Katimoto, K. Haraya, T. Nakane, H. Matsude, N. Koura, T. Sano, Biotechn. Techniques 11 (1997) 921.
[259] H. Kita, 5th International Conference on Inorganic Membranes, International Workshop on Zeolitic Membranes and Films, p. 43.
[260] http://www.japancorp.net/Article.Asp?Art_ID=6795
[261] T. Tomita, K. Nakayama, H. Sakai, Micropor. Mesopor. Mater. 68 (2004) 71.
[262] T. Clark, C. Yoon, H. Deckman, R. Chance, D.M. Ruthven, B. Bonekamp, P. Pex, Proceedings of the 14th Annual Meeting of the North American Membrane Society, Jackson Hole, WY, 2003.
[263] R. Chance, C. Yoon, T. Clark, H. Deckman, E. Thomas, K. Geurts, P. Rubas, T. Tomita, K. Nakayama, H. Sakai, K. Suzuki, Proceedings of the 8th International Congress rInorganic Membranes, Membrane Synthesis and Characterization I-4, CT, 2004.
[264] J.C. Jansen, F. Kapteijn, S. Strous, EP 03078287.4.
[265] p-Xylene production costs lowered significantly with new zeolite membrane, NGK Insulators in: Japan Chemical Week, November 7, 2002.
[266] G.W. Meindersma, A.B. de Haan, Desalination 149 (2002) 29.
[267] S. Tennison, Membrane Technol. 4 (2000).
[268] A. Goldbach, T. Mauer, N. Stroh, Keram. Z. 53 (2001) 223.
[269] http://www.ikts.fhg.de (accessed on July 2009)
[270] http://www.deutscher-zukunftspreis.de/?q=en/content/team-2-35
[271] http://www.separion.com (accessed on July 2009)
[272] http://www.hitk.de (accessed on July 2009)
[273] H. Richter, I. Voigt, G. Fischer, P. Puhlfürß, Sep. Purif. Technol. 32 (2003) 133.
[274] A. Nijmeijer, C. Huiskes, N.G.M. Sibelt, H. Kruidhof, H. Verweij, Am. Ceram. Soc. Bull. 77 (1998) 95.

[275] P.M. Biesheuvel, V. Breedveld, A.P. Higler, H. Verweij, Chem. Eng. Sci. 56 (2001) 3517.
[276] K. Kim, S. Cho, K. Yoon, J. Kim, J. Ha, D. Chun, J. Membr. Sci. 199 (2002) 69.
[277] I.A. Karimi, S. Farooq, Chem. Eng. Sci. 54 (1999) 4111.
[278] S. Farooq, I.A. Karimi, J. Membr. Sci. 186 (2001) 109.
[279] H. Krieg, School for Chemistry and Biochemistry, North-West University, Potchefstroom, South Africa, personal communication.
[280] P. Chu, F.G. Dwyer, V.J. Clarke, EP 358 827, 1990.
[281] A. Arafat, J.C. Jansen, A.R. Ebaid, H. van Bekkum, Zeolites 13 (1993) 162.
[282] J.C. Jansen, A. Arafat, A.K. Barakat, H. Van Bekkum, , in: M.L. Occelli, H. Robson (eds.), Synthesis of Microporous Materials, vol. I, Van Nostrand Reinhold, New York, 1992, p. 507.
[283] I. Girnus, K. Hoffmann, F. Marlow, J. Caro, G. Döring, Micropor. Mater. 2 (1994) 537.
[284] I. Girnus, K. Jancke, R. Vetter, J. Richter-Mendau, J. Caro, Zeolites 15 (1995) 33.
[285] N.N. Greenwood, A. Earnshaw, Chemistry of the Elements, 2nd ed, Butterworth-Heinemann, 1997, p. 623
[286] J. Motuzas, A. Julbe, R.D. Noble, C. Guizard, Z.J. Beresnevicius, D. Cot, Micropor. Mesopor. Mater. 80 (2005) 73.
[287] J. Motuzas, A. Julbe, R.D. Noble, A. van der Lee, Z.J. Beresnevicius, Microp. Mesopor. Mat. 92 (2006) 259.
[288] J. Motuzas, R. Mikutaviciute, R.D. Noble, Z.J. Beresnevicius, A. Julbe, Proceedings of the 9th International Conference on Inorganic Membranes, Lillehammer, Norway, June 25–29, 2006, p. 532.
[289] X. Chen, W. Yang, J. Liu, L. Lin, J. Mater. Sci. 39 (2004) 671.
[290] X. Xu, Y. Bao, C. Song, W. Yang, J. Liu, L. Lin, Micropor. Mesopor. Mater. 75 (2004) 173.
[291] Y.S. Li, J. Liu, W.S. Yang, J. Membr. Sci. 281 (2006) 646.
[292] W. Yang, Y. Li, Proceedings of the 9th International Conference on Inorganic Membranes, Lillehammer, Norway, June 25–29, 2006, p. 273.
[293] J. Hedlund, S. Mintova, J. Sterte, Micropor. Mesopor. Mater. 28 (1999) 185.
[294] Z. Wang, Y. Yan, Chem. Mater. 13 (2001) 1101.
[295] J.H. Koegler, H. van Bekkum, J.C. Jansen, Zeolites 19 (1997) 262.
[296] S.M. Lai, L.T.Y. Au, K.L. Yeung, Micropor. Mesopor. Mater. 54 (2002) 257.
[297] A.J. Bons, P.D. Bons, Micropor. Mesopor. Mater. 62 (2003) 9.
[298] A. Avhale, G.T.P. Mabande, W. Schwieger, T. Stief, R. Dittmeyer, S. Ghosh, Z. Lai, M. Tsapatsis, Proceedings of the 9th International Conference on Inorganic Membranes, Lillehammer, Norway, June 25–29, 2006, p. 279.
[299] M. Noack, J. Caro, in: F. Schüth, K.S.W. Sing, J. Weitkamp (eds.), Handbook of Porous Solids, Wiley-VCH, Weinheim, Germany, 2002, pp. 2433–2507.
[300] V.A. Tuan, S. Li, J.L. Falconer, R.D. Noble, J. Membr. Sci. 196 (2002) 111.
[301] V.A. Tuan, J.L. Falconer, R.D. Noble, Micropor. Mesopor. Mater. 41 (2000) 269.
[302] F. Jareman, J. Hedlund, J. Sterte, Sep. Purif. Technol. 32 (2003) 159.
[303] G.T.P. Mabande, G. Pradhan, W. Schwieger, M. Hanebuth, R. Dittmeyer, T. Selvam, A. Zampieri, H. Baser, R. Herrmann, Micropor. Mesopor. Mater. 75 (2004) 209.
[304] M. Tatlier, S. Tantekin-Ersolmaz, C. Atalay-Oral, A. Erdem-Senatalar, J. Membr. Sci. 182 (2001) 183.
[305] J. Sterte, J. Hedlund, D. Creaser, O. Ohrman, W. Zheng, M. Lassinantti, Q. Li, F. Jareman, Catal. Today 69 (2001) 323.
[306] W. Lai, E. Corcoran, US Patent 6037292, 2000.
[307] Q. Li, J. Hedlund, D. Creaser, J. Sterte, Chem. Commun. 6 (2001) 527.
[308] L. Gora, G. Clet, J. Jansen, Th. Maschmeyer, Stud. Surf. Sci. Catal. 135 (2001) 3145.

[309] Q. Li, J. Hedlund, J. Sterte, D. Creaser, A.-J. Bons, Micropor. Mesopor. Mater. 56 (2002) 291.

[310] G.T.P. Mabande, M. Noack, A. Avhale, P. Kölsch, G. Georgi, W. Schwieger, J. Caro, Micropor. Mesopor. Mater. 98 (2007) 55.

[311] J. Caro, H. Wang, M. Noack, P. Kölsch, F. Kapteijn, N. Kanellopoulos, J. Nolan, EP 06 112 764.3.

[312] (a) N. Stock, T. Bein, Angew. Chem. 116 (2004) 767. (b) N. Stock, T. Bein, Angew. Chem. Int. Ed. 43 (2004) 749.

[313] H. Li, M. Eddaoudi, M. O'Koeffe, O.M. Yaghi, Nature 402 (1999) 276.

[314] O.M. Yaghi, M. O'Koeffe, N.W. Oekwig, H.K. Chae, M. Eddaoudi, J. Kim, Nature 423 (2003) 705.

[315] B. Moulton, M.J. Zaworotko, Chem. Rev. 101 (2001) 1629.

[316] C.J. Janiak, Dalton Trans (2003) 2781.

[317] U. Müller, M. Schubert, F. Teich, H. Puetter, K. Schierle-Arndt, J. Pastré, J. Mater. Chem. 16 (2006) 626.

[318] K. Schlichte, T. Kratzke, S. Kaskel, Micropor. Mesopor. Mater. 73 (2004) 81.

[319] F. Stallmach, S. Gröger, V. Künzel, J. Kärger, O.M. Yaghi, M. Hesse, U. Müller, Angew. Chem. Int. Ed. 45 (2006) 2123.

[320] J. Won, J.S. Seo, J.H. Kim, H.S. Kim, Y.S. Kang, S.J. Kim, Y. Kim, J. Jegal, Adv. Mat. 17 (2005) 80.

[321] A. Car, C. Stropnik, K.-V. Peinemann, Desalination 200 (2006) 424.

[322] S. Hermes, F. Schröder, R. Chelmowski, C. Wöll, R.A. Fischer, J. Am. Chem. Soc. 127 (2005) 13744.

[323] S. Hermes, D. Zacher, A. Baunemann, C. Wöll, R.A. Fischer, Chem. Mater. 19 (2007) 2168.

[324] D. Zacher, A. Baunemann, S. Hermes, R.A. Fischer, J. Mater. Chem. 17 (2007) 2785.

[325] E. Biemmi, C. Scherb, T. Bein, J. Am. Chem. Soc. 129 (2007) 8054.

[326] Y. Yoo, H.-K. Jeong, Chem. Commun. (2008) 2441.

[327] J. Gascon, S. Aguado, F. Kapteijn, Micropor. Mesopor. Mater. 113 (2008) 132.

[328] M. Arnold, P. Kortunov, D.J. Jones, Y. Nedellec, J. Kärger, J. Caro, Europ. J. Inorg. Chem. (2007) 60.

[329] D.N. Dybtsev, H. Chun, S.H. Yoon, D. Kim, K. Kim, J. Am. Chem. Soc. 126 (2004) 32.

[330] R.Q. Zou, L. Jiang, H. Senoh, N. Takeichi, Q. Xu, Chem. Commun. 2 (2005) 3526.

[331] J.Y. Lee, L. Pan, S.P. Kelly, J. Jagiello, T.J. Emge, J. Li, Adv. Mater. 17 (2005) 2703.

[332] T. Ohmura, W. Mori, M. Hasegawa, T. Takei, T. Ikeda, E. Hasegawa, Bull. Chem. Soc. Japan 76 (2003) 1387.

[333] L. Pan, K.M. Adams, H.E. Hernandez, X. Wang, C. Zheng, Y. Hattori, K. Kaneko, J. Am. Chem. Soc. 125 (2003) 3062.

[334] D. Zacher, O. Shekhah, C. Wöll, R.A. Fischer, Chem. Soc. Rev. 38 (2009) 1418.

[335] Y. Yoo, H.-K. Jeong, Chem. Commun. (2008) 2441.

[336] J. Gascon, S. Aguado, F. Kapteijn, Micropor. Mesopor. Mater. 113 (2008) 132.

[337] Y. Liu, Z. Ng, E.A. Khan, H.-K. Jeong, C. Ching, Z. Lai, Micropor. Mesopor. Mater. 118 (2009) 296.

[338] Y. Yoo, Z. Lai, H.-K. Jeong, Micropor. Mesopor. Mater. 123 (2009) 100.

[339] H. Guo, G. Zhu, I.J. Hewitt, S. Qiu, J. Am. Chem. Soc. 131 (2009) 1646.

[340] K.S. Park, Z. Ni, A.P. Côté, J.Y. Choi, R. Huang, F.J. Uribe-Romo, H.K. Chae, M. O'Keeffe, O.M. Yaghi, Proc. Natl. Acad. Sci. 103 (2006) 10186.

[341] X.-C. Huang, Y.-Y. Lin, J.-P. Zhang, X.-M. Chen, Angew. Chem. Int. Ed. 45 (2006) 1557.

[342] R. Banerjee, A. Phan, B. Wang, C. Knobler, H. Furukawa, M. O'Keeffe, O.M. Yaghi, Science 319 (2008) 939.

[343] R. Banerjee, H. Furukawa, D. Britt, C. Knobler, M. O'Keeffe, O.M. Yaghi, J. Am. Chem. Soc. 131 (2009) 3875.
[344] T. Wu, J. Zhang, C. Zhou, L. Wang, X. Bu, P. Feng, J. Am. Chem. Soc. 131 (2009) 6111.
[345] H. Bux, F. Liang, Y. Li, J. Cravillon, M. Wiebcke, J. Caro, J. Am. Chem. Soc., in press.
[346] W. Zhou, H. Wu, T.J. Udovic, J.J. Rush, T.J. Yildirim, Phys. Chem. A 112 (2008) 12602.
[347] M. Fernandez, J. Kärger, D. Freude, A. Pampel, J.M. van Baten, R. Krishna, Micropor. Mesopor. Mater. 105 (2007) 124.
[348] A. Mitra, Z.B. Wang, T.G. Cao, H.T. Wang, L.M. Huang, Y.S. Yan, J. Electrochem. Soc. 149 (2002) B472.
[349] A.M.P. McDonnell, D. Beving, A.J. Wang, W. Chen, Y.S. Yan, Adv. Funct. Mater. 15 (2005) 336.
[350] B.A. Holmberg, S.J. Hwang, M.E. Davis, Y.S. Yan, Micropor. Mesopor. Mater. 80 (2005) 347.
[351] M. Matsukata, M. Ogura, T. Osaki, P.R.H. Prasas Rao, M. Nomura, E. Kikuchi, Top. Catal. 9 (1999) 77.
[352] W. Chaikittisilp, M.E. Davis, T. Okubo, Chem. Mater. 19 (2007) 4120.
[353] W. Chaikittisilp, M.E. Davis, T. Okubo, 15th International Zeolite Conference, Beijing, 2007, Recent Research Reports, p. 59.
[354] S. Eslava, F. Iacopi, M.R. Baklanov, C.E.A. Kirschhock, K. Maex, J.A. Martens, in: R. Xu, Z. Gao, J. Chen, W. Yan (eds.), Stud. Surf. Sci. Catal. 170 (2007) 594.
[355] C.M. Lew, Y.S. Yan, in: R. Xu, Z. Gao, J. Chen, W. Yan, Stud. Surf. Sci. Catal. 170 (2007) 1502.
[356] A.S.T. Chiang, L.-J. Wong, S.-Y. Li, S.-L. Cheng, C.-C. Lee, K.-L. Chen, S.-M. Chen, Y.-J. Lee, in: R. Xu, Z. Gao, J. Chen, W. Yan (eds.), Stud. Surf. Sci. Catal. 170 (2007) 1583.
[357] D. Beving, C.O'Neill, Y.S. Yan, in: R. Xu, Z. Gao, J. Chen, W. Yan (eds.), Stud. Surf. Sci. Catal. 170 (2007) 1629.
[358] X.L. Cheng, Z.B. Wang, Y.S. Yan, Electrochem. Solid State Lett. 4 (2001) B23.
[359] M.R.J. Wyllie, H.W. Patnode, J. Phys. Chem. 54 (1950) 204.
[360] R.M. Barrer, S.D. James, J. Phys. Chem. 64 (1960) 417.
[361] W.L. Robb, Ann. N. Y. Acad. Sci. 146 (1968) 119.
[362] G. Clarizia, C. Algieri, E. Orioli, Polymer 45 (2004) 5671.
[363] T. Moore, W.J. Koros, J. Molec. Struct. 739 (2005) 87.
[364] S.B. Tantekin-Ersolmaz, C. Atalay-Oral, M. Tatler, A. Erdem-Enatalar, B. Schoeman, J. Sterte, J. Membr. Sci. 175 (2000) 285.
[365] W.J. Koros, R. Mahajan, J. Membr. Sci. 181 (2001) 141.
[366] R. Mahajan, W.J. Koros, Ind. Eng. Chem. Res. 39 (2000) 2692.
[367] R. Mahajan, D.Q. Vu, W.J. Koros, J. Chin. Inst. Chem. Eng. 33 (2002) 77.
[368] T.T. Moore, T. Vo, R. Mahajan, S. Kulkarni, D. Hasse, W.J. Koros, J. Appl. Polym. Sci. 90 (2003) 1574.
[369] T.T. Moore, R. Mahajan, D.Q. Vu, W.J. Koros, AIChE J. 50 (2004) 311.
[370] T.-S. Chung, L.Y. Jiang, Y. Li, S. Kulpathipanja, Prog. Polym. Sci. 32 (2007) 483.
[371] L.Y. Jiang, T.S. Chung, C. Cao, Z. Huang, S. Kulprathipanja, J. Membr, Science 252 (2005) 89.
[372] L.Y. Jiang, T.S. Chung, S. Kulprathipanja, AIChE J. 52 (2006) 2898.
[373] L.Y. Jiang, T.S. Chung, S. Kulprathipanja, J. Membr. Sci. 276 (2006) 113.
[374] Y. Li, T.S. Chung, Z. Huang, S. Kulprathipanja, J. Membr. Sci. 277 (2006) 28.
[375] Y. Li, H.-M. Guan, T.S. Chung, S. Kulprathipanja, J. Membr. Sci. 275 (2006) 17.
[376] Y. Li, T.S. Chung, C. Cao, S. Kulprathipanja, J. Membr. Sci. 260 (205) 45.
[377] T.T. Moore, T. Vo, R. Mahajan, S. Kulkarni, D. Hasse, W.J. Koros, A. Appl. Polym. Sci. 90 (2003) 1574.

[378] S.I. Zones, L. Yuen, S.J. Miller, US Patent. 6,709,644, 2004.
[379] S.A. Stern, J. Membr. Sci. 94 (1994) 1.
[380] D.Q. Vu, W.J. Koros, S.J. Miller, J. Membr. Sci. 211 (2003) 311.
[381] B.D. Reid, F.A. Ruiz-Trevino, I.H. Musselmann, K.J. Balkus, J.P. Ferraris, Chem. Mater. 13 (2001) 2366.
[382] Y. Zhang, K.J. Balkus, I.H. Musselmann, J.P. Ferraris, 15th International Zeolite Conference, Beijing, 2007, Recent Research Reports, p. 145.
[383] J.E. Readman, I. Gameson, J.A. Hriljac, P.A. Anderson, Micropor. Mesopor. Mater. 86 (2005) 96.
[384] D.S. Bhange, V. Ramaswamy, Mat. Res. Bull. 41 (2006) 1392.
[385] J.L. Schlenker, J.J. Pluth, J.V. Smith, Mat. Res. Bull. 14 (1979) 751.
[386] P.R. Rudolf, J.M. Garces, Zeolites 14 (1994) 137.
[387] M. Noack, M. Schneider, A. Dittmar, G. Georgi, J. Caro, Micropor. Mesopor. Mater. 117 (2009) 10.
[388] M. Freemantle, C&EN, October 3 (2005) 49.
[389] Y. Li, W. Yang, J. Membr. Sci. 316 (2008) 3.
[390] P.M. Slangen, J.C. Jansen, H. van Bekkum, Micropor. Mater. 9 (1997) 259.
[391] C.S. Cundy, R.J. Plaisted, J.P. Zhao, Chem. Commun. (1998) 1465.
[392] X.C. Xu, W.H. Yang, J. Liu, L.W. Lin, Sep. Purif. Technol. 25 (2001) 241.
[393] X.C. Xu, W.S. Yang, J. Liu, L.W. Lin, Adv. Mater. 12 (2000) 195.
[394] X.B. Chen, W.S. Yang, J. Liu, L.W. Lin, J. Membr. Sci. 255 (2005) 201.

CHAPTER 2

Advances in Aromatics Processing Using Zeolite Catalysts

C. Perego[1,*] *and* P. Pollesel[2]
[1]Instituto Eni Donegani, Eni S.p.A, Via Fauser 4, 28100 Novara, Italy
[2]Eni S.p.A, Refining & Marketing Division, Via Maritano 26, 20097 San Donato Milanese, Italy

Contents

Abstract

Alkylation, transalkylation, isomerization, and disproportionation are important processes for the synthesis and the interconversion of alkylbenzenes. The chemical reactions involved in these processes are catalyzed by acids. During the past three decades, the main innovation in this field has been represented by the introduction of new zeolite catalysts with the aim to improve the overall selectivity of the processes and to comply with the requirements of the more stringent environmental legislation.

In this review, the recent advances in zeolite catalysis for alkylbenzene production are discussed. The processes for xylene, ethylbenzene (EB),

*Corresponding author. Tel.: +39 0321 447252; Fax: +39 0321 447679
E-mail address: carlo.perego@eni.it

Advances in Nanoporous Materials, Volume 1 ISSN 1878-7959, DOI 10.1016/S1878-7959(09)00102-9

cumene, *para*-ethyltoluene, *para*-diethylbenzene (DEB), diisopropylbenzene (DIPB), and cymene, are extensively discussed in relation with the aromatics market evolution and technology key parameters (e.g., feedstock, thermodynamic constraints, isomer selectivities). According to a detailed open and patent literature search, the different technologies are examined, mainly with respect to the zeolite catalysts utilized. The innovation of the zeolite catalysts is discussed under the following different aspects: zeolite structure, zeolite morphology, catalyst formulation, and shaping. A special focus is devoted to the shape selectivity of the different zeolite catalysts, considering both the peculiarity of the zeolite structures and the influence of the various practical post-synthesis modifications.

1. INTRODUCTION

Aromatics, together with light olefins, are the most important building blocks on which a vast petrochemical and organic chemical industry is based. They are important raw materials for many commodity petrochemicals and valuable fine chemicals such as monomers for polyesters, polyamides, and engineering plastics and intermediates for detergents, pharmaceuticals, agricultural products, and explosives [1]. Benzene, toluene, and xylenes (BTX) are the three basic materials for the production of most intermediates of several aromatic derivatives (Fig. 1), most of which are by means of acid catalyzed reactions: alkylation, transalkylation, isomerization, and disproportionation [2]. In fact, ethylbenzene (EB), cumene, *para*-diethylbenzene (*para*-DEB), *para*- and *meta*-diisopropylbenzene (*para*- and *meta*-DIPB), C_{10}–C_{14} linear alkylbenzenes (LAB), *para*-ethyltoluene, and *para*- and *meta*-cymene are some of the chemical intermediates obtained by alkylation on the aromatic ring of benzene or toluene [1]. Other important alkylbenzenes are obtained by isomerization (e.g., xylene isomerization), by disproportionation (e.g., toluene disproportionation to *para*-xylene and benzene), and by transalkylation (e.g., polyethylbenzenes transalkylation with benzene to EB). Process development in alkylaromatics production by alkylation or alkylaromatic isomerization, disproportionation, and transalkylation is therefore an important research task with great industrial demands.

Several different driving forces have been pushing the process innovation in this field. In addition to the economically relevant variables such as market demand, feedstock availability, and operating cost, legislative aspects such as environmental laws and new reformulated gasoline specifications also played an important role. For instance, the US Clean Air Act, which came into effect in 1995, changed the aromatics supply picture. It stipulates that benzene in gasoline must be decreased to 1% and the remaining aromatics content must be no greater than 25%. In 1990, gasoline contained as much as 3% benzene and 36% total aromatics [3]. The modifications involved

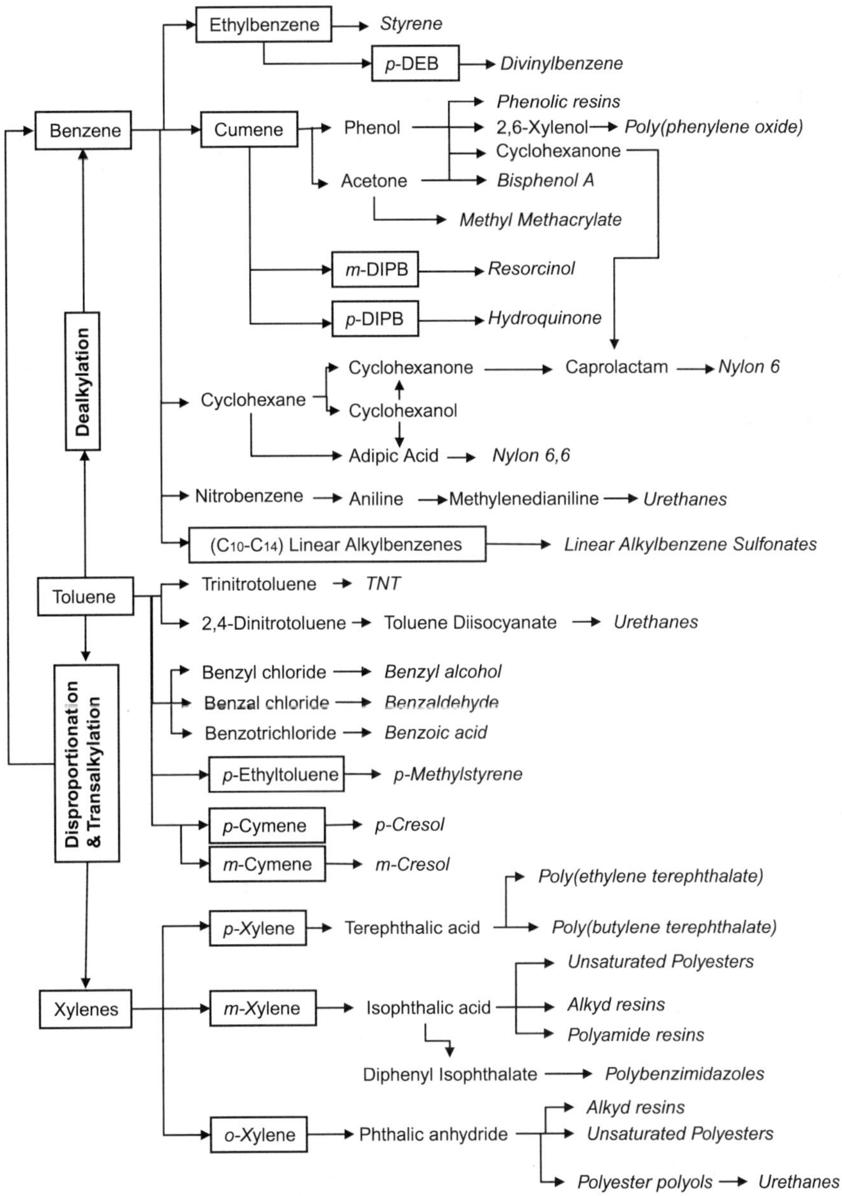

Figure 1 Derivatives from BTX (benzene, toluene, and xylenes). Products obtained by isomerization, disproportionation, and transalkylation are highlighted in boxes.

switching the application of aromatics from gasoline to petrochemicals, especially to benzene and xylenes.

Catalytic reforming and naphtha pyrolysis are the main sources of BTX production. The product yields of those processes are normally controlled by thermodynamics and hence result in a substantial mismatch between the supply and the actual market demands. The ratio for B:T:X obtainable from the two above-mentioned processes is 32:36:32 [2]. This is in contrast to their global market demand that was roughly 47:20:33 for the year 2004. In the United States, most benzene comes from the catalytic reforming of naphtha, which yields a mixture of BTX. In Europe and Japan, pyrolysis gasoline is the major source of benzene and toluene. It results from the steam cracking of virgin naphtha and/or gas oil. Xylenes also occur in pyrolysis gasoline but are not easily isolated because of a high concentration of EB. In the United States, only 24% of benzene comes from this source.

The total demand for benzene in 2004 was 36.4×10^6 t/a. In Fig. 2, the primary sources of benzene are represented [4]. Seventy-nine percent of benzene comes from naphtha reforming and pygas (pyrolysis gasoline) from naphtha cracking. Most of the world's benzene is produced as a by-product from the production of ethylene and *para*-xylene. Hydrodealkylation (HDA) had been the swing capacity that came on stream as necessary to satisfy the market's needs. However, over the past decade, 2.3×10^6 t/a of benzene capacity has been permanently taken out of service. Much of this was HDA swing capacity. This reduction in HDA swing capacity is one of the main reasons for the current tightness in the benzene market. Benzene consumption subdivided by final use and for different geographical regions is reported in Table 1.

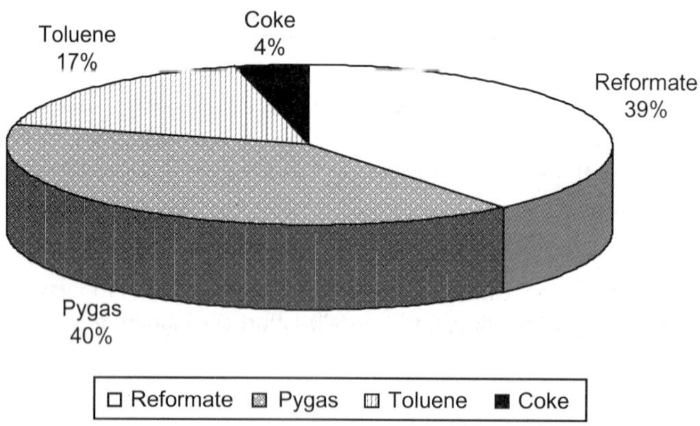

Figure 2 Sources of benzene supply in 2004 (adapted from Ref. [4]).

Table 1 Breakdown of benzene consumption by final use, for the year 2004 (10^3 t/a)

	Ethylbenzene	Cumene	Cyclohexane	Nitrobenzene	Others	Total
Western Europe	4461	2054	993	1080	320	8908
Africa	0	0	0	3	76	79
Asia/Pacific	7770	1982	1758	474	1686	13,670
East. Europe (incl. CIS)	720	304	442	120	283	1869
Middle East	1047	0	233	0	72	1352
North America	5114	2424	1060	825	299	9722
South America	388	123	143	34	142	830
Total	19500	6887	4629	2536	2878	36,430
Percentage of total	53.5	18.9	12.7	7.0	7.9	100

Toluene, which has the lowest market demand, is always in surplus from the production of reformate and pyrolysis gasoline, whereas benzene and xylenes are in strong demand with the average annual growth rates of around 4% and 6%. As a result of demand and supply, the price for toluene is always lower than for the other aromatics. The conversion of dispensable toluene into the more valuable aromatics therefore has an economic incentive. A serious discrepancy between production and market demand was also found for most dialkylbenzene isomers, among which the *para*-isomers apparently have the greatest market demand.

In response to market situation and legislation changes, Tsai et al. identified several areas of new aromatics process innovations [2]. Among them, it is worthwhile citing:

1. Conversion of surplus toluene,
2. Upgrading of heavy aromatics to more valuable products, and
3. Selective production of *para*-dialkylbenzene isomers against thermodynamic equilibrium, such as *para*-xylene and *para*-DEB.

In response to the worldwide environmental awareness, there are active programs to search for clean processes. In fact, most of the eldest technologies, mainly the alkylation/transalkylation ones, are based on catalysts having drawbacks. Often, such catalysts are strong mineral acids or Lewis acids (e.g., HF, H_2SO_4, $AlCl_3$). These acids are highly toxic and corrosive. They are dangerous to handle and to transport as they corrode storage and disposal containers. Besides, because the reaction products are mixed with acids, the separation at the end of the reaction is often a difficult and energy-consuming process. Very frequently, these acids are neutralized at the end of the reaction, and therefore, the corresponding salts have to be disposed. To avoid these problems, many efforts have been devoted to the search for solid acids, which are more selective, safer, and more environmentally friendly. Among the different solid acids, zeolites have been extensively evaluated for such a purpose. Comprehensive reviews have been published on the use of zeolites as catalysts in these fields (see, e.g., Refs. [2,5–9]). The present review is aimed to summarize the recent advances and perspectives of zeolite-based technology for aromatic hydrocarbon production in the petrochemical industry.

2. ZEOLITE CATALYSIS FOR REACTIONS INVOLVING AROMATIC HYDROCARBONS

Today, the majority of solid acid catalysts used in refineries and in the petrochemical industry are zeolites. The use of zeolite catalysts in industrial applications evolved from the pioneering studies of R.M. Barrer at Imperial College, UK, and those started in 1948 at the Union Carbide laboratories, USA, by R.M. Milton (i.e., discovery of A and X zeolites), D.W. Breck (i.e., discovery of Y zeolite and of the unique cracking activity of X zeolite), and J.A. Rabo (i.e., discovery of catalytic active ingredients used worldwide in catalytic cracking) [10]. Since then, the use of zeolite-based catalysts had a relevant growth in the area of oil refining and bulk chemical production. In 2001, the world consumption of synthetic zeolites amounted to 1.46×10^6 t/a, of which 81% were used for detergent builders, 13% for catalysts, and 6% for adsorbents/desiccants. The total market value was around 2×10^9 US$, with catalysts representing around 55% of the overall value. It is also interesting to observe that about 95% of the zeolite catalyst market is related to the production of fluidized catalytic cracking (FCC) catalysts. With regard to chemical commodities, zeolites have demonstrated to provide a promising new class of catalysts, and therefore, we should expect to see an expansion of their application for petrochemical processes. The application of zeolites in petrochemical processes occurs in two main areas: the production of olefins and derived products and the production of aromatics [6].

2.1 Zeolite structures and acid sites

Zeolites are porous crystalline aluminosilicates built from SiO_4 and AlO_4 tetrahedra. These tetrahedra are linked to each other through oxygen ions. In almost all natural zeolites, aluminum or silicon occupies all the tetrahedra. In some synthetic molecular sieves, boron, gallium, germanium, iron, titanium, phosphorus, or other heteroatoms may substitute for aluminum or silicon. Exchangeable cations allow the introduction of different cations with various catalytic properties. When cationic sites are exchanged with H^+, a high number of strong acid sites are obtained.

According to Cejka and Wichterlova [5], the main advantages of zeolites and zeolite-like catalysts compared to conventional solid acids include the following:

1. Well-defined inorganic crystalline structures, with a variety of structures differing in channel diameters, geometry, and dimensionality;
2. A precisely defined inner void volume providing high surface area;
3. The ability to adsorb and transform molecules in the inner volume;
4. The possibility for isomorphous substitution of some trivalent cations into the silicate framework enabling tuning of the strength and concentration of the acid sites;

5. Shape selectivity, given by the ratio of the kinetic diameters of the reactants, intermediates, and products to the dimensions of the channels and/or cavities; and
6. Environmental tolerance.

The Brønsted acid sites are associated with framework aluminum or other trivalent ions (e.g., B, Ga, Fe). The number of the Brønsted acid sites is directly proportional to the concentration of these framework trivalent ions. However, as a completely isolated Al tetrahedron will create the strongest type of Brønsted acid site, the strength of the acid site is usually inversely proportional to the concentration of framework aluminum, up to about a silica/alumina ratio of 10. Changing the framework silica/alumina ratio, either by direct synthesis or by postchemical synthesis, will change the acid strength.

The acid strength depends on the type of the heteroatom: gallium or iron zeolites are much less acidic than aluminum zeolites. Boron-substituted zeolites have very weak acidity.

Lewis sites in zeolites are uslly associated with extraframework Al. The simultaneous presence of Brønsted and Lewis sites could enhance the acid activity of zeolites. However, extraframework Al affects different reactions differently and therefore could affect selectivities.

Pore diameters depend on the number of tetrahedra in the ring around the pores. According to their pore dimensions, zeolites are grouped as follows:

1. Small-pore zeolites (8-membered rings (MRs) up to 0.43 nm) – for example, Erionite;
2. Medium-pore zeolites (10-MR up to 0.55 nm) – for example, ZSM-5, ZSM-11, ferrierite, MCM-22 (possessing 12-MR and large intracrystalline cavities pockets on the crystal surface);
3. large-pore zeolites (12-MRs up to 0.75 nm) – zeolites Y, Beta, mordenite, ZSM-12; and
4. extra-large-pore zeolites (14 or more MRs – for example, CIT-5, UTD-1, ECR-34, ITQ-33 and ITQ-37.

The orientation of the plane of the ring also affects pore diameters. Besides, the pores may be straight or zigzag-like, and the pore system may be one-, two-, or three-dimensional. In some cases, the channel system can consist of channels of different diameters (ferrierite, mordenite), or it possesses two independent channel systems (MCM-22).

In principle, it is possible to select the most appropriate acid strength, dimensionality, pore size, and pore shape from a large variety of zeolites to catalyze almost any reaction. The most utilized structure types of zeolites for synthesis and transformations of mono- and dialkylbenzenes are listed in Table 2.

2.2 Shape selectivity of zeolite catalysts

Shape selectivity is a very important property of zeolite catalysts. Shape selective catalysis differentiates between reactants, products, or reaction intermediates according to their shape and size. Only molecules whose dimensions are less than a critical size can enter the pores, have access to internal catalytic sites, and react there. Furthermore, only molecules that can leave appear in the final product. According to Degnan [11], few concepts have had a greater impact on the design and development of novel catalytic processes for petroleum refining and petrochemical manufacture than that of shape selective catalysis. First proposed by Weisz and Fridette in 1960 [12], the concept of shape selectivity is today the basis of several commercial processes. The initial concepts proposed by Weisz and extended by Csicsery have provided the foundation for much of the theory upon which shape selective catalysis is built. Insightful papers by Weisz and co-workers of Mobil Research and Development [13–15] and Csicsery [16–19] are particularly relevant. Other prominent contributions include Refs. [11,20–27].

There are three well-accepted types of shape selectivity:

1. In reactant selectivity, some of the molecules in a reactant mixture are too large to diffuse through the catalyst pores.
2. In product selectivity, some of the products formed within the pores are too bulky to diffuse out as observed products.
3. In restricted transition-state selectivity, certain reactions are prevented because the corresponding transition state would require more space than available in the cavities or pores. Reactions requiring smaller transition states proceed unhindered [18,28]. Advantages of shape selective catalysts are that they may favor the formation of desirable isomers over less desirable ones, crack undesirable molecules to smaller fragments that are easily removed by distillation, and avoid undesirable competing reactions such as coking or polymerization.

An example of product shape selectivity is the *para*-selectivity of ZSM-5 in the alkylation of monoalkylaromatics, that is, among the other positions of the aromatic ring (*ortho* and *meta*), *para* is preferred for steric reasons.

The occurrence of restricted transition-state selectivity may affect the reaction mechanisms. For instance, many reactions may proceed by both mono- and bimolecular mechanisms. The bimolecular reaction has a lower activation energy than the monomolecular one. However, when the space

Table 2 Structure characteristics for zeolites relevant to the synthesis and transformation of aromatic hydrocarbons

Zeolite	IZA code[a]	Channel dimensionality	Channel type	Pore dimensions (nm)
Beta	*BEA[b]	3D	12	0.66 × 0.67
			12	0.56 × 0.56
ITQ-7	ISV	3D	12	0.61 × 0.65
			12	0.59 × 0.66
$AlPO_4$-5,	AFI	1D	12	0.73 × 0.73
SAPO-5				
NU-87	NES	2D	10	0.48 × 0.57
SSZ-33	CON	3D	12	0.59 × 0.70
			10	0.45 × 0.51
			12	0.64 × 0.70
CIT-5	CFI	1D	14	0.72 × 0.75
L	LTL	1D	12	0.71 × 0.71
Mordenite	MOR	1D	12[c]	0.70 × 0.65
MCM-22,	MWW[d]	2D	10	0.40 × 0.55
ERB-1,			10	0.41 × 0.51
MCM-56				
NCL-1	–	1D	12	0.57 × 0.86
Y	FAU	3D	12[e]	0.74 × 0.74
Omega	MAZ	1D	12	0.74 × 0.74
ZSM-5	MFI	3D	10	0.53 × 0.56
			10	0.51 × 0.55
ZSM-11	MEL	3D	10	0.53 × 0.54
ZSM-12	MTW	1D	10	0.56 × 0.60

Table 2 *(Continued)*

Zeolite	IZA code[a]	Channel dimensionality	Channel type	Pore dimensions (nm)
ZSM-22	TON	1D	10	0.46 × 0.57
ZSM-23	MTT	1D	10	0.45 × 0.52
ZSM-48	–	1D	10	0.53 × 0.56
ZSM-21	FER	1D	10[f]	0.42 × 0.54
EU-1	EUO	1D	10[g]	0.41 × 0.54

* BEA indicates the Polymorph or Polytype A of Beta zeolite (BEA) according to International Zeolite Association nomenclature.
[a] International Zeolite Association.
[b] Polymorph A, tetragonal.
[c] Interconnected with two 8-membered ring channel systems with dimensions 0.34 × 0.48 and 0.26 × 0.57 nm.
[d] Two independent channel systems, one of them intersects large cylindrical cavities (diameter 0.71 nm, height 1.8 nm); surface pockets located on the (001) crystal surface (dimensions 0.71 × 0.71 nm)
[e] Interconnecting large cages.
[f] Interconnected with a 8-membered ring channel system with dimensions 0.35 × 0.48 nm.
[g] With large side pockets.

around an active site is insufficient to accommodate the bimolecular transition state, the reaction proceeds through the monomolecular route. Alkylaromatics may isomerize either by monomolecular or by bimolecular mechanisms. The monomolecular reaction proceeds through consecutive 1,2-shifts. Diphenylmethane intermediates are involved in the latter. For instance, the ratio of the monomolecular/bimolecular reaction paths in EB disproportionation depends on acid site density and zeolite pore structure. In medium-pore zeolites like ZSM-5, the monomolecular mechanism is proposed, while with large-pore zeolites like Y, the bimolecular is generally accepted [29].

Contributions of shape selectivity to the synthesis and transformation of aromatic hydrocarbons will be discussed in the following section.

3. XYLENE

3.1 Xylene market

The xylene isomers are important chemical intermediates. *ortho*-Xylene is oxidized to make phthalic anhydride, which is mainly used to produce phthalate plasticizers. *meta*-Xylene is converted into isophthalic acid, which is used in the production of unsaturated polyesther resins. However, *para*-xylene has by far the largest market of the three isomers. The largest use of *para*-xylene is in its oxidation to synthesize terephthalic acid. Terephthalic acid is used in turn to synthesize important polymers such as polyethylene terephthalate (PET) and polybutylene terephthalate (PBT). PET in particular is one of the largest volume polymers in the world, and the demand for *para*-xylene is several times that for *meta*- and *ortho*-xylene.

Principal sources of xylene isomer mixtures are catalytically reformed naphthas and pyrolysis distillates [1], with the distribution of xylene isomers being approximately 50%–60% of *meta*-xylene and 20%–25% *ortho*- and *para*-xylenes. *para*-Xylene holds about 80% of the xylenes market share [4]. Global *para*-xylene demand in 2004 was 21.6×10^6 t/a with an estimated average growth rate of 6.2% per year until 2009. The growth will be drawn by the fiber sector, mainly in Asia, while the PET bottle grade will increase slowly. *ortho*-Xylene demand was 3.4×10^6 t/a in 2004. Its market has been quite stable in the past years: the growth rate is relatively slow (2.3% annual). Significant capacity expansions are expected only in Iran (Borzoye PC) and China (BASF-YPC, Nanjing).

3.2 Xylene production

To meet the requirements of xylene demand, different processes are available, which are based on zeolite catalysis. The reactions involved in these processes include isomerization of xylenes, disproportionation of toluene, and transalkylation of trimethylbenzenes with toluene. Less relevant from an industrial point of view is toluene alkylation with methanol.

3.2.1 Xylene isomerization

The xylene isomerization process converts *meta*- and *ortho*-xylenes to *para*-xylene. To meet the *para*-xylene demand, the much less used *ortho*- and *meta*-xylenes are converted through the xylene isomerization reaction, a major industrial process for this aromatic [1]. Xylene isomerization is a thermodynamic equilibrium controlled reaction, and therefore, total conversion is impossible under conventional conditions. Equilibrium product distributions in the standard state (atmospheric pressure) for temperatures from 0 to 700 °C are reported in Fig. 3 [1].

Separation of *para*-xylene from its isomers is essential but is difficult due to their close boiling points. *ortho*-Xylene has the highest boiling point ($T_b = 144.4$ °C) and can be separated from the other isomers by distillation in a so-called xylene column. From the bottom of this column, *ortho*-xylene plus C_9 aromatics are recovered. *para*-Xylene and *meta*-xylene (plus EB) are recovered from the top and then sent to the isomerization step. After

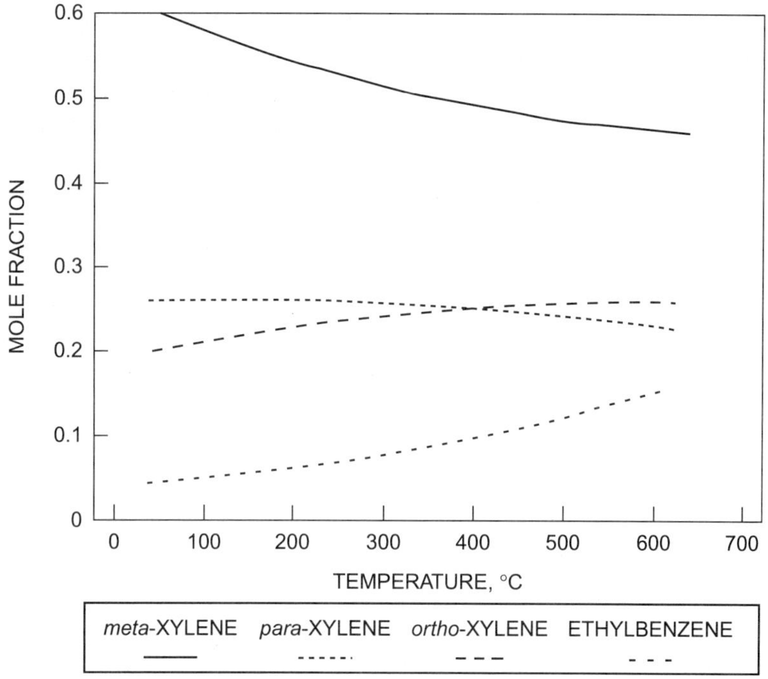

Figure 3 C_8 aromatics equilibrium composition versus temperature [1].

isomerization, *para*-xylene (T_b = 138.3 °C) and *meta*-xylene (T_b = 139.1 °C) cannot be separated by fractional distillation since the difference in their boiling points is too small. Crystallization was initially the most important method of producing *para*-xylene. Several technologies (*Isofining* – Exxon, *Antar* – HRI) and proprietary processes have been developed (Krupp, Maruzen, ARCO) [2]. For typical thermodynamic equilibrium xylene compositions, the degree of *para*-xylene recovery obtained by crystallization is around 65%. Low recovery degree and low crystallization temperatures have pushed toward the use of adsorption methods rather than crystallization for *para*-xylene purification. Adsorption technologies prevalently utilize modified faujasite zeolites as selective adsorbent. Well-known technologies in this context are the *Parex* (UOP), the *Aromax* (Toray), and the *Eluxyl* (IFP) process. Crystallization has been recently revived by advances in disproportionation processes that produce a C_8 mixture that allows a *para*-xylene recovery above 90%.

In the past, dual-functional catalysts (e.g., in *Octafining*) and various monofunctional acid catalysts (e.g., silica-aluminas, Lewis acids) were used to isomerize xylenes. The Octafining process was developed by Atlantic

Richfield in the 1960s. Isomerization takes place at an initial temperature of 425 °C, in the presence of hydrogen, on a platinum/aluminosilicate catalyst. Only a small amount of EB is converted, since its initial concentration in the feedstock is usually well below thermodynamic equilibrium levels [1]. Processes based on monofunctional catalysts (*Isomar* – UOP, *Isarom* – IFP) are usually characterized by short service life. In the early 1970s, Mobil developed zeolite catalysts for the isomerization of xylene with improved selectivities. Zeolites ZSM-5, ZSM-12, and ZSM-21 have been proposed for this purpose [30]. As a whole, selectivities are much better and coking is much less over shape selective ZSM-5, modified ZSM-5, and related other "pentasil" catalysts than in the older systems or in large-pore zeolites. Trialkylbenzene formation and coking are inhibited in the ZSM-5 pore system because there is not enough space to accommodate the large, bimolecular transition states of this reaction (restricted transition state selectivity) [18]. There is a strong increase in the ratio between the rates of isomerization and disproportionation when zeolite pores become smaller. This ratio increases from 20 to 70 up to 1000 for faujasite, mordenite, and ZSM-5, respectively [31]. Less coking allows much longer operation between regenerations than it was ever possible with the old amorphous catalysts.

The isomerization may proceed by either a monomolecular or a bimolecular mechanism. Diphenylmethane intermediates are involved in the bimolecular reaction. The monomolecular reaction proceeds through consecutive 1,2-shifts [32]. Zeolites with strong acid sites isomerize xylenes by the monomolecular pathway, whereas mesoporous molecular sieves with weaker acid sites catalyze mostly the bimolecular reaction.

In ZSM-5, at 200 °C, the reaction kinetics controls the rate of isomerization. At and above 300 °C, the rate is diffusion controlled. The rate of diffusion of *para*-xylene is higher than that of *ortho*-xylene. *meta*-Xylene has the lowest diffusion rate.

Xylene isomerization feedstocks usually contain more than 10% EB. If not removed, EB will accumulate in the recycle streams. Shape selective zeolite catalysts offer the additional advantage that they can convert EB to easily removable products. For example, EB could be converted simultaneously through disproportionation to benzene and DEB. Distillation can remove both. Or the EB may be dealkylated, and the ethylene formed could be hydrogenated over a hydrogenation component (e.g., platinum) under moderate hydrogen pressure. This makes the de-ethylation irreversible.

In the mid-1980s, Indian Petrochemical Corporation Limited (IPCL), in association with National Chemical Laboratory (NCL, Pune), developed a composite catalyst containing a high-silica ZSM-5-type metallosilicate, called "Encilite," with unique catalytic activity for *meta*-xylene isomerization. This led to the development of India's first petrochemicals catalyst and process (*Xylofining*) for the isomerization of *meta*-xylene to *ortho*-and *para*-xylene with conversion of EB to benzene. Subsequently, based on extensive fundamental studies, like changes in the composition of the high-silica zeolite, support modification and impregnation with active metal ingredients, IPCL developed an improved new generation catalyst of its own, known as "Encilite-501." The performance of the newly developed Encilite-501 catalyst, a conventional C_8 aromatics isomerization catalyst, and the Encilite catalyst is presented in Table 3. The zeolite-based Encilite-501 has shown excellent stability and life for more than 10 years in commercial operation for the production of *para*- and *ortho*-xylene [33].

Shape selectivity also affects the disproportionation/isomerization rates. The rate of disproportionation relative to that of isomerization decreases with decreasing zeolite pore diameter. The catalyst pore size and the size of pore mouth can be tailored to minimize disproportionation and to optimize *para*-xylene selectivities. Catalyst pre-coking and silanation are the most commonly used procedures for isomerization catalyst modification. There

Table 3 Xylene isomerization processes: typical process parameters [33]

	Conventional	Encilite catalyst	Encilite-501 catalyst.
Approach to equilibrium (%)			
ortho-Xylene	85–90	>90	>90
para-Xylene	95–100	>100	>100
H_2/hydrocarbon ratio (mol/mol)	6	2	1
WHSV (h^{-1})	3	10	12
Pressure (kg cm^{-2})	12–16	8–12	8–12
Desirable product	Nil	Benzene	Benzene
Flexibility to *ortho*/*para*-xylene	–	Possible	Possible
Life cycle (years)	1	1–3	3

are at least three incentives for catalyst modification:

1. The suppression of xylene losses by disproportionation to toluene and trimethylbenzenes,
2. The enhancement of EB conversion, and
3. The increase in *para*-xylene selectivity.

Systematic studies have compared ZSM-5 pre-coking and silanation. It has been observed [34] that pre-coking is more effective in inactivating strong acid sites on the external surface, where rate-limiting steric constraints do not exist, and hence, bimolecular disproportionation reactions can more easily occur. Therefore, pre-coking is more efficient in suppressing xylene losses and to some extent also favors EB conversion. Silanation with tetraethyl *ortho*-silicate (TEOS) does not passivate efficiently the external surface, but it results in a pore narrowing of zeolite ZSM-5, enhancing *para*-selectivity of the silanated catalyst.

The maximization of EB conversion during xylene isomerization is an important point to be considered for process improvements. Mordenite and ZSM-5 have been compared in this respect. A platinum-containing ZSM-5 extrudate catalyst allows high EB conversion (more than 80%) through dealkylation with, at the same time, very low xylene loss (Fig. 4). The

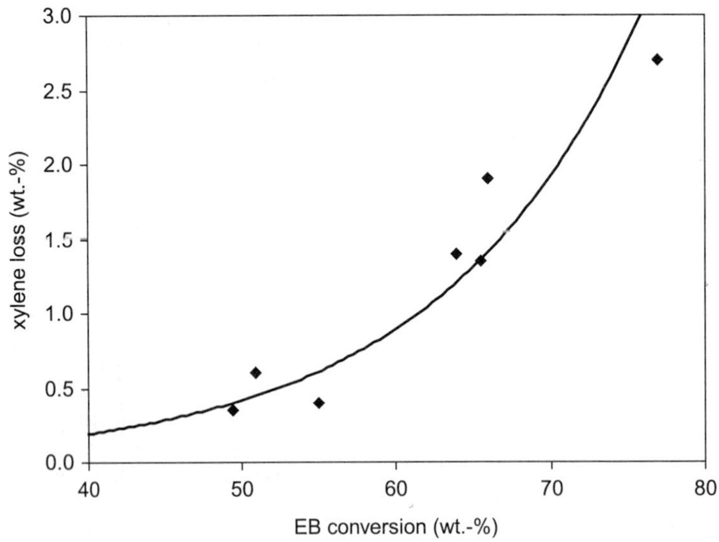

Figure 4 Xylene loss and ethylbenzene (EB) conversion in xylene isomerization with a ZSM-5-based catalyst [35].

Table 4 *Para-ortho* ratio (*p/o*), isomerization/disproportionation ratio (i/d), and trimethylbenzene (TMB) selectivity at low conversion of *meta*-xylene for different zeolites

	Conversion (% mol)	p/o ratio	i/d ratio	1,2,4 TMB selectivity (% normalized)
ZSM-5	10.0	2.3	34.3	100
	21.1	2.1	55.0	100
NU-87	26.9	1.3	3.4	92.5
	30.7	1.3	2.4	93.3
Beta	19.2	1.1	2.9	65.8
	34.8	1.1	2.3	64.4
Equilibrium (350 °C)		1.0		68.0

Source: Adapted from [36].

addition of iron or molybdenum oxide also improves selectivity with less C_9–C_{10} aromatics formation [35]. The mordenite-based catalyst needs a higher platinum loading and mainly converts EB (50%–60% maximum) to xylenes through bifunctional isomerization.

Recently, a zeolite with a unique pore topology, NU-87, has also been considered for xylene isomerization. NU-87 has a structure with 10-MR channels intersected by perpendicular 12-MR cavities. This pore arrangement can be interesting in catalysis because it may combine high conversion with shape selectivity features. For this application, however, NU-87 does not seem to perform in the desirable way, as can be seen from Table 4 [36]. The *para/ortho* ratio is considerably lower with respect to ZSM-5, and it is only slightly better than for zeolite Beta. The isomerization/disproportionation ratio is again similar as for zeolite Beta and one order of magnitude lower than for ZSM-5. The potential enrichment in *para*-xylene due to diffusional restrictions in the 10-MR pores of NU-87 is almost completely neutralized in the 12-MR cavities. In the large cavities, there is also time and space for *meta*-xylene to react and disproportionate: the isomerization/disproportionation ratio is again slightly better than for Beta but one order of magnitude lower than for ZSM-5. A diffusional effect can be revealed in NU-87 for trimethylbenzenes, among which the 1,2,4-isomer is clearly predominant.

MFI-based materials remain the most effective catalyst type for xylene isomerization processes. Today, more than one-half of the Western World's xylene isomerization plants use one of Mobil's xylene isomerization processes. Since 1973, 21 plants have been licensed for use throughout the Americas, Europe, Middle East, and Asia, with the ExxonMobil isomerization technologies *XyMax* or *Advanced MHAI*.

3.2.2 *para*-Xylene processes from toluene

The economic incentive to convert toluene to *para*-xylene is that toluene is less valuable than either benzene or xylenes. As stated before, there is an unbalanced market demand for BTX aromatics from conventional aromatics production processes; in particular, toluene production exceeds market demand (plants operating factor in 2004 was 73%), while the demand for benzene and *para*-xylene is strong and still growing. This unbalancement between market needs and chemical process yields can be partially corrected through aromatics interconversion processes. Disproportionation-transalkylation is the main route for increasing benzene and dialkylbenzenes production starting from toluene. Toluene with its high octane value can be sold as a component in the gasoline pool. However, to gain the benefits of higher prices, toluene is frequently converted to benzene and xylenes. Surplus toluene can be converted into other aromatics through methyl group transfer or HDA reactions. HDA converts toluene into benzene and methane in the presence of hydrogen. HDA processes were quite common in the United States, but recently, disproportionation demand has grown strongly. The strong growth in demand for xylene derivatives and the impact of benzene limitation in gasoline give more relevance to processes that convert toluene to xylenes, together with benzene. Moreover, compared to methyl group transfer reactions, HDA occurs normally at much higher reaction temperature and requires higher operation and capital costs.

Toluene may be converted to *para*-xylene also by alkylation with methanol. This route has so far not been very much exploited, but it is recently getting more consideration.

Toluene disproportionation-transalkylation. Disproportionation processes have been available for licensing since the early 1970s; main processes are UOP-*Tatoray* and ExxonMobil-*MTDP-3*. Most of these processes operate with high toluene conversion efficiency, typically from 42% to 48% [37]. The main drawback, however, is the production of an equilibrium xylene mixture, which does not correspond to market demand for isomers.

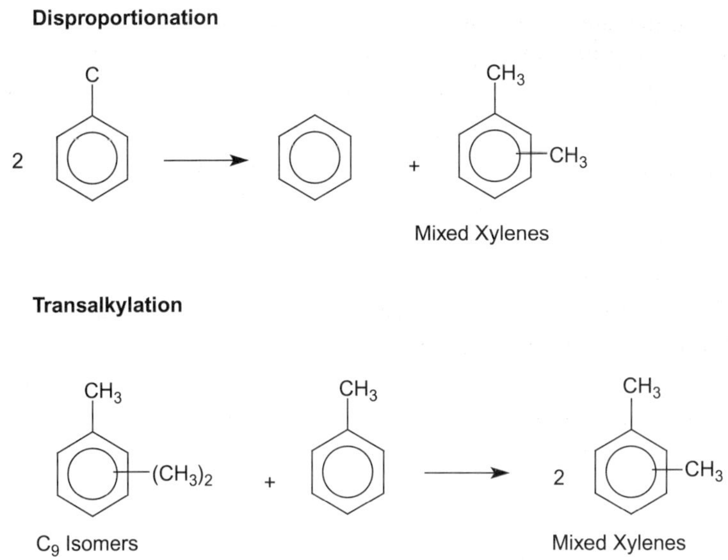

Scheme 1 Toluene disproportionation and transalkylation reactions.

Mobil's toluene disproportionation process converts two moles of toluene to one mole each of xylenes and benzene. Here, too, restricted transition state selectivity minimizes coking and the formation of higher-molecular-weight hydrocarbons. Additional advantages of disproportionation processes are that separation costs are lower than those of xylene isomerization and that the benzene produced is very pure (99.99%, pure enough for EB manufacturing). The first plant went on stream in 1975 in Naples, Italy [38]. The catalyst was ZSM-5.

With some implementations, the process can also accept C_9 and C_{10} aromatic feedstocks. In that case, the reactions for toluene feedstocks (disproportionation) is accompanied by the reactions for C_9–C_{10} feedstocks (transalkylation) as shown in Scheme 1.

ExxonMobil's *MTDP-3* can handle C_9 aromatics on a limited basis. UOP has licensed a total of 48 *Tatoray* units; feedstocks range from 100 wt.% toluene to a mixture of 30 wt.% toluene and 70 wt.% C_{9+}, with up to 10% C_{10} aromatics in the feed. ExxonMobil has also commercialized a heavy aromatics transalkylation process, *TransPlus*, which can treat, in principle, up to 100% C_{9+} with up to 25% C_{10} aromatics [37]. When C_{9+} molecules are present in the feed, *TransPlus* process works by first dealkylating ethyl and higher alkyl groups leaving primarily methyl aromatics behind. The remaining feed components are then converted through disproportionation and transalkylation to produce a near-equilibrium product.

During the late 1980s, Mobil developed a selective toluene disproportionation process (MSTDP) that uses shape selective catalysis to maximize *para*-xylene in the produced xylene stream [11]. As already underlined, in conventional toluene disproportionation, an equilibrium mixture of xylenes is expected (*p:m:o* = 24:50:26 at 300 °C). *para*-Xylene, formed as primary product, should rapidly equilibrate with the other two isomers since the intrinsic isomerization rate constant, k_I, is normally at least an order of magnitude larger than k_D, the intrinsic disproportionation rate constant. In fact, Olson and Haag [39] found that (k_I/k_D) intrinsic was greater than 7000 for MFI molecular sieves. The objective of STDP is to use shape selectivity to direct the primary product to be highly *para*-selective and to inhibit the secondary isomerization of the primary *para*-xylene product (i.e., k_I/k_D observed significantly less than unity). Olson and Haag evidenced the dependence of *para*-selectivity on isomer diffusivities and zeolite crystallite radius. As the diffusivity of *para*-xylene in ZSM-5 is three orders of magnitude larger than for *ortho*- and *meta*-xylene, by a proper increase of the crystallite radius, it is possible to minimize the isomerization so to produce *para*-xylene in excess of equilibrium. Olson and Haag also examined the effects of selectively modifying the diffusion properties through selective incorporation of phosphorus, magnesium, boron, silicon, antimony, and coke. Ultimately, it was determined that the deposition of coke on the exterior of the MFI molecular sieve was preferred since it increased the diffusion path length and covered all of the nonselective external sites of the zeolite crystallites without affecting the number of internal sites. MSTDP (i.e., Mobil's STDP process) is an excellent example of the application of shape selectivity to vastly exceed equilibrium yields of preferred products by playing off rates of diffusion versus rates of reaction [11]. Beside coke deposition, silanization may be considered to increase *para*-selectivity in toluene disproportionation [40].

More recently, ExxonMobil has developed a newer toluene dispoportionation process referred to as PxMax, which was first commercialized at Chalmette Refining's Louisiana refinery in 1996 and subsequently applied in 1997 at Mobil Chemical's grass-roots unit in Beaumont, Texas. Five additional units have been licensed since the PxMax technology was made available. The process uses a catalyst named MTPX, based on a silica-modified HZSM-5 zeolite (5%–10% SiO_2/HZSM-5). The selectivity to *para*-xylene can reach 98% at toluene conversions of about 20–25 wt.% (*para*-selectivity decreases as toluene conversion increases). Compared to its precursor, MSTDP it operates at lower temperatures and reduced H_2/hydrocarbon ratios. The MTPX catalyst has also been modified to reduce by-products formation, especially EB. This is done by incorporating a hydrogenating/dehydrogenating function in the catalyst, that is, platinum. Patent examples show that EB formation can be reduced by a factor of 3–4 and C_9^+ aromatics by a factor of 3 by adding 0.025% platinum on a 10% SiO_2/HZSM-5 catalyst formulation [41].

Similarly, UOP has developed a toluene disproportionation process *PX-Plus*, which is much more selective to *para*-xylene compared to its previous Tatoray process. Details about formulation and selectivation procedure for the PX-Plus catalyst are not disclosed. Because the PX-Plus process operates at lower conversions and higher recycle toluene levels than the Tatoray process, the destruction of aromatic rings must be minimized by the proper tuning of catalyst acidity and process conditions. At 30% toluene conversion, the selectivity to *para*-xylene is around 90%, with less then 2% light by-products per pass [42]. Typical product yields are shown in Fig. 5 [43].

For transalkylation of toluene with C_9 aromatic streams, large-pore zeolites can also be used, namely, mordenite has been used on an industrial scale [33]. Moreover, zeolites Beta, NU-87, and ZSM-12 have also shown interesting performances [44,45].

The selection between conventional toluene disproportionation-transalkylation (TDP) and STDP processes is driven by their peculiarities. STDP processes provide a highly concentrated *para*-xylene stream (greater than

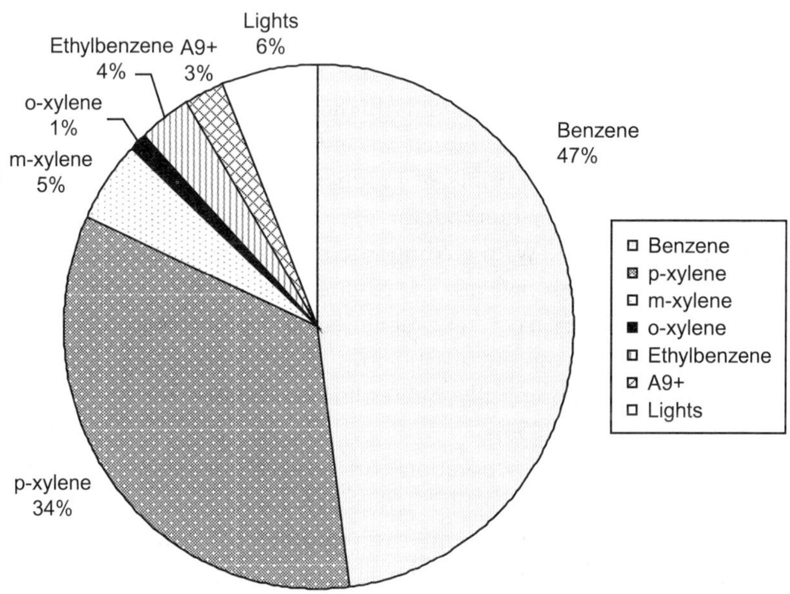

Figure 5 PX-Plus: typical product distribution in wt.% [43].

80%) and a large benzene by-production but requires a toluene feed. The conventional TDP technologies can process a C_9 aromatics stream, along with toluene but produce an equilibrium mixture of xylenes (approximately 20%–25% of *para*-isomer) with a smaller benzene by-production.

Toluene alkylation with methanol. Toluene methylation to produce xylenes has been the subject of considerable research. Compared to the toluene disproportionation process, the alkylation has some advantages:

1. Co-production of undesired benzene by-product is negligible,
2. Cheap and abundant methanol feedstock is used, and
3. Reduced toluene requirement per unit of *para*-xylene produced.

No processes have been developed at a commercial level so far, mainly because of the low conversion level of toluene that can be maintained (8%–10%) to achieve high *para*-xylene selectivity (more than 85%). Also, for this reaction, the selectivity to the *para*-isomer is affected by thermodynamic equilibrium. The *ortho*- and *para*-isomers formed by primary alkylation will then isomerize to the more stable *meta*-isomer. A similar strategy as in the case of selective toluene disproportionation has been adopted to improve *para*-selectivity. The catalyst, usually a MFI-type zeolite, can be modified by external surface passivation and pore size control. ZSM-5 treated with Mg, P, and B yields *para*-xylene preferentially. It is probable that inside the pores all three xylene isomers are made at equilibrium concentrations. However, the diffusivity of *para*-xylene in this system is several orders of magnitude higher than those of the other two isomers. As a consequence, product *para*-xylene concentrations may be as high as 90%. This is an example of product shape selectivity [38,46]. Das and Halgeri developed a method for pore size regulation of ZSM-5 by silanization with TEOS. With a composite catalyst based on this pore size–regulated ZSM-5, *para*-xylene selectivities above 85% with toluene conversions up to 28% have been obtained in a pilot plant. The catalyst has shown stability for more than 500 h on-stream, and these performances could provide an economically interesting technology for a stand-alone toluene conversion unit [33].

GTC Technology Corporation has started marketing a toluene methylation process (GT-TolAlk^{SM}) that uses a propietary zeolite catalyst [47]. In the process, the xylenes are easily purified to chemical-grade *para*-xylene in a simple crystallization unit. There is virtually no benzene by-product from this reaction, so that the *para*-xylene yield on toluene feed basis is the highest among all toluene conversion processes (Table 5).

Table 5 *para*-Xylene yield from toluene processes

Use of toluene	Tons *para*-xylene/100 tons toluene
TDP/STDP	40
C_7/C_9 transalkylation	70
Alkylation with methanol	100

STDP, selective toluene disproportionation process; TDP, toluene disproportionation-transalkylation.

4. ALKYLBENZENES BY ALKYLATION-TRANSALKYLATION REACTIONS

EB, cumene, *para*-DEB, *para*-DIPB, C_{10}–C_{14} LAB, *para*-ethyltoluene, and cymene are some of the chemical intermediates obtained by acid-catalyzed alkylation on the aromatic ring of benzene or toluene.

Scheme 2 summarizes several aromatic alkylations industrially applied for the preparation of important chemical intermediates. These reactions include the most important aromatic substrates, BTX, and different olefins. Besides, they include two different kinds of alkylation: the electrophilic alkylation on the aromatic ring catalyzed by acids and the side-chain alkylation catalyzed by bases.

From the point of view of production volumes, the acid-catalyzed alkylations of benzene are the most important: in 2004, about 75% of the benzene produced in the world (36.4×10^6 t/a) was alkylated with olefins, 72% of which was to produce EB and cumene (Table 1).

The alkylation reaction of aromatic hydrocarbons can be performed using various alkylating agents: alkyl halides, alcohols, alkyl sulphates, and olefins. Olefins are the most extensively used alkylating agents in the petrochemical industry. The alkylation reaction of the aromatic ring with olefins is an exothermic reaction and thus is favored, from the thermodynamic point of view, at low temperatures. The enthalpy of reaction at 25 °C in the gaseous state is -105.51 kJ mole^{-1} and -99.65 kJ mole^{-1}, for the formation of EB and cumene, respectively. Thus, the formation of these alkyl aromatics is accompanied by a release of energy, in the form of heat, which needs to be taken into account in the design of production plants.

4.1 Acid catalysis

Different types of acids are used as catalysts for the alkylation of aromatic hydrocarbons: (a) metal halides, such as aluminum and gallium chloride and borium fluoride; (b) mixed oxides and zeolites; (c) protonic acids such as

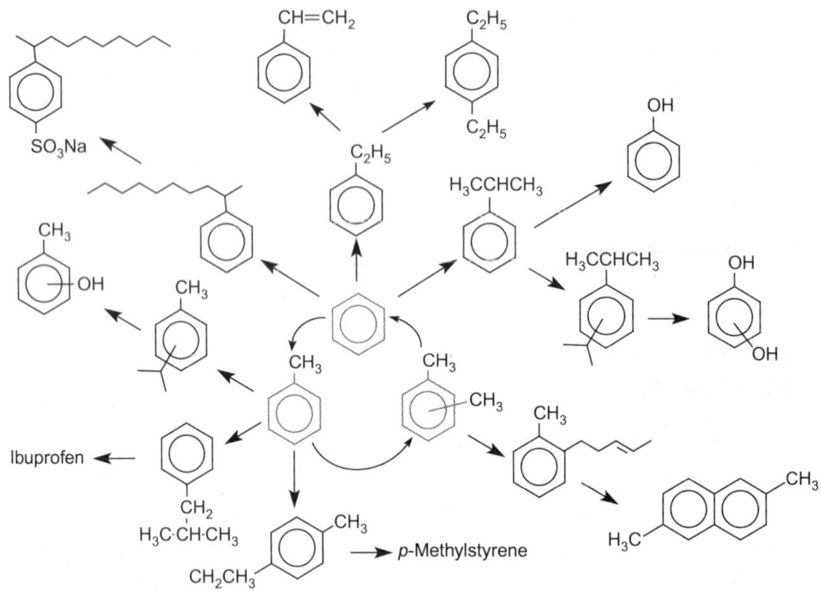

Scheme 2 Main aromatic hydrocarbon alkylations with olefins of industrial interest.

sulfuric acid, hydrofluoric acid, and phosphoric acid; and (d) sulfonic resins. The most active ones are the Brønsted acids, which contain an acidic proton. The metal halides, which are Lewis acids, are not very active alkylation catalysts if used as such and must be activated by means of adding small quantities of a co-catalyst such as a hydrohalic acid. The co-catalyst reacts with the Lewis acid, thus generating a Brønsted acid. These catalysts are also known as Friedel–Crafts catalysts and are still extensively used in alkylation, even if the new processes use solid acid calalysts.

The zeolites of interest for the alkylation of aromatics are predominantly of medium- and large-pore size (see Table 2).

4.1.1 Reaction mechanism

The alkylation mechanism first involves the formation of an electrophile E^+ by means of the interaction between the olefin and the acid, followed by electrophilic attack on the aromatic ring (Ar–H) with the formation of an intermediate $[E\text{–}Ar\text{–}H]^+$, known also as the Wheland intermediate or arenium ion. From this intermediate, an alkyl aromatic is formed by elimination of H^+, that is, of a proton. In the case of alkylation of benzene with ethylene catalyzed by a generic HA acid, the activated species $CH_3\text{–}CH_2^+A^-$ is often represented as a free carbocation $CH_3\text{–}CH_2^+$. This is, however, a qualitative representation, and in fact, the completely free carbocation is never obtained: the protonation of ethylene produces a single primary carbocation. In the case of higher olefins, the protonation can form two different types of carbocations. For example, from propylene, the secondary carbocation (*i*-propyl, $CH_3CH^+CH_3$) and the primary one (*n*-propyl, $CH_3CH_2CH_2^+$) may be generated. The relative stability of the carbocations increases in the order primary < secondary < tertiary and influences the rate and selectivity of the alkylation reaction.

4.1.2 Kinetics

The nature of the alkyl group affects the reaction rate. For example, the isopropylation of toluene is approximately 1460 times faster than its ethylation, using $GaBr_3$ as the catalyst and the alkyl bromide as the alkylating agent [48]. Similar data were reported for the alkylation of benzene with olefins, with a zeolite catalyst (REY, rare earth exchanged Y zeolite): at 100 °C, propylation is about 300 times faster than ethylation [49]. These data agree with the protonic affinity [50], that is, the tendency of an olefin to be protonated, thus creating the corresponding carbocation:

Ethylene to ethyl	672 kJ mole^{-1}
Propylene to *i*-propyl	755 kJ mole^{-1}
i-Butene to *t*-butyl	810 kJ mole^{-1}

4.1.3 Selectivity

The products of alkylations using acid catalysts are those derived from the most stable carbocations. For example, in the case of the alkylation with propylene, the *i*-propyl benzene (cumene) is practically the only product of monoalkylation, because the formation of the *i*-propyl carbocation is favored with respect to the *n*-propyl carbocation: the difference in the enthalpies of formation between the two carbocations in the gas phase is 67 kJ mole^{-1}. In addition to the alkylation of the aromatic ring, the olefin can also follow other reaction paths: it can react with itself forming higher oligomers, as in the case of the production of cumene, which is accompanied by the formation of propylene oligomers (e.g., nonene), or it can isomerize forming other olefins that, in their turn, can produce other alkylation products. An example for this case is the alkylation of benzene with 1-dodecene where, together with 2-phenyl dodecane, other isomers are also formed (3-, 4-, 5-, 6-phenyl dodecane).

Oligomerization and isomerization (of the olefin) are parallel reactions to alkylation and reduce its selectivity. Both are reactions that are catalyzed by acids and can be limited through the use of very selective alkylation catalysts or by means of suitable operating conditions. The selectivity is also affected by consecutive reactions. After the first alkylation, the aromatic substrate can undergo successive alkylations, forming polyalkylated by-products. The presence of alkyl substituents on the aromatic ring increases its reactivity, due to their ability to favor the delocalization of the positive charge on the Wheland intermediate. In fact, the presence of alkyl substituents such as ethyl or *i*-propyl on the aromatic ring increases the rate of Friedel–Crafts alkylation by a factor of 1.4 to 3.2 with respect to unsubstituted benzene [51]. Various methods are employed to maximize the yield of the monoalkylate. The most obvious is to operate with a large molar excess of the aromatic, namely, with a high aromatic/olefin ratio. This also makes it possible to limit the by-production of oligomers. The disadvantage is constituted by the necessity of separating the excess aromatics and recycling them, with elevated energy costs. The use of very selective catalysts such as zeolites allows to reduce the formation of the bigger polyalkylates and to favor the formation of the monoalkylates. For example, in the alkylation of benzene with propylene catalyzed by zeolite Beta, the ratio between the rate constants for the formation of the di- and monoalkylated products is equal to 0.54 [52].

4.2 Ethylbenzene

In 2004, the world production of EB was about 26×10^6 t/a, with the demand growing on average at a rate of 4%–5% per year. Almost all EB is used for the production of styrene, a raw material for thermoplastic polymers and elastomers.

The alkylation of benzene with ethylene catalyzed by acids proceeds according to the pathways reported in Scheme 3 [9]. The first stage is represented by the formation of the ethyl carbocation, which then follows one of two principal reaction paths. Either it reacts with benzene to give EB, following which, by successive alkylations, forms DEB and triethyl benzene (TEB), or it reacts with another molecule of ethylene to form a C_4 carbocation, which can then successively undergo alkylation, oligomerization, isomerization, and cracking to yield other alkyl benzenes and olefins. To a very limited extent, EB can alkylate benzene to 1,1-diphenylethane. It is important to emphasize that DEB and TEB can react with benzene to give EB by transalkylation. Transalkylation is an equilibrium reaction and under suitable conditions can already take place during alkylation.

The traditional process for the production of EB was developed around 1930. The catalyst used was $AlCl_3$-HCl, and all the operations were carried out in an agitated reactor, under somewhat gentle conditions: 170 °C and 0.7 MPa. On leaving the reactor, after separation, the polyethyl benzenes (mainly DEB and TEB) were recycled in the alkylation reactor where they

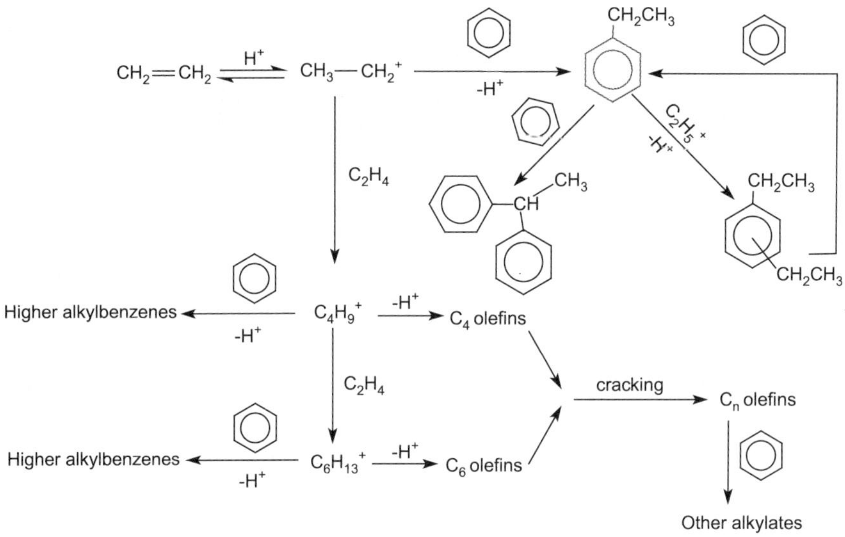

Scheme 3 Reaction pathway for benzene alkylation with ethylene.

were converted, in the presence of an excess of benzene, by transalkylation. This way, a conversion near to the thermodynamic equilibrium is achieved. The equilibrium composition is a function of the ratio of ethylene/benzene; this ratio is typically in the range between 0.35 and 0.55 [1]. The reaction is performed in the liquid phase and, because of the corrosive effect of the catalyst, the process is carried out in enamelled or glass-lined reactors. As a possible solution to the corrosion problems, supported catalysts have been proposed in the 1960s by UOP for liquid-phase (BF_3/Al_2O_3 in the Alkar^{TM} process) and gas-phase (Kieselguhr supported phosphoric acid, SPA) operations. However, due to partial release of the acids, corrosion problems were not completely avoided. Besides, these catalysts are not active in the polyethylbenzenes transalkylation and cannot be regenerated. For these reasons, such processes did not obtain large industrial interest.

4.2.1 EB processes based on zeolitic catalysis

To overcome the above problems, starting in the mid-1960s, various zeolitic catalysts were tested for this reaction. In 1976, Mobil-Badger started up the first industrial plant for the production of EB in the gas phase, with a fixed-bed reactor, loaded with a catalyst based on ZSM-5. The reactor was operated under high temperature (390–450 °C) and pressure (1.5–2 MPa) conditions. As in the process with $AlCl_3$, after separation, the polyalkylates were recycled to the reactor for transalkylation. Due to the deactivation related to the deposition of carbon-containing residues (coke) in the zeolitic pores, the catalyst had to be regenerated every 40–60 days. The regeneration was performed *in situ*, by blowing in air to allow the combustion of the coke. The high frequency of this operation made it necessary to have two reactors, one for the regeneration and one for the reaction, to guarantee continuous production. This process, used commercially from 1980 onwards, was successively improved by the addition of a reactor dedicated to the transalkylation of the polyethylbenzenes, thus obtaining an improvement both in the yield and in the catalyst life [53].

Another process was jointly developed by the NCL of India and by Hindustan Polymers Ltd., still using a ZSM-5-like zeolite that contains iron instead of aluminum in its framework [54]. This process, named Albene, is aimed to alkylate benzene with ethanol in a single-step reaction. The Albene process is operated at 350–450 °C, with a molar benzene/ethanol ratio of 4 and a weight hourly space velocity (WHSV) of 6 h^{-1}. Any DEB formed was transalkylated with benzene in a separate reactor to form more EB. The Albene EB process used dilute ethanol instead of ethylene because of the availability of ethanol from molasses in India.

A considerable improvement with respect to the processes described above was later obtained by UOP/Lummus/Unocal with the development of a process working in the liquid phase. The advantage of the liquid phase is represented by better thermal control that is reflected in an extension of the catalyst life. In this way, the regenerations are less frequent and can be conducted on the catalyst, which is discharged from the reactor and put into dedicated ovens. Due to problems related to diffusion control, medium-pore zeolites such as ZSM-5 were not suitable for the use in the liquid phase. For this reason, in the new process, the large-pore Y zeolite was used. The use of zeolite Y in the alkylation of benzene in the liquid phase was patented by Unocal in 1979 [55]. The process was commercially used for the first time in Japan in 1990 [56].

Other large-pore zeolites have been shown to be suitable catalysts for the liquid-phase alkylation of benzene with ethylene (e.g., L, Omega, ZSM-12, Beta). As a result of these investigations in the early 1990s, zeolite Beta was claimed to show the best performance [57,58]. In particular, zeolite Beta has turned out to be more selective than an Ultra-Stabilised Y zeolite (USY), with a global selectivity (EB + DEB + TEB) of 99.3% against 91.1% for zeolite Y [59]. Both zeolites have a three-dimensional system of channels, but the presence of large cavities (1.2 nm in diameter) at the intersections of the channels (in zeolite Y) is probably the cause of the formation of a large amount of by-products that, apart from reducing the selectivity, produce a faster deactivation of the catalyst [59].

Very interesting results have also been obtained with MCM-22, a zeolite of medium-pore size, characterized by two systems of channels independent of each other, one of which containing large cavities opened to the exterior on the external surface, with openings formed by 12 tetrahedra and with a diameter of 0.71 nm. Thanks to this peculiarity, MCM-22 demonstrates a catalytic activity comparable to USY, but inferior to that of Beta. However, the selectivity is higher, with respect to both USY and Beta, as the formation of DEB and TEB is particularly reduced [60]. MCM-22 was applied as catalyst for a liquid-phase process called EBMax^{TM}, which has been commercialized by ExxonMobil since 1995. In this new technology, the alkylation is performed in the liquid phase, while the transalkylation is still operated in the gas phase with a ZSM-5-based catalyst. The first commercial plant, at Chiba Styrene Monomer Co., started up in 1995 in Chiba, Japan. More recently, a new zeolite catalyst (trade name TRANS-4) was developed so that the transalkylation step is carried out in the liquid phase too. The nature of the zeolite was not disclosed, but according to a recent patent, it could be a TEA-mordenite [61].

Zeolite Beta is the process catalyst developed by Polimeri Europa for the production of EB. After some years of evaluation in a pilot plant, the catalyst was evaluated in an industrial reactor of an existing EB plant, starting in 2001. The catalyst demonstrated exceptional performance, from the point of view of both the consumption of raw materials and the quality of the EB produced. The flow diagram of the Polimeri Europa process is shown in Fig. 6 [62]. The catalyst is distributed in the reactor over several beds, and the supply of ethylene is directed over these in such a way as to create a much higher local ratio of benzene/ethylene with respect to the global ratio, for the reasons described above. This arrangement is frequently used in alkylation processes using reactors with catalytic beds. The first catalyst developed for industrial EB synthesis is the proprietary zeolite Beta-based PBE-1 catalyst, which is equally effective for alkylation and transalkylation to EB. More recently, the new proprietary zeolite-based catalyst PBE-2 has been specifically developed for industrial DEB transalkylation [62]. The nature of the zeolite used in PBE-2 has not been disclosed. In a recent patent, a catalytic composition based on a Y-type zeolite, characterized by a peculiar extra-zeolite porosity (i.e., the porosity of the binder) is reported to have an outstanding performance in DEB transalkylation to EB [63]

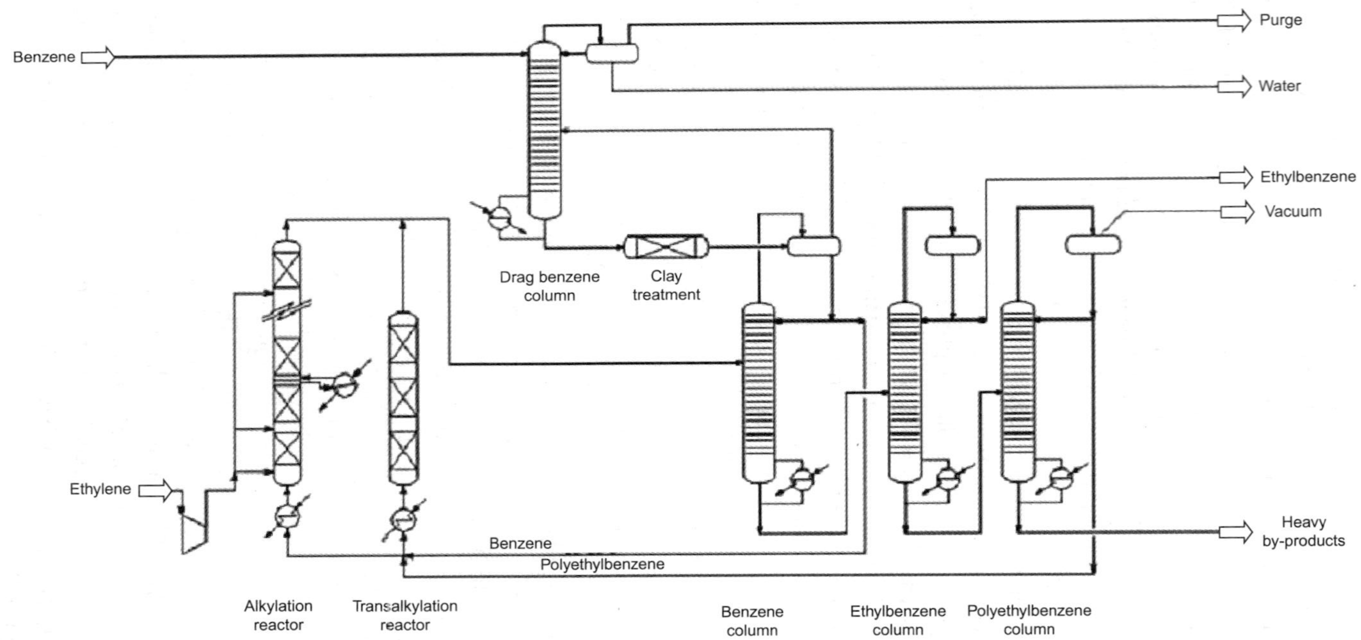

Figure 6 Process flow diagram of the Polimeri Europa ethylbenzene (EB) process [62].

Since 1990, UOP/Lummus has improved its own liquid-phase process [56], which now is used commercially under the name *EBOne*. The catalyst, originally a Y-type zeolite, now consists of a modified zeolite Beta. Two different catalysts are used in alkylation and transalkylation (UOP trade names EBZ-500 and EBZ-100, respectively). The improvements made to the catalyst have also been extended to the process developed by CDTECH (a consortium between ABB Lummus and Chemical Research and Licensing), based on catalytic distillation that combines the reaction and distillation in a single operation. The catalyst, packaged in "bales," is positioned on the plates of the distillation column. On each individual plate, as a result of the resistance to mass transfer and to the liquid-vapour equilibrium, the ratio of benzene/ethylene proves to be very high (>1000), with the resulting advantages in terms of selectivity. In addition, the heat of reaction is exploited to distill the benzene, thus making an energy saving. This process, known as CDTECH EB, is particularly suitable for diluted streams of ethylene [64].

Though less important than in gas-phase operation, deactivation also affects zeolite catalysts operating in the liquid phase, asking for a regeneration process. As already reported, in the case of liquid-phase operation, the regeneration is usually performed off-site, namely, on the spent catalyst discharged from the reactor. Several studies on the deactivation of zeolite catalysts during aromatics alkylation with olefins have been reported (see, e.g., for zeolites Beta [65,66], Y and mordenite [67], MCM-22 [68]). Coking is recognized to be the main reason of deactivation, and regeneration is usually performed by burning off the coke. A commercial experience concerning the regeneration of a batch of a spent catalyst (e.g., UOP EBZ-500 catalyst) has been reported [69]. The spent catalyst, originating from a European petrochemical site, has been successfully regenerated by Eurecat, France. The catalyst was deactivated by coke as well as nearly 0.2 wt.% nitrogen, which inhibited the zeolitic acid function. The oxidative treatment enabled complete carbon and nitrogen removal without any damage of the zeolite structure. Commercially regenerated catalyst performance in the customer unit is as good as for the fresh catalyst. Also, the mechanical strength of the EBZ-500 catalyst and the smoothness of the industrial Eurecat equipment avoided the formation of large quantity of fines during processing: only a low 0.5 wt.% fines generation occurred during regeneration. This allows a satisfactory lifetime duration of the regenerated catalyst.

Of over 70 industrial EB plants in the world in 2004, only 21% still used non-zeolite catalysts (mainly, $AlCl_3$–HCl); the others were using zeolitic catalysts: 22% in gas phase and 57% in the liquid phase [70].

4.2.2 Zeolite catalyst formulation

The alkylation catalyst is a composite material usually obtained by mixing the selected zeolite with a binder (e.g., γ-alumina or silica) and shaping into pellets or beads. Therefore, the final catalyst is characterized by:

1. Zeolite structure;
2. Zeolite crystallite morphology; and
3. Extrazeolite catalyst texture, that is, binder domain porosity.

Zeolite structure. It has already been reported that liquid-phase operations ask for zeolite structures characterized by large pores (e.g., Y, Beta) or cavities opened to the exterior of the zeolite crystallites (e.g., MCM-22). Several attempts have been reported to improve the performance of a given zeolite structure, with respect to not only the crystallite size and morphology (see later) but also to take advantage from structural peculiarities.

This is the case for MCM-22. This medium-pore zeolite has a peculiar character: the as-synthesized, uncalcined precursor possesses a layered structure. Only during calcinations, oxygen bridges are formed between layers, giving rise to the three-dimensional structure. The latter is characterized by a complex porous structure, formed by two non-interconnected pore systems, both having 10-MR pore openings. One-pore system is constituted by large supercages (with cylinder-like shape, 1.8 nm long and 0.71 nm wide), which are interconnected by slightly elliptical openings. The other pore system consists of sinusoidal channels, also with a slightly elliptical cross-section. To account for the interesting performance of MCM-22 in EB synthesis, it is assumed that a sensible number of the large cavities that characterize its structure are opened to the exterior at the termination of the crystallites. Following this model, it is assumed that on the [001] surface of the platelet-like crystallites, 12-MR hemisupercages are present, having a free diameter of approximately 0.71 nm. In this case, the EB formation should occur in these cavities, practically without diffusion barriers [60,71].

To further increase the number of these hemisupercages, two different approaches were suggested. Researchers from Mobil [72] have pillared the layered precursor, and a new MCM-36 structure has been developed in which the 1.8 × 0.7 nm cages are not closed but have the two halves separated by the pillars (Fig. 7). MCM-36 has been claimed as catalyst for EB production [73]. Corma et al. suggested to delaminate the MCM-22 precursors to obtain a new material called ITQ-2, where the channel system is still preserved but with the cavities being more accessible (see. Fig. 7) [74]. The new material was claimed to behave better than MCM-22 in the EB production (Fig. 8) [75].

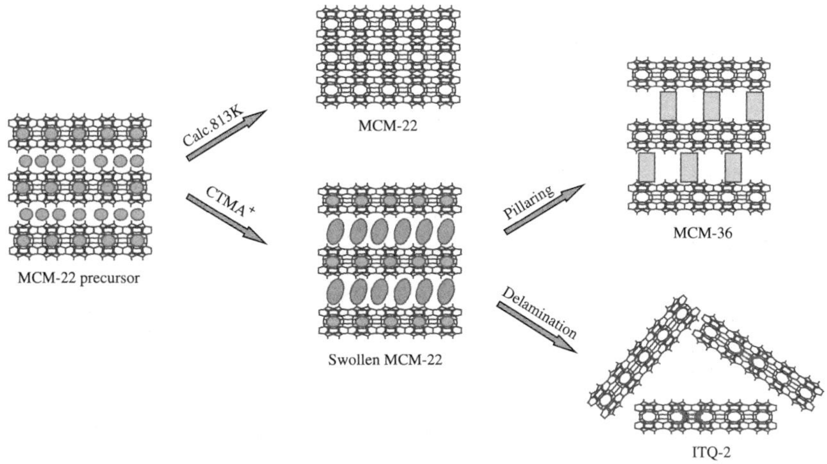

Figure 7 Scheme for the preparation of the different materials obtained from a MCM-22 precursor [74].

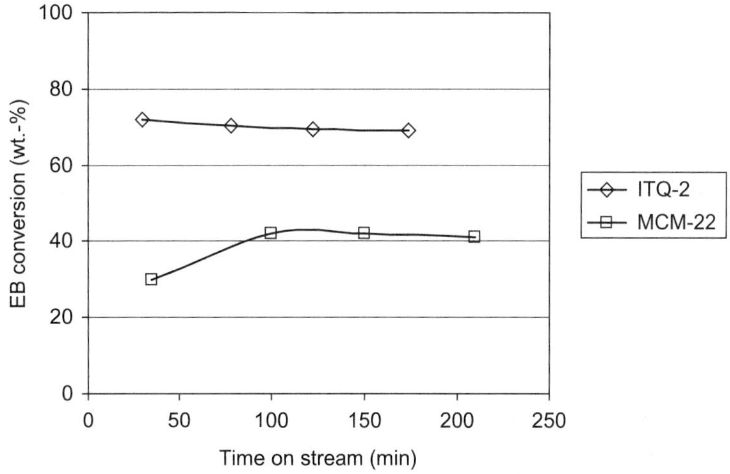

Figure 8 Comparison of the catalytic behavior of ITQ-2 and MCM-22 in ethylbenzene (EB) synthesis (T = 240 °C; P = 3.5 MPa; benzene/ethylene = 8) [75].

Zeolite crystallite morphology. Bellussi et al. [59] reported a significant decrease of the catalytic activity of zeolite Beta by increasing the size of crystallite conglomerates from 0.2 to 1 μm. Such a behavior was attributed to the influence of intracrystalline diffusion. To overcome the intracrystalline diffusion limitations, beside the reduction of the crystallite dimensions [59],

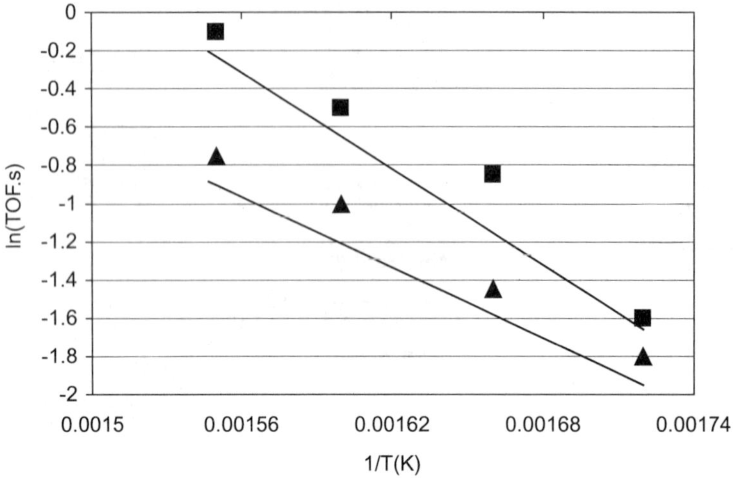

Figure 9 Arrhenius plot illustrating the activity difference between conventional and mesoporous zeolite catalysts in the alkylation of benzene with ethylene: mesoporous (■) and conventional (▲) MFI. The activities are expressed as turn-over frequency (TOF) [76].

other different approaches have been suggested. Christensen et al. [76] reported a new family of zeolite catalysts, that is, the mesoporous zeolite single crystals, exhibiting significantly improved catalytic activities and selectivities as compared to conventional zeolite catalysts in the alkylation of benzene with ethene. A classical synthesis of ZSM-5 was modified by introducing carbon materials as pore former. The mesoporous zeolite single crystals so obtained combine in each individual zeolite crystallite an intracrystalline micropore system typical of zeolites with an intracrystalline mesopore system. This dramatically improves the mass transport in the zeolite crystallites. In the gas-phase alkylation of benzene with ethylene, a significant improvement of the turn-over frequencies with respect to the parent ZSM-5 has been reported (Fig. 9). The increased activity is ascribed to the improved mass transport in the mesoporous zeolites, which is indicated by an increase of the apparent activation energy from 59 to 77 kJ mole^{-1}. Attributing the higher activity to improved mass transport is further substantiated by the differences in selectivities observed for the two catalysts. The higher selectivity to EB is explained by Christensen et al. in the following way: whenever an EB molecule is formed, it can either be transported into the product stream or undergo further alkylation. However, in a mesoporous zeolite, the diffusion path is significantly shorter than in the conventional zeolite and further alkylation is suppressed.

Another approach to overcome intracrystalline limitations was proposed by Lummus researchers [77]. By a so called dry synthesis, ultra-small crystallite of zeolites Beta, Y, and ZSM-5 were grown on a preformed silica-alumina particle having mesopores within the solid oxide. The composite so obtained shows higher activity with respect to a conventional zeolite: for the alkylation of benzene with ethylene, Dautzenberg and Angevine reported a more than double-catalytic activity [78].

Extrazeolite catalyst texture. A typical shaping procedure is the extrusion of dough obtained by mulling the zeolite powder with the binder. Starting from the same zeolite, it is possible to obtain catalysts with different textural properties (i.e., pore volume, pore size distribution, specific surface area). According to Girotti et al. [62], an important feature of a catalyst pellet is the extrazeolite porosity that can be tailored by a proper extrusion procedure. Increasing the extrazeolite porosity resulted in an increase of the catalyst life. Another critical parameter of the zeolite catalyst is the pellet size. Ercan et al. [79] reported the influence of the size by comparing zeolite Y extrudates (1.6 mm diameter) as such and crushed to smaller granules. This way, the catalyst utilization, which was limited due to pore diffusion limitation, could be increased from 11% to 38% for EB synthesis. Again, an improvement of the catalytic performance could be achieved by decreasing the pellet size, but in turn, this is limited by the pressure drop of the industrial reactor. To overcome the mass transfer limitations, a different formulation approach was proposed by Jansen et al. [80]. This new approach is based on the coatings of shaped support with zeolites. The advantages of these structured catalysts are to increase the activity per weight of zeolite, to reduce the pressure drop, and to improve the heat transfer. Coatings of α-alumina supports (i.e., extrudates and spheres) with zeolite Beta were prepared and tested for the transalkylation of DEB and in EB synthesis. These catalysts may be usefully applied in catalytic distillation units.

4.3 Isopropyl benzene (cumene)

The world production of cumene in 2004 was 9.5×10^6 t/a, with an installed capacity of around 10.5×10^6 t/a. It is almost exclusively used for the production of acetone and phenol, and a growth of 5% per year is predicted for the consumption of phenol.

The alkylation reaction of benzene with propylene is very similar to that with ethylene (Scheme 4) [8]: the *i*-propyl carbocation reacts with benzene to give cumene and, by successive alkylations, di- and triisopropylbenzenes. Di- and triisopropylbenzenes (TIPBs) can transalkylate to

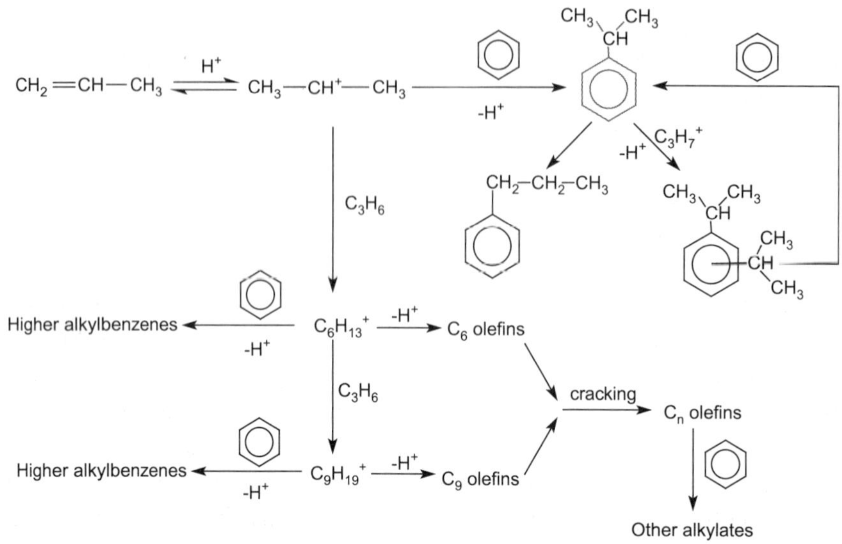

Scheme 4 Reaction pathway for benzene alkylation with propylene.

cumene in the presence of an excess of benzene. The carbocation can, in addition, react with propylene producing C_6 carbocations, which may react further through oligomerization, cracking, and alkylation to give higher oligomers and other alkyl benzenes. Small quantities of *n*-propyl benzene are also obtained by isomerization of cumene. This represents a very critical aspect, as *n*-propyl benzene cannot be separated by simple distillation, and thus, its formation affects the final quality of the cumene.

The demand for cumene as a high-octane additive for military aeroplanes in the Second World War led to the development of the first process based on the use of sulfuric acid. The problems related to the use of a free acid were overcome in the 1940s with the introduction, by UOP, of a catalyst based on SPA. This technology is still widely used today: the catalyst is loaded in a fixed-bed reactor operating in the liquid phase (180–240 °C; 3–4 MPa). The formation of polyalkylates, which are not transalkylated by SPA, and oligomers of propylene is minimized by operating with a high ratio of benzene/propylene (from 5 to 10). Also in this case, the ratio is further increased by splitting the propylene feed into individual catalytic beds. The SPA catalyst, even though supported, nevertheless generates problems of corrosion, due to a release of free acid. In addition, at the end of its life cycle, it cannot be regenerated.

In the 1970s, Monsanto-Lummus introduced a new technology based on the use of $AlCl_3$–HCl, very similar to that used for EB synthesis. The advantage of $AlCl_3$ resides in its ability to catalyze the transalkylation of the polyalkylates that can then be recycled to the alkylation reactor, in contrast to the SPA catalyst. However, this technology has been implemented only in a few plants.

4.3.1 Cumene processes based on zeolitic catalysts

The search for zeolitic catalysts for the production of cumene has been in many ways similar to that for EB, even if many more years were necessary to arrive at a significant result than with EB. This is mainly due to the fact that zeolite ZSM-5, the catalyst for Mobil's EB process, demonstrated a major limitation in the cumene reaction, represented by the elevated level of co-production of *n*-propyl benzene. On the contrary, ZSM-5, being of medium-pore size, is not sufficiently active in the liquid phase [59]. Also for cumene, a noticeable improvement was thus obtained by operating in the liquid phase with large-pore zeolites. Based on these zeolites, new commercial processes or industrial test runs were announced in the 1990s by Dow-Kellogg, Mobil-Raytheon, CDTech, EniChem, and UOP [81]. In all cases, it was a matter of improvements made in existing plants through the substitution of SPA with a zeolitic catalyst. Fig. 10 reports company names, process names, and zeolite types in use.

Mobil-Raytheon, EniChem (now Polimeri Europa), and UOP independently started up separate industrial zeolite-based cumene plants in 1996. Cumene yields are above 99.5% in all zeolite-based processes. Product purity is as high as 99.95%.

Fig. 11 compares the quality of cumene obtained with the Polimeri Europa process (i.e., PBE-1, zeolite Beta catalyst) and with the traditional SPA catalyst in terms of the concentration of impurities. After years of operation, PBE-1 has shown a high level of stability, reaching a production of

Company	Mobil - Raytheon	UOP	CD-Tech	Dow - Kellog	Polimeri Europa
Process		Q-max	CDCumene®	3-DDM	
Zeolite	MCM-22	Beta	Y	Mordenite	Beta
	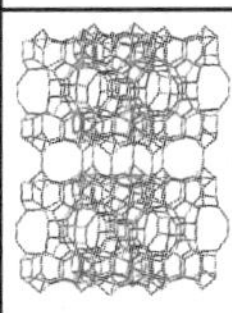	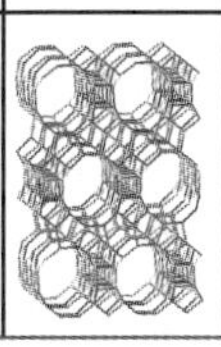	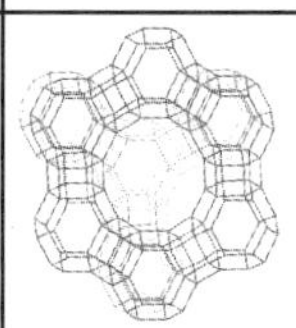	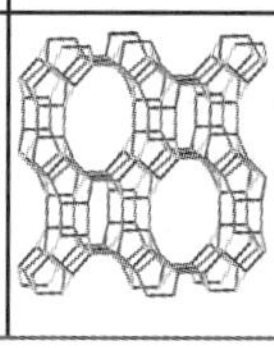	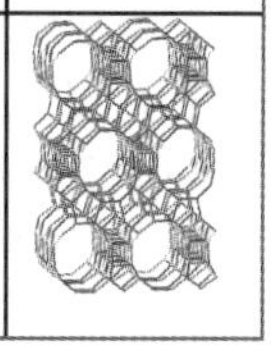

Figure 10 Processes for the production of cumene with zeolite catalysts.

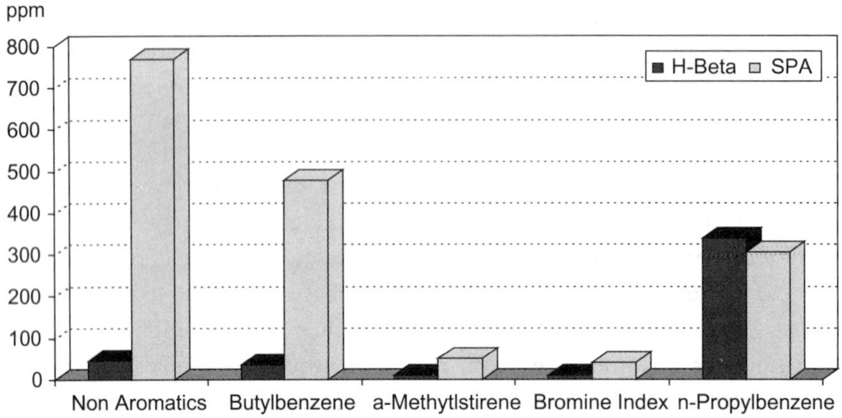

Figure 11 Comparison of catalyst performance between zeolite Beta and supported phosphoric acid (SPA) as catalyst in the production of cumene [62].

more than 30,000 tons of cumene per ton of catalyst, compared with the 1500–2000 tons obtained per ton of SPA. The flow diagram of the Polimeri Europa cumene production process is shown in Fig. 12 [62]. In 2001, already 14 cumene plants out of around 48 existing in the world were operating with a zeolitic catalyst [7]. According to Chemical Week, in 2004, about 50% of the total cumene production units use zeolite catalysts, as producers have switched from older phosphoric acid–based processes [82].

Progress in the zeolite catalyst. A number of different structure types of medium- and large-pore zeolites have been evaluated in benzene alkylation with propylene. Medium-pore zeolites (e.g., ZSM-5, ZSM-11, EU-1, and MCM-22) and large-pore zeolites (e.g., Y, mordenite, ZSM-12, Beta, and NCL-1) have been studied for this reaction. The subject has been extensively reviewed in the past decade by some comprehensive overviews [5,7,8,11,81,83]. Despite the large number of structures considered, only few of them at the end of the story have been really applied.

Mordenite has been applied by Dow-Kellog in so-called 3-DDM cumene process. 3-DDM stands for three-dimensional dealuminated mordenite [84]. In fact, considering only the 12-ring channels, mordenite is a "one-dimensional" zeolite. Dealumination converts the material into a novel catalyst by introducing some mesoporosity, resulting in a pseudo three-dimensional structure that provides optimal performance and stability. This 3-DDM catalyst has been industrially applied by Dow-Kellog as transalkylation catalyst in the cumene plant of Terneuzen since 1992.

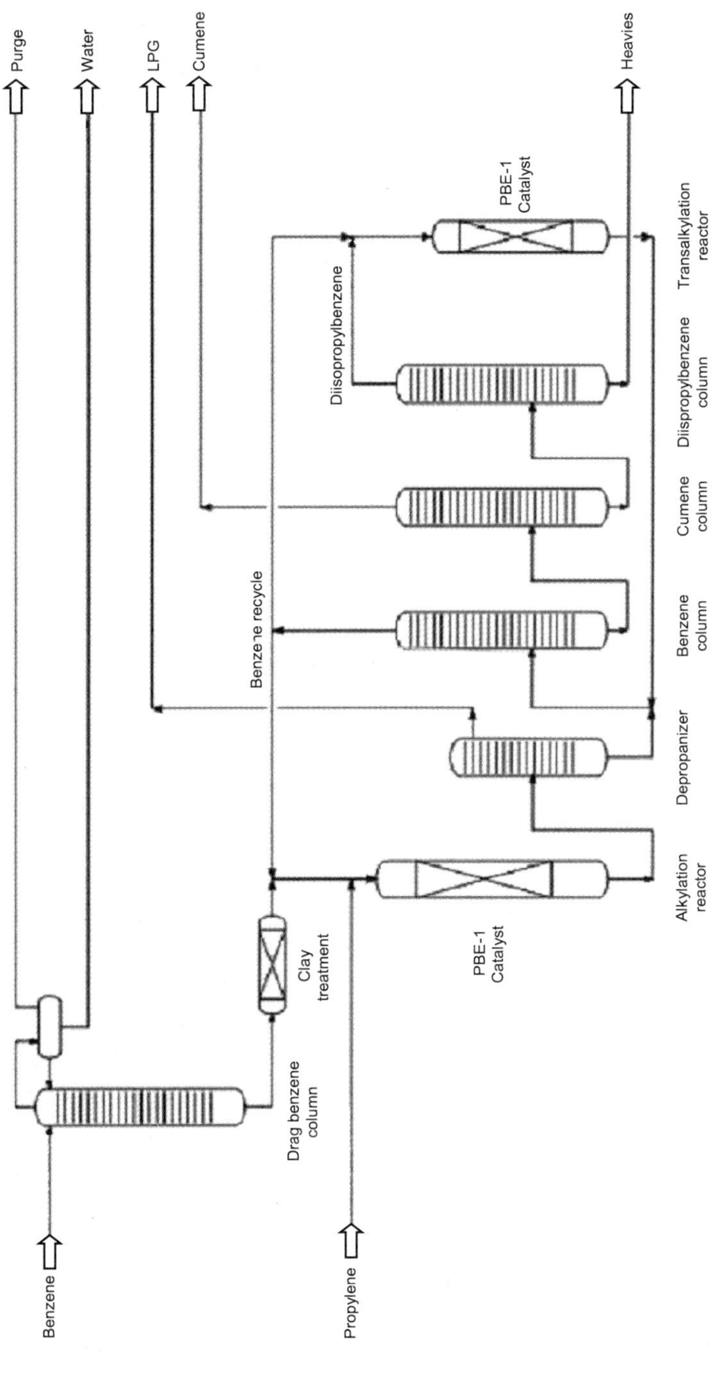

Figure 12 Process flow diagram of Polimeri Europa cumene process [62].

Table 6 Comparison of MCM-22 and MCM-56 in the liquid-phase alkylation of benzene with propylene [85]

Parameters	Catalyst MCM-22	Catalyst MCM-56
Temperature (°C)	112	113
Propylene WHSV (h^{-1})	1.3	10.0
Propylene conversion (%)	98.0	95.4
Selectivity (%)		
Cumene	84.35	84.98
DIPB	11.30	13.20
TIPB	2.06	1.28
C_3 oligomers	1.8	0.52
n-PBZ (ppm)	70	90

DIPB, diisopropylbenzene; *n*-PBZ, *n*-propyl benzene; TIPB, triisopropylbenzene.

The other zeolites used in the current processes for cumene production are similar to that used for EB, that is, zeolites Beta and MCM-22. Over the past decade, great progress has been made in improving and optimizing catalyst formulations for the use in cumene production. For example, as further development of the MCM-22-based catalyst, Mobil researchers have patented MCM-56 with improved catalytic activity for both EB and cumene production [85]. Table 6 compares the catalytic activity of MCM-22 and MCM-56 in the liquid-phase alkylation of benzene with propylene. Though the space velocity is one order of magnitude higher in the case of MCM-56, conversion and selectivity of both catalysts are comparable. MCM-56 and MCM-22 are synthesized using the same structure directing agent, that is, hexamethyleneimine, and share the same MWW topology. The interesting feature of MCM-56 is that the XRD pattern of calcined MCM-56 has very broad peaks suggesting that it is formed of very thin layers [86]. As the structure of MCM-56 is similar to that of MCM-22, then a larger number of the 12 MR hemisupercages would be exposed to the crystal exterior, making MCM-56 more active as solid acid catalyst. This behavior was further confirmed by Corma et al. [87], who measured the kinetics of TIPB cracking: MCM-56 is more active than MCM-22. Extending the comparison to ITQ-2, Corma et al. evidenced a better activity of the latter with respect to both MCM-22 (as already pointed out

Table 7 Kinetic rate constants (s^{-1}) for cracking of triisopropylbenzene (TIPB) on different zeolite samples [87]

Sample	k (s^{-1}) × 10^2
MCM-22	8.6
MCM-56	9.9
ITQ-2	25.1

for EB production) and MCM-56. In Table 7, the kinetic constants measured for all mentioned zeolites are reported.

The possibility to synthesize zeolite Beta in a wide range of SiO_2/Al_2O_3 ratios has given catalyst designers the opportunity to tailor the zeolite into a form that optimizes activity and selectivity. A parametric study on the effects of SiO_2/Al_2O_3 ratio on activity and selectivity was published by Bellussi et al. [59]. In this work, it was found that as the SiO_2/Al_2O_3 ratio was increased from 28 to 70; there was a decrease in both activity and selectivity toward isopropylbenzenes (cumene, di- and TIPBs). Additionally, the less active catalysts had a greater tendency toward oligomerization and were more prone to coking. According to Schmidt [88], this study parallels work performed at UOP where it has been found that with a SiO_2/Al_2O_3 ratio of 25, the catalyst maintains sufficient activity to achieve polyalkylate equilibrium (e.g., DIPB equilibrium) and, at the same time, minimizes formation of heavier diphenyl compounds (and hence maximizes yield) in cumene synthesis. This work [88] also underlines the importance to minimize the Lewis acidity of the zeolite Beta and at the same time maintains high Brønsted acidity. In fact, olefin oligomerization is reported to be directly related to the Lewis acid function. As the oligomerization reactions can lead to the formation of heavy compounds (coke-type precursors), the minimization of the Lewis acid character of Beta should lead to a catalyst with high stability. Schmidt also reports the historical development of zeolite Beta for aromatics alkylation, showing that early versions of zeolite Beta catalyst demonstrated non-optimum performance when compared to today's state-of-the-art formulation. Fig. 13 is a plot of the relative stability of zeolite Beta as a function of SiO_2/Al_2O_3 ratio in the framework in which the dominating influence of this parameter is evident. UOP has applied this knowledge to the development of the new generation cumene alkylation catalyst QZ-2001 [88] with respect to the former QZ-2000 [89]. The new formulation exhibits as much as two times

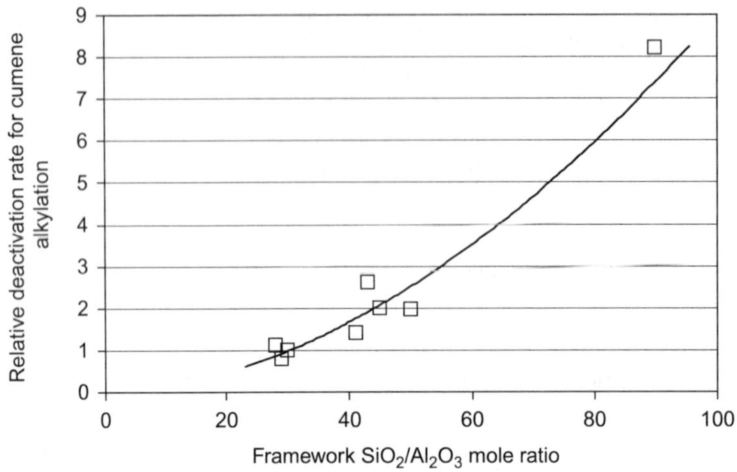

Figure 13 Relative stability of zeolite Beta as a function of the SiO_2/Al_2O_3 ratio [88].

the time-on-stream stability when compared to the older formulation. The importance of the Lewis and Brønsted acidity of zeolite Beta has been also underlined by a very recent patent of Polimeri Europa [90]. This patent claims the synthesis procedure of a zeolite Beta with a molar ratio of Lewis sites/Brønsted sites equal to or higher than 1.5. Such a zeolite behaves better than samples with lower ratio in the alkylation/transalkylation of benzene to cumene.

4.4 *para*-Ethyltoluene and *para*-diethylbenzene

para-Ethyltoluene and *para*-DEB are two important dialkylaromatic intermediates. The first one is used for the production of an important monomer, *para*-methylstyrene, by dehydrogenation. The worldwide production of this monomer accounted for 17,000 tons in 2002. *para*-DEB is used as desorbent in adsorptive separation processes, for example, in the Parex SM (UOP) and the EluxylSM (IFP) process [33,91]. According to Tsai et al. [2], the worldwide annual demand for *para*-DEB was estimated to 12,000 tons in 1999. These two dialkylbenzenes can be synthesized by the following routes:

1. Ethylation of toluene and EB by ethylene or ethanol,
2. Transalkylation of toluene and DEB, and
3. Disproportionation of EB.

The main goal in all these reactions is the *para*-selectivity, that is, the capability of the catalytic system to selectively favor the formation of the *para*-isomer. This subject has been extensively studied by several authors. A comprehensive review of the subject was published quite recently by Cejka and Wichterlova [5].

The report of Kaeding et al. [92] on the alkylation of toluene with ethylene showed the ability of zeolite H-ZSM-5 to catalyze the formation of dialkylbenzenes with good selectivities. In an attempt to increase the zeolite selectivity for *para*-ethyltoluene, Kaeding et al. modified H-ZSM-5 with phosphorus and by impregnation with Mn, Mg, and B salts [92,93]. A dramatic increase in the *para*-selectivity up to 98% was observed and ascribed mainly to spatial restrictions in the narrowed zeolite pores rather than differences in the zeolite acidity. Based on this zeolite catalyst, Mobil developed an ethyltoluene process that was industrially applied in Baton Rouge, Louisiana, in 1982.

In addition to the modification mentioned above, selective silanation of the surface sites of H-ZSM-5 by TEOS has been shown to increase the zeolite *para*-selectivity in the methylation, ethylation, or propylation of toluene and in the ethylation of EB [40]. Table 8 compares the performance of silanized zeolite for the above-mentioned four alkylation reactions. Alkylbenzene conversion is higher in ethylation than methylation as the latter needs higher reaction temperature for an appreciable conversion. The *para*-dialkylbenzene selectivities were 100%, 95%, 93%, and 88%, respectively, for *para*-cymene, *para*-DEB, *para*-ethyltoluene, and *para*-xylene. This is in the order of the number of carbon atoms in the side chain of the dialkylbenzenes, which in turn is related to diffusivity differences between the *para* and the other isomers of dialkylbenzenes [40]. TEOS cannot penetrate the narrow pores of ZSM-5 and therefore only affects the surface. It narrows the pore openings, coats the external surface, and eliminates external acid sites without affecting the internal ones. This treatment is called chemical vapor deposition (CVD). Thus, the primary *para* product could not isomerise on the non-shape selective outer surface of the zeolite crystallites to the other isomers. The CVD treatment increased the *para*-DEB selectivity from 49.4% to 97.1%. The silanation of ZSM-5 has been successfully applied in the *para*-DEB production plant operated by Indian Petrochemical Company since 1994 with a capacity of 1200 t/a [33].

para-DEB can also be produced by EB disproportionation. A variety of different zeolitic catalysts, either medium-pore zeolites (e.g., ZSM-5, ZSM-22, MCM-22) or large-pore zeolites (e.g. ZSM-12, ZSM-34, Y, EMT, L, Beta) have been studied [94–96]. The products obtained in the disproportionation reaction normally give the thermodynamic equilibrium composition and therefore require energy intensive operations for

Table 8 Performance comparison of silylated ZSM-5 for different alkylbenzene alkylation reactions [40]

	Reactions			
	1	2	3	4
Monoalkylbenzene conversion (wt.%)	6.12	19.51	19.27	8.03
Selectivity to products (wt.%)				
Cymene				
meta	0.00			
para	28.98			
ortho	0.00			
DEB				
meta		3.28		
para		70.72		
ortho		0.00		
Ethyltoluene				
meta			5.45	
para			83.55	
ortho			0.00	
Xylene				
meta				5.16
para				63.88
ortho				3.54
Others	71.02	26.00	10.04	27.42
Dialkylbenzenes isomer				
meta	0.00	4.43	6.12	7.11
para	100.0	95.57	93.88	88.02
ortho	0.00	0.00	0.00	4.87

DEB, diethylbenzene.Conditions: temperature = 350 °C, H_2/HC = 3, alkylbenzene/alcohol = 5 and WHSV = 5.2 h^{-1}, silanation period = 180 min (16% silica), 1: toluene isopropylation, 2: ethylbenzene ethylation, 3: toluene ethylation, 4: toluene methylation.

para-DEB recovery [2]. Thus, the technical challenge in producing *para*-DEB by EB disproportionation is technology development to enhance *para*-selectivity. Wang et al. [97,98] applied the surface silanation by CVD to modify ZSM-5, by which they were able to obtain *para*-DEB with product selectivities up to 99%. These authors also concluded that, among the various types of orthosilicates investigated, silanation by TEOS yielded the best *para*-DEB selectivity. The above *para*-DEB selectivation process using modifed ZSM-5 by Si-CVD surface silanation method was successfully commercialized by Taiwan Styrene Monomer Company (TSMC) in 1988. The commercial plant that produced *para*-DEB with 96% purity went on stream in 1990 with a capacity of approximately 4000 t/a. Recently, TSMC further upgraded their *para*-DEB production purity to 99%. The TSMC process has a cycle length of more than six months and its catalyst is fully regenerable [2].

4.5 Diisopropylbenzene

DIPBs are industrially used in the hydroperoxidation technology for the production of dihydroxybenzenes. This technology is analogous to the production of phenol from cumene (i.e., the Hock process): acetone is a by-product in both cases. Three successive reactions are involved: the alkylation of benzene with propylene to DIPBs, oxidation of DIPBs with air to the corresponding dihydroperoxide, and the acid cleavage of the latter to dihydroxybenzene and acetone. The dihydroperoxide step requires purified individual DIPB isomers. The boiling points of the *ortho*-, *meta*-, and *para*-isomers are 203.8, 203.2, and 210.4 °C, respectively. Therefore, while it is feasible to obtain *para*-DIPB by fractional distillation, it is not practically possible to separate the *ortho*- and *meta*-isomers. During alkylation, for steric reasons, the amount of *ortho*-DIPB formed is lower than those of *meta*- and *para*-DIPBs [9] and can be limited by a proper selection of the zeolite catalyst. In fact, due to the different steric hindrance of the isomers, the barriers for diffusion inside the zeolite channels increases in the order *para* < *meta* < *ortho* [99]. The alkylation of benzene to DIPBs has been already discussed in Section 4.3. According to Scheme 4, first benzene reacts with propylene to cumene, which undergoes further alkylation to DIPBs. The isopropylation of cumene is an electrophilic substitution. Isopropyl is an *ortho-para* orienting group. Therefore, first, *ortho*- and *para*-isomers are formed, followed by isomerization to the *meta*-isomer, which is thermodynamically more stable. As for the transalkylation, the isomerization of polyalkylbenzenes requires a catalyst with stronger acidity than the alkylation. Fig. 14 compares the *meta*/*para* DIPB ratio observed in benzene alkylation with propylene for different

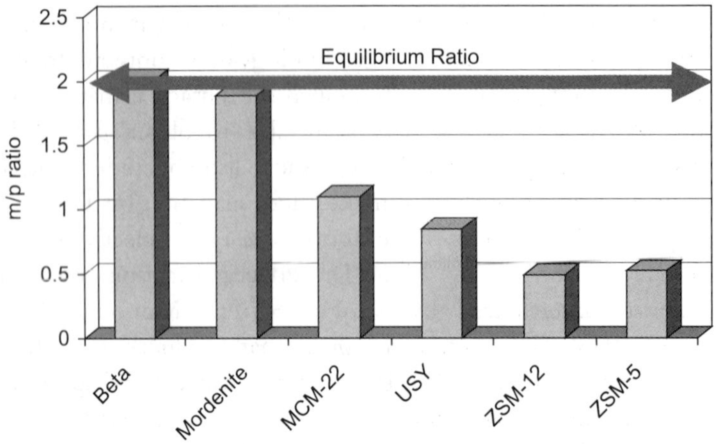

Figure 14 Alkylation of benzene with propylene (T = 150 °C; P = 3.5 MPa; benzene/propylene = 7; WHSV = 3 h^{-1}): *meta-/para*-diisopropylbenzene ratio [96,99].

zeolites (as reported in [96,99]) with the values of the thermodynamic equilibrium. A high *meta/para* ratio, close to the equilibrium value, indicates a high isomerization activity of zeolites Beta and mordenite. In contrast, the low *meta/para* ratio obtained with ZSM-5 and ZSM-12 can be explained by a shape selectivity effect. As a matter of fact, the *para*-selectivity of ZSM-5 and ZSM-12 in the isopropylation of cumene was already reported [100,101]. In the case of USY, because of the presence of supercages, shape selectivity effects should be excluded; therefore, the low *meta/para* ratio indicates a lower isomerization activity, that is, a milder acidity with respect to zeolites Beta and mordenite. MCM-22 shows an intermediate behavior, probably because of a combination of low isomerization activity and shape selectivity.

If the desired product is hydroquinone (i.e., *para*-dihydroxybenzene), *para*-DIPB is recovered by distillation of the alkylate. When the desired product is resorcinol (i.e., *meta*-dihydroxybenzene), the alkylate, after addition of recycled *para*-DIPB and TIPB is subjected to isomerization/transalkylation. During this step, most of the *ortho*- and *para*-DIPBs and TIPB are converted to the more stable *meta*-DIPB. The reaction mixture is separated by distillation into three fractions: benzene/cumene, *meta*-DIPB, and *para*-DIPB/TIPB. During the following autoxidation step, the residual presence of *ortho*-DIPB is not critical as it is resistant to oxidation.

Companies that produce resorcinol and/or hydroquinone through the DIPB peroxidation process include Sumitomo Chemical Company, Mitsui Chemicals, The Goodyear Tire & Rubber Company, and Eastman Kodak

[102,103]. The total capacity of the corresponding plants in terms of DIPBs transformation is around 56,000 t/a.

As for DEB, also DIPBs are obtained by disproportionation of the correspondent monoalkylbenzene, that is, cumene [104]. In this case, however, the overall selectivity is affected by a competing side reaction, namely, the isomerization of cumene to *n*-propylbenzene, followed by the possible formation of nine dipropylbenzene (DPB) isomers. DPB has three skeletal isomers, namely, DIPB, *n*-propylisopropylbenzene (NIPB), and di-*n*-propylbenzene (DNPB), which differ by the branching of the propyl group. Each of the three skeletal isomers can be further divided into three ring positional (*para*, *meta*, and *ortho*) isomers [2]. Tsai et al. [104] compared the catalytic performance of various zeolites in cumene disproportionation. It was demonstrated that among zeolite Y, mordenite, ZSM-12, and Beta, the last one had the greatest activity and disproportionation selectivity. In addition to selectivity, catalytic stability is the other major concern in the development of an industrial IPB disproportionation process. In a recent paper, Tsai et al. [105] reported that a mordenite-type catalyst maintains a cumene conversion of 50% (which is the approximate equilibrium conversion) with a DIPB selectivity of nearly 100% for 300 h without aging.

4.6 Cymenes (methylisopropylbenzenes)

Cymene (methylisopropylbenzene) production is commercially carried out by alkylation of toluene with propylene. *meta*-Cymene and *para*-cymene are intermediates for the production of *meta*- and *para*-cresol through oxidation and acid cleavage (Scheme 5). Although the demand for cymene is much lower than for cumene, some commercial units are operating with an installed capacity of around 40,000 t/a [8]. The alkylation produces a mixture of cymene isomers (i.e., *ortho*, *meta*, and *para*). The most preferred isomer distribution requires a low *ortho*-cymene content, since *ortho*-cymene is difficult to oxidize and inhibits the oxidation of the other isomers. The lowest *ortho*-cymene content is obtained from an isomer mixture with the

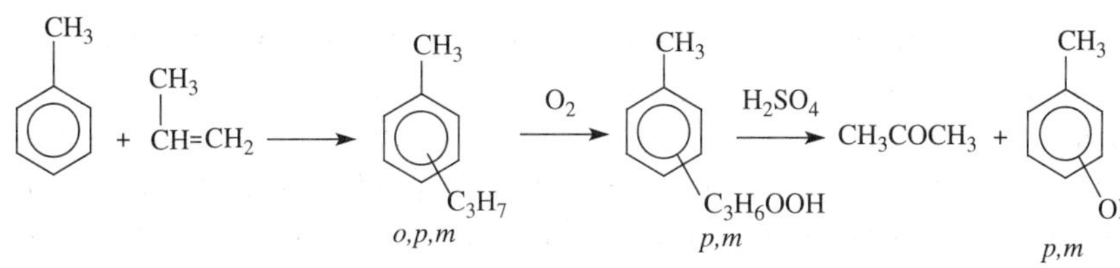

Scheme 5 Cresols production from toluene through alkylation/oxidation and acid cleavage.

thermodynamic equilibrium (i.e., 4.5% *ortho*-, 64.51% *meta*-, and 30.99% *para*-cymene at 200 °C [106]).

The isopropylation of toluene is an electrophilic substitution on the aromatic ring, activated on the *ortho* and the *para* positions by the presence of the methyl group. The *para* position is favored because of steric hindrance of the methyl group. *meta*-Cymene is thermodynamically more stable and can be obtained by either direct isopropylation or isomerization. Hence, the final isomer distribution depends on the combination of alkylation versus isomerization rates.

As for cumene, two catalytic technologies are applied for the production of cymenes, based on $AlCl_3$–HCl and SPA. Using $AlCl_3$–HCl (liquid-phase, $T = 60-80\,°C$), an isomer ratio close to the equilibrium one is obtained (namely, 3% *ortho*-, 64% *meta*-, and 33% *para*-cymene) [1]. After oxidation to cymene hydroperoxide, the excess of cymene, containing more *ortho*-cymene than the feed, is recycled to the alkylation step, so that the *ortho*-cymene content can again be lowered through isomerization on $AlCl_3$. The SPA process differs from the $AlCl_3$ process with respect to the isomer distribution obtainable, which is far from the equilibrium one (40% *ortho*-, 25% *meta*-, and 35% *para*-cymene) [107]. The process has a separation unit (Cymex, UOP), based on a 13X-type molecular sieve, for the separation of *meta*- and *para*-isomers, which allows the production of the pure corresponding cresols [108]. The *ortho*-isomer is then transferred to an isomerization unit.

4.6.1 Zeolite catalysts

The isopropylation of toluene with propylene or isopropanol was largely studied using different zeolite catalysts. Fraenkel and Levy reported that, in the gas phase alkylation of toluene with isopropanol ($T = 250\,°C$), ZSM-5 is more *para*-selective than large pore zeolites like mordenite and zeolite Y (*para*-cymene = 97.2% vs. 46.0% and 37.6%, respectively) [109]. On the contrary, ZSM-5 catalyzes the production of *n*-propyltoluene in non-negligible amounts (9.9 wt.%), while mordenite and Y produce only cymene. The mechanism of *n*-propyltoluene formation was elucidated by Wichterlova and Cejka [110]: the cymene isomers formed in the first alkylation step subsequently react with toluene with a bimolecular mechanism, forming *n*-propyltoluene. The ability to produce *n*-propyltoluene in the gas-phase alkylation of toluene with isopropanol increases in the order mordenite < Y < ZSM-5. The relative concentration of *n*-propyltoluene to cymene increased substantially by increasing the temperature. ZSM-5 produces more *n*-propyltoluene than zeolite Y, as reported in Table 9.

Table 9 *n*-Propyltoluene/cymene ratio in the alkylation of toluene with isopropanol (9.6 molar ratio; WHSV = $10\,h^{-1}$, atmospheric pressure) [110]

Zeolite	247 °C	297 °C	347 °C
ZSM-5	0.13	2.5	2.5
Y	0	0.23	1

Better selectivities to cymene in the gas-phase alkylation of toluene with isopropanol were reported for zeolite Beta with respect to ZSM-5 [111] and to other large-pore zeolites (e.g., mordenite and ZSM-12) [112] at a lower temperature (180–220 °C).

In the liquid-phase alkylation of toluene with propylene (T = 180 °C, P = 3.9 MPa) [113], zeolite Beta produces a low concentration of *ortho*-cymene and a high *meta*/*para* ratio, close to the thermodynamic equilibrium (4.3% *ortho*-, 68.3% *meta*-, and 27.5% *para*-cymene vs. 4.1% *ortho*-, 64.82% *meta*-, and 31.08% *para*-cymene at 180 °C [106]). Also, amorphous mesoporous silica-aluminas (i.e., MCM-41 and MSA) were considered for the liquid-phase alkylation of toluene with propylene [113]. MSA and MCM-41 show an alkylation activity comparable to zeolite Beta, but lower cymene selectivity, because of the larger formation of polyalkylates. The distribution of cymene isomers is far from equilibrium (namely, 34.0% *ortho*-, 28.1% *meta*-, and 37.9% *para*-cymene with MSA and 36.7% *ortho*-, 24.8% *meta*-, and 38.6% *para*-cymene with MCM-41) and similar to the one reported for SPA. In conclusion, zeolite Beta is the best candidate to substitute $AlCl_3$–HCl for the liquid-phase alkylation of toluene with propene, while MSA and MCM-41 can be considered as a possible alternative to the SPA catalyst.

An alternative route to produce cymene was proposed by Bandyopadhyay et al. [114]. Accordingly, cymene was produced by transalkylation of toluene with cumene and 1,4-DIPB. Among the large-pore zeolites considered, Beta was more active and selective than zeolites Y and ZSM-12.

5. CONCLUSIONS

The growing demand for selected alkylbenzenes, the characteristics of the aromatic feedstock available, and the stringent environmental legislation have pushed the innovation in zeolite catalysis for alkylbenzene production. Alkylation of aromatics with olefins and alkylbenzene interconversion (isomerizations, disproportionations, transalkylations) have gained advantages from the achievements in zeolite science and catalysis.

The processes based on benzene alkylation for both EB and cumene production have been characterized by an impressive improvement since the application of liquid-phase operation instead of the gas-phase mode. To such a purpose, the use of large-pore zeolites (e.g., Beta and MCM-22) has played a major role for the successful development of new processes. Starting from these catalysts, new advances have been then announced either by the development of new materials (e.g. ITQ-2, MCM-36, mesoporous MFI zeolite) or by improved catalyst formulations.

The selectivation of suitable zeolites for the maximization of *para*-alkyl aromatics (e.g., *para*-xylene, *para*-ethyltoluene, *para*-DEB, *para*-DIPB) by means of post-synthesis treatment technologies (pre-coking, silanation) have been advantageously exploited.

In the near future, zeolites with extra-large-pore openings (i.e., larger than 12-MRs) and new specially tailored mesoporous materials could be introduced for the same reactions (e.g., alkylation/transalkylation) applied to bulkier molecules (i.e., alkylnaphthalene, dialkylbiphenyl, alkyldibenzothiophene).

REFERENCES

[1] H.G. Franck, J.W. Stadelhofer, Industrial Aromatic Chemistry, Springer-Verlag, Berlin-Heidelberg, 1988.

[2] T.C. Tsai, S.B. Liu, I. Wang, Appl. Catal. A: Gen. 181 (1999) 355.

[3] H.A. Wittcoff, B.G. Reuben, J.S. Plotkin, Industrial Organic Chemicals, 2nd edn, John Wiley & Sons, Inc., 2004, (Chapter 7).

[4] V. Zukaskas, CMAI – 2005 World Petrochemical Conference, p. 298.

[5] J. Cejka, B. Wichterlova, Catal. Rev.-Sci. Eng. 44 (2002) 375.

[6] G. Bellussi, Stud. Surf. Sci. Catal. 154A (2004) 53.

[7] T.F. Degnan, C.M. Smith, C.R. Venkat, Appl. Catal. A: Gen. 221 (2001) 283.

[8] C. Perego, P. Ingallina, Catal. Today 73 (2002) 3.

[9] C. Perego, P. Ingallina, Green Chem. 6 (2004) 274.

[10] R.M. Milton, Zeolite synthesis, ACS Symp. Ser. 398 (1989) 1.

[11] T.F. Degnan, J. Catal. 216 (2003) 32.

[12] P.B. Weisz, V.J. Frilette, J. Phys. Chem. 64 (1960) 382.

[13] P.B. Weisz, Chem. Tech. 3 (1973) 498.

[14] P.B. Weisz, V.J. Frilette, R.W. Maatman, E.B. Mower, J. Catal. 1 (1962) 307.

[15] P.B. Weisz, Ind. Eng. Chem. Res. 34 (1995) 2692.

[16] S.M. Csicsery, Zeolite chemistry and catalysis, in: J.A. Rabo (ed.), ACS Monograph, vol. 171, Am. Chem. Society, Washington, DC, 1976, p. 680.

[17] S.M. Csicsery, Zeolites 4 (1984) 202.

[18] S.M. Csicsery, Pure Appl. Chem. 58 (1986) 841.

[19] S.M. Csicsery, Stud. Surf. Sci. Catal. 94 (1995) 1.

[20] P.B. Venuto, P.S. Landis, Adv. Catal. 18 (1968) 259.

[21] H. Heinemann, Catal. Rev.-Sci. Eng. 23 (1981) 315.

[22] N.Y. Chen, W.E. Garwood, Catal. Rev.-Sci. Eng. 28 (1986) 185.

[23] N.Y. Chen, W.E. Garwood, F.G. Dwyer, Shape Selective Catalysis in Industrial Applications, Chemical Industry Series 136, Marcel Dekker, New York and Basel, 1989
[24] F.G. Dwyer, Stud. Surf. Sci. Catal. 67 (1991) 179.
[25] W.O. Haag, Stud. Surf. Sci. Catal. 84 (1994) 1375.
[26] P.B. Venuto, Micropor. Mater. 3 (1994) 297.
[27] C.R. Marcilly, Top. Catal. 13 (2000) 357.
[28] S.M. Csicsery, J. Catal. 108 (1987) 433.
[29] N. Arsenova-Hartel, H. Bludau, R. Schumacher, W.O. Haag, H.G. Karge, E. Brunner, U. Wild, J. Catal. 191 (2000) 326.
[30] G. Burress, US Patent 3,856,873 to Mobil Oil Corp., 1974.
[31] H. Heinemann, Catal. Rev.-Sci. Eng. 23 (1981).
[32] A. Corma, J. Catal. 216 (2003) 298.
[33] J. Das, A.B. Halgeri, Catalysis Survey from Asia. 7(1) January (2003) 3.
[34] F. Bauer, W.-H. Chen, H. Ernst, S.-J. Huang, A. Freyer, S.-B. Liu, Micropos. Mesopor. Mater. 72 (2004) 81.
[35] H.H. John, H.D. Neubauer, P. Birke, Catal. Today 49 (1999) 211.
[36] F.J. Llopis, G. Sastre, A. Corma, J. Catal. 227 (2004) 227.
[37] Xylenes, ChemSystems PERP Report01/02-7, June, 2002.
[38] K.W. Smith, W.C. Starr, N.Y. Chen, Oil Gas J. 78 (1980) 75.
[39] D.H. Olson, W.O. Haag, ACS Symp. Ser. 248 (1984) 275.
[40] A.B. Halgeri, J. Das, Catal. Today 73 (2002) 65.
[41] G.D. Mohr, J.P. Verduijn, US Patent 6,039,864 to Exxon Chemical Patent Inc., 2000.
[42] J.A. Johnson, C.M. Roeseler, T.J. Stoodt, DeWitt Petrochemical Review, Houston, March 18 20, 1997.
[43] http://www.uop.com/aromatics/3050.html
[44] T.C. Tsai, W.H. Chen, S.B. Lin, C.H. Tsai, I. Wang, Catal. Today 73 (2002) 39.
[45] J.M. Serra, E. Guillon, A. Corma, J. Catal. 227 (2004) 459.
[46] N.Y. Chen, W.W. Kaeding, F.G. Dwyer, J. Am. Chem. Soc. 101 (1979) 6783.
[47] European Chemical News, April 10–16, 2000.
[48] R.H. Allen, L.D. Yats, J. Am. Chem. Soc. 83 (1961) 2799.
[49] J.S. Beck, W.O. Haag, Alkylation of aromatics, in: Handbook of Heterogeneous Catalysis, VCH, Weinheim, 1997, p. 2123.
[50] P. Vogel, Carbocation Chemistry, Amsterdam, Elsevier, 1985.
[51] G.A. Olah, Friedel-Crafts Chemistry, New York, John Wiley & Sons, 1973.
[52] C. Perego, S. Amarilli, G. Bellussi, O. Cappellazzo, G. Girotti, Proceedings of the 12th International Zeolite Conference on Materials Research Society, Vol. 1, Warrendale, Pennsylvania, 1999, p. 575.
[53] S.H. Wang, Styrene, PEP Report 33C, Supplement C, Process Economics Program, Menlo Park, California, SRI International, 1993.
[54] P. Ratnasamy, R. Kumar, Catal. Today 9 (1991) 329.
[55] C.G. Wight, US Patent 4,169,111 to Union Oil Company of California, 1979.
[56] F. Narsolis, G. Woodle, G. Gajda, D. Gandhi, Petrol. Technol. Q., Summer (1997) 77.
[57] R.A. Innes, S.I. Zones, G.J. Nacamuli, US Patent 4,891,458 to Chevron USA Inc., 1990.
[58] F. Cavani, V. Arrigoni, G. Bellussi, European Patent 0432814A1 to Eniricerche, EniChem, Snamprogetti, 1991.
[59] G. Bellussi, G. Pazzuconi, C. Perego, G. Girotti, G. Terzoni, J. Catal. 157 (1995) 227.
[60] J.C. Cheng, T.F. Degnan, J.S. Beck, Y.Y. Huang, M. Kalyanaraman, J.A. Kowalasky, C.A. Loehr, D.N. Mazzone, Stud. Surf. Sci. Catal. 121 (1999) 53.
[61] W.J. Roth, B.J. Ratigan, D.N. Mazzone, US 6,984,764 to ExxonMobil Oil Corp., 2006.

[62] G. Girotti, F. Rivetti, F. Assandri, Hydrocarb. Eng., November (2004).
[63] E. Bencini, G. Girotti, WO 2004/056475 to Polimeri Europa, 2004.
[64] S. Cho, W. Zhu, Proceedings of ERTC Petrochemical Conference, Paris, France, March 3–5, 2003.
[65] C. Flego, G. Pazzuconi, E. Bencini, C. Perego, Stud. Surf. Sci. Catal. 126 (1999) 461.
[66] R. Millini, C. Perego, W.O. Parker, C. Flego, G. Girotti, Stud. Surf. Sci. Catal. 154B (2004) 1214.
[67] A.R. Pradhan, B.S. Rao, J. Catal. 132 (1991) 79.
[68] J. Rigoreau, S. Laforge, N.S. Gnep, M. Guisnet, J. Catal. 236 (2005) 45.
[69] P. Dufrense, M. Meens, G. Mendakis, D. Kauff, Proceedings of Styrene Conference 2000, Louisville, USA, August 13–17, 2000.
[70] B. Maerz, Proceedings of CRI Catalyst Styrene Global Conference 2004, Estoril, Portugal, 2004.
[71] A. Corma, V. Martinez-Soria, E. Schnoeveld, J. Catal. 192 (2000) 163.
[72] W.J. Roth, C.T. Kresge, J.C. Vartuli, M.E. Leonowicz, A.S. Fung, S.B. McCullen, Stud. Surf. Sci. Catal. 94 (1995) 301.
[73] C.T. Kresge, Q.N. Le, W.J. Roth, R.T. Thomson, US 5,258,565 to Mobil Oil Corp., 1993.
[74] A. Corma, V. Fornes, J. Martınez-Triguero, S.B. Pergher, J. Catal. 186 (1999) 57.
[75] P.J. Van den Brink, A. Corma, E.J. Creyghton, V. Fornes, V. Martines-Soria, WO 01/21562 to CSIC – Univ. Politecnica Valencia, 2001.
[76] C.H. Christensen, K. Johannsen, I. Schmidt, C.H. Christensen, J. Am. Chem. Soc. 125 (2003) 13370.
[77] L.L. Murrell, R.A. Overbeek, Y.-F. Chang, N. van der Puil, C.Y. Yeh, US Patent 6,004,527 to ABB Lummus Global Inc., 1999.
[78] F.M. Dautzenberg, P.J. Angevine, Catal. Today 93–95 (2004) 3.
[79] C. Ercan, F.M. Dautzenberg, C.Y. Yeh, H.E. Barner, Ind. Eng.Chem. Res. 37 (1998) 1724.
[80] J.C. Jansen, J.H. Koegler, H. van Bekkum, H.P.A. Calis, C.M. van den Bleek, F. Kapteijn, J.A. Moulijn, E.R. Geus, N. van der Puil, Micropos. Mesopor. Mater. 21 (1998) 213.
[81] G.R. Meima, CATTECH. June (1998) 5.
[82] Chemical Week, April 7/14, 2004, p. 41.
[83] S. Siffert, L. Gaillard, B.L. Su, J. Mol. Catal. A: Chemical 153 (2000) 267.
[84] G.R. Meima, M.J.M. van der Aalst, M.S.U. Samson, J.M. Garces, J.G. Lee, in: R. von Ballmoos et al. (eds.), Proceedings of 9th International Zeolite Conference, Reed Publishing, 1993, p. 327.
[85] J. C. Cheng, A.S. Fung, D.J. Klocke, S.L. Lawton, D.L. Lissy, W.J. Roth, C.M. Smith, D.E. Walsh, US 5,453,554 to Mobil oil Corp., 1995.
[86] G.G. Juttu, R.F. Lobo, Micropos. Mesopor. Mater. 40 (2000) 9.
[87] A. Corma, U. Diaz, V. Fornes, J.M. Guil, J. Martınez-Triguero, E.J. Creyghton, J. Catal. 191 (2000) 218.
[88] R.J. Schmidt, Appl. Catal. A: Gen. 280 (2005) 89.
[89] M.F. Bentham, G.J. Gajda, R.H. Jensen, H.A. Zinnen, Proceedings of the DGMK Conference on Catalysis on Solid Acids and Bases, Berlin, Germany, March 14–15, 1996, p. 155.
[90] G. Spanò, S. Ramello, G. Girotti, F. Rivetti, A. Carati, WO 2006/002805 A1 to Polimeri Europa, 2006.
[91] R. Koon-Ling, I. Yoji, Styrene, Chemical Economics Handbook, report 694.3000 A, SRI International, September 2002.
[92] W.W. Kaeding, L.B. Young, C. Chu, J. Catal. 89 (1984) 267.
[93] W.W. Kaeding, L.B. Young, A.G. Prapas, Chemtech 12 (1982) 556.

[94] J. Weitkamp, S. Ernst, P.A. Jacobs, H.K. Karge, Erdöl und Kohle-Erdgas-Petrochemie 39 (1986) 13.
[95] H.K. Karge, S. Ernst, M. Weihe, U. Weiss, J. Weitkamp, Stud. Surf. Sci. Catal. 84 (1994) 1805.
[96] C. Perego, M. Margotti, L. Carluccio, L. Zanibelli, G. Bellussi, Stud. Surf. Sci. Catal. 135 (2001) 178.
[97] I. Wang, C.L. Ay, B.J. Lee, M.H. Chen, Appl. Catal. 54 (1989) 257.
[98] I. Wang, B.J. Lee, M.H. Chen, US 4,950,835 to Taiwan Styrene Monomer Corp., 1990.
[99] C. Perego, S. Amarilli, R. Millini, G. Bellussi, G. Girotti, G. Terzoni G., Micropors. Mesopor. Mater. 6 (1996) 395.
[100] W.W. Kaeding, R.E. Holland, J. Catal. 109 (1988) 212.
[101] W.W. Kaeding, J. Catal. 120 (1989) 409.
[102] J.G. Lacson, Hydroquinone, Chemical Economics Handbook, report 664.8000, SRI International, March 2003.
[103] F. Haiduk, A. Kishi, Resorcinol, Chemical Economics Handbook, Report 691.7000, SRI International, September 2001.
[104] T.C. Tsai, C.L. Ay, I. Wang, Appl. Catal. 77 (1991) 199.
[105] T.C. Tsai, S.Y. Chang, I. Wang, Ind. Eng. Chem. Res. 42 (2003) 6053.
[106] R. Brzozowski, J.C. Dobrowolski, M.H. Jamròz, Catal. Commun. 3 (2002) 141.
[107] M.J. Schlatter, R.D. Clark, J. Am. Chem. Soc. 75 (1953) 361.
[108] H. Fiege, Ullmann's Encyclopedia of Industrial Chemistry A8 (1987) 25.
[109] D. Fraenkel, M. Levy, J. Catal. 118 (1989) 10.
[110] B. Wichterlova, J. Cejka, J. Catal. 146 (1994) 523.
[111] A.B. Halgeri, J. Das, Appl. Catal. A: Gen. 181 (1999) 347.
[112] K.S.N. Reddy, B.S. Rao, P.V. Shiralkar, Appl. Catal. A 121 (1995) 191.
[113] C. Perego, S. Amarilli, A. Carati, C. Flego, G. Pazzuconi, C. Rizzo, G. Bellussi, Micropos. Mesopor. Mater. 27 (1999) 345.
[114] R. Bandyopadhyay, P.S. Singh, R.A. Shaikh, Appl. Catal. A 135 (1996) 249.

CHAPTER 3

Mesoporous Non-Siliceous Materials and Their Functions

Ajayan Vinu*

International Center for Materials Nanoarchitectonics, World Premier International (WPI) Research Center, National Institute for Materials Science, 1-1, Namiki, Tsukuba, 305-0044, Japan

Contents

1. INTRODUCTION

Porous materials with regular geometries have been recently paid much attention owing to their scientific importance and great potentials in practical applications such as catalysis, adsorption, separation, sensing,

*Corresponding author. Tel.: +81-29-860-4563; Fax: +81-29-860-4667
E-mail address: vinu.ajayan@nims.go.jp

Advances in Nanoporous Materials, Volume 1 ISSN 1878-7959, DOI 10.1016/S1878-7959(09)00103-0

medical usage, ecology, and nanotechnology [1–14]. Depending on the predominant pore size, the porous materials are classified by IUPAC into three classes: microporous (pore diameter in the size range of 0.2–2.0 nm), mesoporous (pore diameter in the size range of 2.0–50.0 nm), and macroporous (pore sizes exceeding 50.0 nm). Microporous and macroporous materials are widely used as adsorbents, but they are incapable of selectively adsorbing a broad spectrum of large organic molecules of technical interest because of the pore size distribution. Pore sizes of mesoporous materials allow not only an easy accessibility for molecules with sizes up to a certain range but also a possible controllability in functions depending on the pore geometries.

Well-ordered mesoporous silica or aluminosilicate materials have attracted significant attention in recent years because of their excellent textural characteristics such as high surface area, large pore volume, uniform pore size distribution, and high thermal and hydrothermal stability, which are highly critical for several applications ranging from catalysis and biomolecules adsorption to advanced nanotechnologies with integrated nanosystems [1–27]. These materials can be prepared by using either a cationic or an anionic or a neutral surfactant as a structure directing agent [1–10]. There are numerous reports that deal with the preparation of various types of one- (1D) and three-dimensional (3D) mesoporous materials such as MCM-41, MCM-48, SBA-1, SBA-15, AMS, HMS, and MSU [2–13]. Mesoporous silica materials with different macroscopic morphologies, for example, spheres [28], helicoids [29], tops [30], 1D objects as well as fibers [30], and thin films [31], have been also reported. By controlling the synthesis conditions, namely, the temperature, solution pH, surfactant-to-silica ratio, nature of the surfactants, and nature of the silica sources, the morphology of the mesoporous silica materials can be fine tuned. Also, mesostructured silica tubes have been prepared using oil–aqueous interfaces as were mesoporous silica tube-inside-silica tubes. Although the mesoporous silica materials possess interesting adsorption properties, they have some limitations that are due to the amorphous character of the pore walls and the neutral silica framework that does not provide acid sites and due to the chemical nature of the surface hydroxyl groups. To provide the proper function on the surface, the surface silanol groups require *in-situ* or post-synthetic modification with different organic functionalities. Moreover, these materials suffer from their poor thermal and hydrothermal stabilities that limit their widespread use in many applications. Recently, much progress has been made in the fabrication of mesoporous materials other than silica, namely, non-siliceous mesoporous materials such as transition metal oxides, carbons, polymers, nitrides, metals, and semiconductors. The fabrication of such materials leads to the production of more advanced nanoscale devices, sensors, microelectronic devices, catalysts, and adsorbents. Members of this novel class of materials are prepared either by soft-templating approach using organic surfactant molecules as templates or through a hard-templating approach using mesoporous inorganic metal oxides as templates. In this contribution, we will describe the recent developments in the fabrication and the application of various non-siliceous mesoporous materials including metal oxides, metals, polymers, semiconductors, carbons, and nitrides.

2. PREPARATION OF MESOPOROUS NON-SILICEOUS METAL OXIDES

As mesoporous silica materials exhibit neutral frameworks, they cannot provide acid sites for various acid-catalyzed reactions. To introduce acid sites into the silica framework of the mesoporous materials, the framework structure has to be modified with the inclusion of di- or trivalent metal ions that can allow for the formation of acid sites. Moreover, the chemistry of non-siliceous materials is much more diverse than that of the siliceous ordered mesoporous oxides, and thus, the synthesis strategies for their production need to be more diverse. With this in mind, several researchers have been actively working on metal-substituted mesoporous materials that resulted in several mesoporous metallosilicate materials including Cu-, Zn-, Al-, B-, Ga-, Fe-, Cr-, Ti-, V-, and Sn-substituted mesoporous materials [13,15,26,27,32–43]. Furthermore, synthesis of mesoporous structures of materials other than silica is possible, that is, template synthesis adopted for mainly silica-based mesophases can be extended to the preparation of various mesoporous metal oxides such as TiO_2, Ta_2O_5, Nb_2O_5, ZrO_2, Al_2O_3, and V_2O_5 [18,44–66]. Synthesis of mesoporous aluminophosphates was similarly reported [67–71] as well as preparation of mesoporous metals and related materials [72,73]. In this section, several examples from recent results on the preparation of non-siliceous metal oxide mesoporous materials through the soft- and hard-templating approach are presented.

The preparation of ordered mesoporous crystalline networks and mesostructured nano-island single layers, composed of multicationic metal oxides having perovskite, tetragonal, or ilmenite structures, was successfully achieved by Sanchez and coworkers [74]. Evaporation-induced self-assembly (EISA) was also applied to synthesize nanocrystalline mesoporous films with controlled wall thickness (10–20 nm) of dielectric $SrTiO_3$,

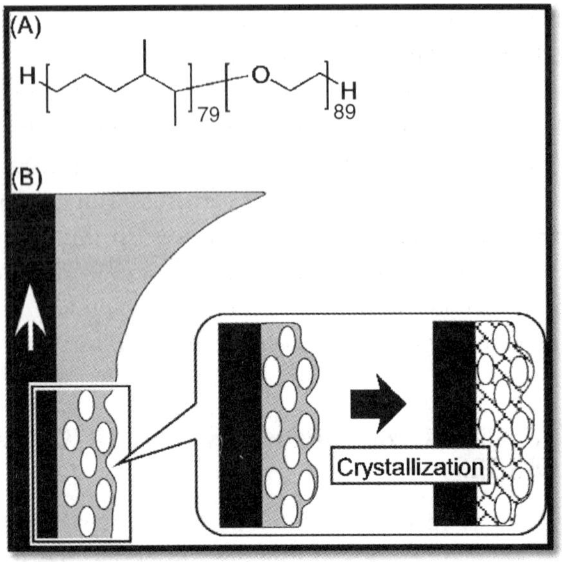

Figure 1 Preparation of ordered mesoporous crystalline networks composed of multicationic metal oxides: (a) template used in this research and (b) formation of mesoporous film followed by crystallization.

photoactive $MgTa_2O_6$, or ferromagnetic semi-conducting $Co_xTi_{1-x}O_{2-x}$ using a non-ionic triblock-copolymer template (Fig. 1). Solutions containing a stoichiometric ratio of different metallic precursors and a certain volume fraction of the triblock co-polymer KLE3739 (Fig. 1a) were dissolved in a highly volatile medium. Films with a mesostructure were prepared by dip coating the silicon wafer substrate in a controlled atmosphere. The thickness, quality, and the defects in the nanocrystalline films can be controlled by varying the withdrawal rate. Furthermore, thermal consolidation (dehydration) at 300 °C under air for 30 min was performed to stabilize the network before crystallization of the metal oxide frameworks occurred. Nanocrystallization was achieved by heating with a ramp of 8 °C min^{-1} under air up to 500 °C for anatase $Co_{0.15}Ti_{0.85}O_{1.85}$, 610 °C for $SrTiO_3$, and 760 °C for $MgTa_2O_6$. As pioneering work, Yang et al. also demonstrated a general method for the synthesis of mesoporous metal oxides including TiO_2, ZrO_2, Al_2O_3, Nb_2O_5, Ta_2O_5, WO_3, HfO_2, SnO_2, and mixed oxides such as $SiAlO_{3.5}$, $SiTiO_4$, $ZrTiO_4$, Al_2TiO_5, and ZrW_2O_8, with semicrystalline frameworks [75]. The structure and the pore diameter of the mesoporous metal oxides can be controlled by varying the synthesis conditions or the addition of swelling agents to the reaction mixture. Unlike the synthesis of mesoporous silica, the structural order and the textural parameters of the mesoporous metal oxide may be affected upon varying the synthesis conditions [75].

Titania is one of the fascinating semi-conducting materials and has been given much attention in the recent years owing to its strong oxidizing and reducing ability under UV light irradiation [76–79]. These exciting properties could find the material in various industrial applications including the removal of hazardous organic substances and dye-sensitized solar cells. Specific surface area and crystallinity are the most important factors affecting its photocatalytic activity. Therefore, several research groups tried to introduce mesoporosity into the titania materials as it can provide a large specific surface area that can help to improve the properties and the performance of these materials. Hence, mesoporous titania with ordered pore structure is highly expected to be useful as a photocatalyst and support for wet-type solar cells. Sakai and coworkers succeeded in synthesizing crystalline mesoporous titania [80]. Direct synthesis of crystalline mesoporous titania without any supports requires the deposition of very fine crystalline titania particles on the nanoscale surface of molecular surfactant assemblies. The combination of a $TiOSO_4$ precursor and a cetyltrimethylammonium bromide (CTAB) template with quaternary ammonium cation affects the formation rate of anatase-type crystal nuclei at the surface of the assemblies. Crystalline anatase particles, while maintaining the hexagonal pore structure, were obtained by optimization of the composition of these two components and the temperature. Poly(ethylene oxide)-based organics were also used as templates for the preparation of mesoporous titania. This way, Sanchez and coworkers could produce highly organized and oriented mesoporous titania thin films [81,82].

Dai and coworkers proposed a unique method to immobilize titania layers at the inner pore walls of mesoporous silica [83]. They adopted a nonhydrolytic surface sol–gel process for layer-by-layer modification of powdered mesoporous silica (Fig. 2). Multi-layered titania was prepared on SBA-15 channels through the direct condensation between titanium tetrachloride ($TiCl_4$) and titanium tetraisopropoxide [$Ti(OiPr)_4$] in anhydrous chloroform. The first-layer growth of titania was achieved by the reaction of titanium precursors with the surface silanols in chloroform under an atomic-level thickness control, followed by alternate addition of $TiCl_4$ and titanium tetraisopropoxide for further layer growth. A similar synthetic pathway is expected to enable the preparation of layers of other metal oxides.

To prepare highly stable mesoporous metal oxides after the removal of the surfactant, Kim and coworkers proposed the so-called nano-propping

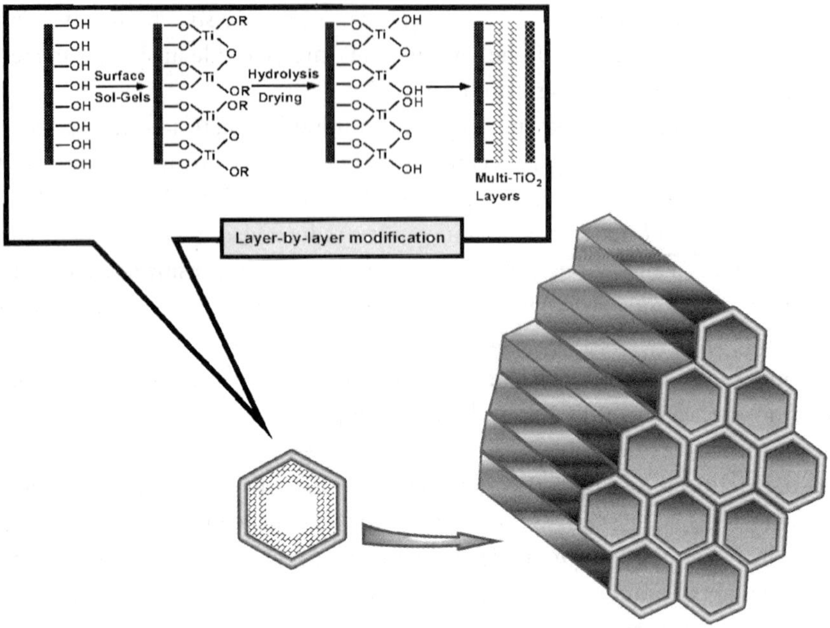

Figure 2 Layer-by-layer modification of the inner pore walls of mesoporous silica through nonhydrolytic surface sol–gel process. Reprinted with permission from Ref. [83] © 2005, American Chemical Society.

pathway. This method uses a tailored gemini surfactant containing a siloxane moiety for the preparation of mesoporous zirconium oxide (Fig. 3) [84]. From X-ray diffraction (XRD) patterns, it was found that the obtained mesoporous zirconium oxide exhibited an ordered structure even after calcination at 550 °C. Generally, mesoporous structures constructed from zirconium oxide are often collapsed or transferred to a poorly ordered structure by the thermal damage of their structural integrity. Unfortunately, mesoporous zirconia with highly ordered structure and good textural parameters cannot be obtained when a gemini surfactant without the siloxane moiety was used as the template and applying the same procedure mentioned above. In this case, the structure of the materials completely collapsed after calcination at 500 °C. These results indicate that the siloxane moiety introduced into the gemini surfactants is highly necessary to stabilize the mesostructure that can also retain its excellent textural parameters even after the calcination of the materials at high temperature.

Alumina is one of the most popular metal oxides, and one may anticipate the improvement in the properties of γ-alumina with exciting properties through the formation of mesostructures. Zang and Pinnavaia

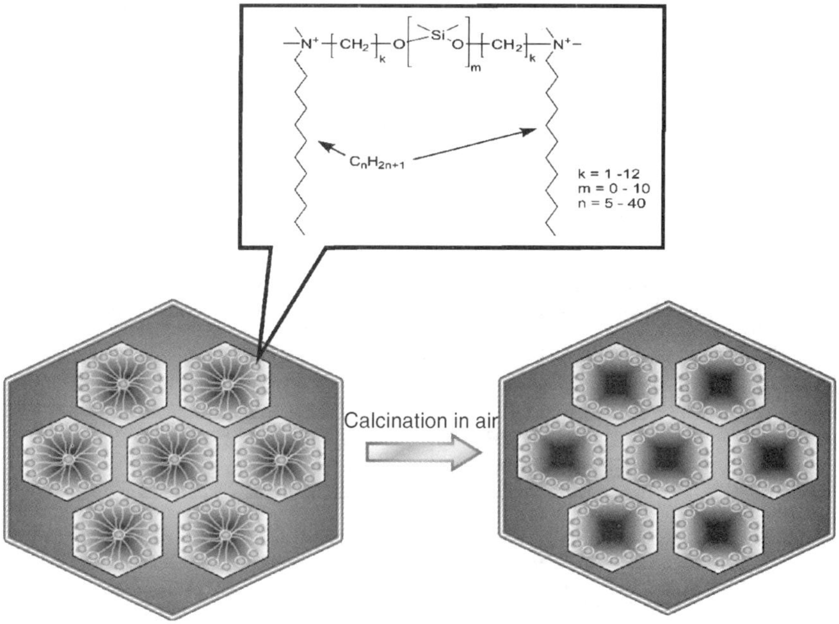

Figure 3 The nano-propping method for the preparation of highly stable mesoporous materials. Reprinted with permission from Ref. [84]. © 2004, American Chemical Society.

elucidated the stepwise reaction process leading to the formation of mesostructured forms of γ-Al_2O_3 with lath-like particle morphologies from aluminum salts in the presence of both triblock and diblock surfactants as the structure-directing reagents (Fig. 4) [85]. Mesostructured MSU-X-type (MSU denotes Michigan State University) alumina with a wormhole framework was formed in the first reaction step through a supramolecular assembly process. However, in this material, the walls of the calcined mesostructures are essentially amorphous. The MSU-X-type alumina became a novel type of precursor for a subsequent transformation to mesostructured MSU-S/B surfactant-boehmite mesophases that can be further transformed upon calcination into the mesostructured MSU-γ alumina with pore walls made of lathlike γ-Al_2O_3 nanoparticles.

Mesoporous structures from mixed oxides belong to another class of exciting materials in the family of non-siliceous mesoporous metal oxides that can provide different functions at the same time. Mixed titanium-vanadium oxides were fabricated using co-deposition from aqueous solutions by Shyue and De Guire [86]. Before the removal of the template, the solid materials showed a (Ti, V)$O_2^{+\delta}$ phase with the anatase structure.

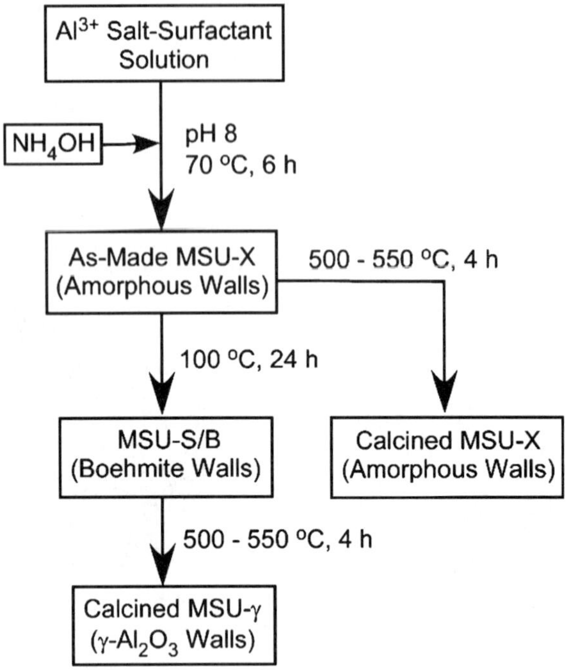

Figure 4 Reaction sequence for the synthesis of mesoporous materials with various alumina walls. Reprinted with permission from Ref. [85]. © 2002, American Chemical Society.

The catalytic behavior of these solids was studied in the oxidation of lactic acid to pyruvic acid using hydrogen peroxide as oxidising agent under liquid-phase conditions. The activity of the mesoporous structures prepared through co-deposition was significantly higher than that of the materials obtained through wet impregnation. This improved performance is attributed to the existence of nanoscale porosity in these solids. The presence of different metal sites in the pore walls of the materials, which offer different active sites for a given reaction, is also responsible for their high activity.

Domen and coworkers reported mesoporous niobium oxide with 3D structure and excellent textural parameters [87]. It is believed that the mesoporous niobium oxide could find useful applications, for example, in electronic and magnetic devices, biotechnology, and nanotechnology, where 3D structures are having much advantage over 1D structures. Moreover, it was found that the initial wormhole-like mesoporous structure was dramatically changed to highly ordered 3D arrays by the addition of cations under precisely controlled conditions. Nazar and coworkers first reported the synthesis of mesoporous transparent conducting oxides that combine both a high surface area and regular pore geometry with the walls composed of an appreciably electronically conductive and stable indium-tin oxide (ITO) framework [88]. This material was prepared by dissolving indium acetate and tin isopropoxide as the source for indium and tin respectively, in an excess of triethanolamine, using CTAB as a surfactant. The mesoporous ITO can be used for selective gas sensors and micro-electrochemical reactor chambers for specific electrocatalytic transformations, while transparency provides the opportunity for tunable display devices.

Bhaumik and Inagaki reported a synthesis procedure and detailed structures of novel mesoporous titanium phosphates [89]. Such ordered mesoporous titanium phosphate materials have many potential uses, for example, as ion exchangers, acid–base catalysts, photocatalysts, and liquid-phase oxidation catalysts. Zhao and coworkers realized this opportunity and successfully synthesized various mesoscopically ordered, large-pore, homogeneous, and stable metal phosphates with tailored morphologies using an acid–base pair concept [90]. In this acid–base pair concept, various metallic and non-metallic sources are sorted according to their relative acidity and alkalinity on solvation. The larger acidity or alkalinity difference between the metallic and/or non-metallic sources leads to appropriate pH of the desired sol–gel reactions during the entire preparation without using extra reagents such as aqueous HCl and NH_3. Various types of mesoporous materials, 2D hexagonal materials, 3D body-centred cubic materials, 3D hexagonal materials, 3D bicontinuous cubic materials, and lamellar materials were successfully prepared using the acid–base pair concept. However, in most cases, the mesoporous metal oxides prepared through a direct templating approach exhibit poor structural order and textural parameters. This is mainly due to the fact that the hydrolysis and the polymerization of the metal ions are too difficult to control and the polymer-metal ions assembly was not robust enough to maintain the mesostructure upon the formation of the metal oxide. Moreover, the mesostructured metal oxides prepared from the soft-templating approach are composed of amorphous inorganic walls, which limit their applications to catalysis and microdevices and electronic, magnetic, and magnetoelectric applications.

Mesoporous silica materials with 2D structures prepared through the microwave digestion method were used as template for the preparation of various mesoporous metal oxides. The microwave digestion method can produce template molecules with a lot of silanol groups on the surface, which can help to attain a high loading of the metal precursor inside the channels of the mesoporous silica template and can render it as the perfect template for the nanocasting synthesis. At the beginning, the mesopores of the templates are filled with an alcoholic solution containing a metal salt. The subsequent evaporation of the solvent followed by the heat treatment at high temperature in air and the subsequent removal of the silica template yield a highly ordered mesoporous metal oxide with excellent textural parameters. Zhao et al. successfully prepared a series of highly ordered metal oxides including Cr_2O_3, Mn_2O_3, Fe_2O_3, Co_3O_4, NiO, CuO, WO_3, CeO_2, and In_2O_3 [91,92]. Mesoporous metal oxides with different structural order can also be obtained by varying the space group of the mesoporous silica template. Accordingly, mesoporous metal oxides with 1D to 3D structures were prepared by using SBA-15, FDU-5, KIT-6, and SBA-16.

Mesoporous carbons were also utilized for the preparation of mesoporous metal oxides other than silica. Roggenbuck and Tiemann first realized the approach and successfully prepared mesoporous magnesium oxide using the hexagonally ordered mesoporous carbon CMK-3, which is composed of ordered carbon rods that were originally formed inside the cylindrical mesopores of SBA-15 template [93]. These primary carbon rods are rigidly interconnected by smaller carbon rods that are formed inside the micropores between the main cylindrical pores of SBA-15. A detailed description of the mesoporous carbons can be found in Section 7. In their approach, magnesium nitrate was impregnated into the mesopores of CMK-3, and the subsequent heating of the composites to 300 °C in air resulted in the conversion of magnesium nitrate to magnesium oxide within the pores. Finally, the carbon was removed upon oxidation by raising the temperature to 800 °C in the presence of air, resulting in mesoporous magnesium oxide with high structural ordering.

Organic functionalized mesoporous silica was also employed as template for the preparation of mesoporous metal oxides through a nanocasting technique. These materials have a lot of advantages over microwave-digestive mesoporous silica template. The organic functional groups on the surface of the materials help to adsorb large amounts of metal ions due to the strong coordination ability of the functional groups such as –SH or NH_2 on the porous surface of the template. This high precursor loading strategy was first followed by Zhu et al., who successfully prepared various mesoporous metal oxides using organic functionalized mesoporous silica as template [94]. The preparation process involves the adsorption of inorganic precursors onto the amine functionalized mesoporous silica materials, followed by the thermal decomposition of the precursor to assemble the individual metal hydroxide nanoparticles in the mesochannels and the subsequent removal of the amine functionalized silica by hydrofluoric acid (HF). Various mesoporous metal oxides such as Cr_2O_3, V_2O_5, WO_3, and Fe_2O_3 with ordered structure have been synthesized in this manner by choosing appropriate metal precursor solutions [94–97]. A similar strategy was also followed for the preparation of mesoporous In_2O_3 using metal-doped mesoporous monoliths as template, in which the calcination treatment and the polymers in the monolith help to reduce the indium nanoparticles. The indium nanoparticles are successfully converted into In_2O_3 single crystal nanowires inside the mesochannels of the monolith upon further oxidation at 500 °C. The obtained materials possess excellent structural ordering with a specific surface area of $170\,m^2\,g^{-1}$.

Although several mesoporous metal oxides could be prepared using the nanocasting technique, still the exact replica of the parent template could not be achieved as the structural order of most of the replica materials is lower than that of the parent template. Thus, it is a big challenge for material scientists to find a way to create well-ordered mesoporous metal oxide materials with structural features similar to those of mesoporous silica materials by a simple one-step approach.

3. MESOPOROUS METALS

After one-and-a-half decades of extensive activity in mesoporous silica materials research, more and more attention was focused on the preparation of mesoporous metals in the past few years. Mesoporous metals analogous to those of mesoporous ceramic oxides are of considerable interest for several applications including catalysis, batteries, fuel cells, capacitors, and sensors, owing to their metallic frameworks. However, it is very difficult to prepare mesoporous metals using the soft-templating approach developed by Beck et al. [1], where relatively low concentrations of surfactants are employed. It is believed that the low concentration of the surfactants helps to enhance the interactions with the inorganic silica nanostructures leading to the formation of materials with regular arrays of mesopores. Unfortunately, mesoporous metals cannot be prepared using the organic surfactant molecules in low concentration (<25 wt.%). In this section, the various synthetic pathways used for the preparation of mesoporous metals, including the soft- and hard-templating approach, are reviewed. Moreover, an overview over their structural characterization and the applications in sensors and electrochemical reactions is given. Attard et al. made an historic attempt to prepare various mesostructures through the soft-templating approach by using organic structure directing agents such as surfactants in very high concentrations. They introduced a novel strategy for the fabrication of mesoporous materials, namely, a much higher concentration of the surfactant (>50 wt.%) is employed with the aim of producing the lyotropic liquid crystalline phase ("lyotropic liquid crystal templating" (LLCT)), which acts as nanoscale mold and directs the formation of the mesoporous materials [98]. This approach set up an excellent platform for the production of various mesoporous metals as well as related nanomaterials. The LLCT method has been used for the first time by Tanon et al. for the generation of metal nanoparticles in surfactant solutions or lamellar lyotropic liquid crystals (LLC). The nanoparticles possessed a regular morphology with a high degree of control over shape and size; however, well-ordered mesoporous structures could not be obtained [99].

Later, Attard et al. introduced a combinatorial approach in which both, the liquid LLCT and the chemical reduction processes, were employed for the synthesis of mesoporous platinum [100]. The synthesis process involves the reduction of metal salts dissolved in the aqueous domains of the liquid-crystalline hexagonal mesophases of oligoethylene oxide surfactants. Mesoporous platinum is a coarse granular material and possesses cylindrical pores with a diameter of approximately 3.0 nm separated by platinum walls of 3.0 nm thickness, with an average pore-to-pore distance of 6.0 nm. It is interesting to note that the wall thickness of the mesoporous platinum is much larger than that of the corresponding mesoporous silica (1.2 nm) prepared from the LLCT method. The specific surface area of the mesoporous platinum amounts to approximately 23 $m^2 g^{-1}$, which is almost similar to that of platinum black. It was also reported that the concentration of the surfactant, the type of the reducing agent, and the pH of the synthesis medium significantly influence the structural order of the resulting mesoporous platinum. Mesoporous platinum films can also be fabricated by using the LLCT method with the help of an electrodeposition technique [72]. The same group also tried the challenge and successfully demonstrated the preparation of mesoporous platinum films using an electrodeposition method, which involves the electrodeposition of platinum metals from appropriate salts dissolved in the aqueous domains of the lyotropic liquid crystalline phases of non-ionic surfactants onto highly polished gold electrodes (Fig. 5) [72]. The resulting mesoporous platinum film possesses a well-ordered mesoporous structure with a high specific surface area,

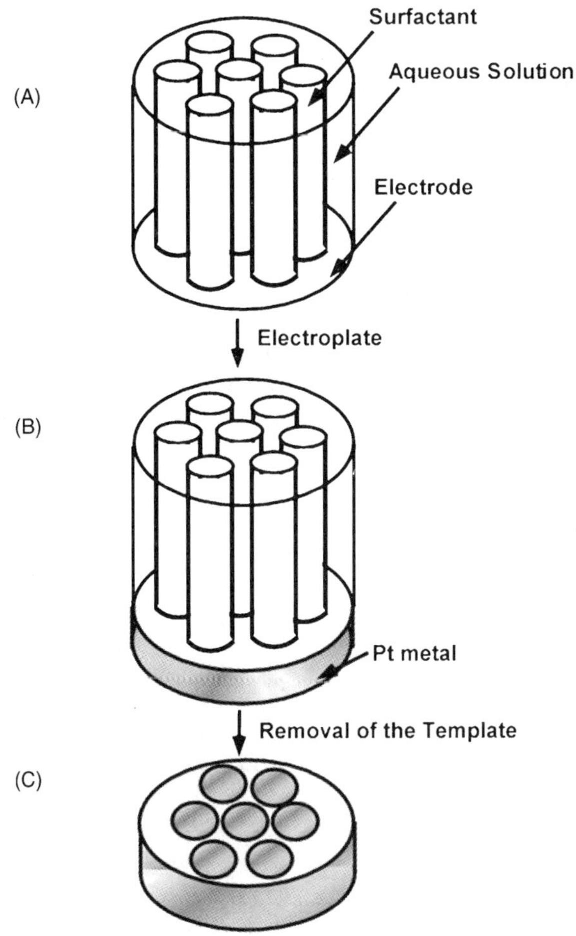

Figure 5 Schematic representation of the templating process used to deposit mesoporous platinum films. The cylinders represent the micellar rods in the lyotropic liquid crystalline phase: A) before platinum deposition, B) after platinum deposition, and C) mesoporous platinum films after removing the template. Reprinted with permission from Ref. [72].

electrical conductivity, fast electrolyte diffusion, and good mechanical and electrochemical stability. All this is highly advantageous for various applications including catalysis and fuel cells.

Mesoporous platinum metals with lamellar (Fig. 6), 2D (*p6mm*) and 3D cubic structures (*Ia*3*d*) can also be prepared by varying the composition of the reaction bath, which controls the architecture of the liquid crystalline phase, the electrodeposition potential, and the electrolyte concentration (Fig. 7) [101–103]. Boo et al. reported that the ionic strength of the

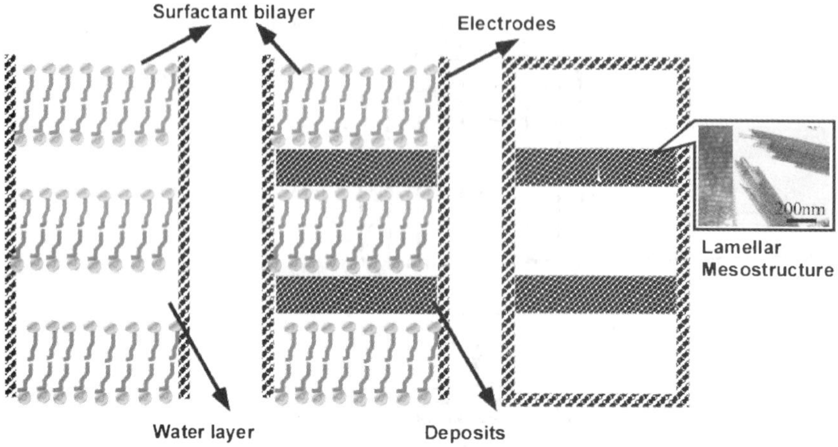

Figure 6 Schematic description of the electrodeposition of a lamellar platinum mesostructure templated by lyotropic liquid crystals.

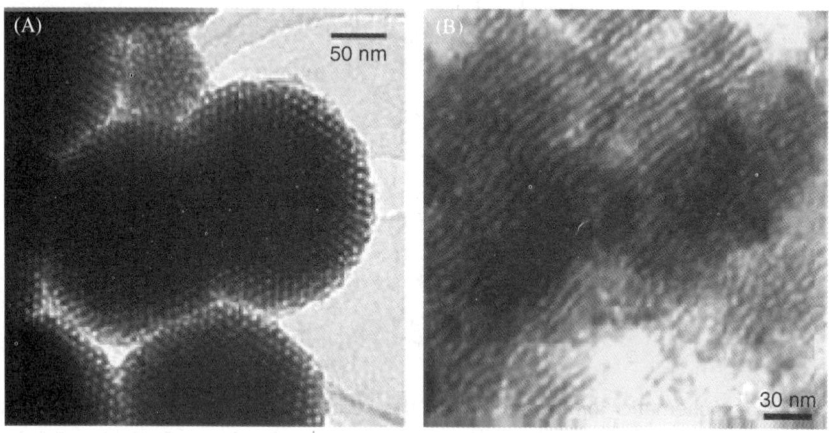

Figure 7 TEM images of lamellar mesoporous metal structures: (a) mesoporous Ni particles and (b) lamellar mesoporous platinum. Reprinted with permission from Ref. [103]. © 2004, Japan Chemical Society.

electrolyte solution can control the interface between the electrical double layer and the bulk solution, and the faradaic current at the mesoporous platinum can be tuned by varying the electrolyte concentration [102]. The pore diameter and the thickness of the platinum walls of the mesoporous platinum can also be controlled by simple adjustment of the length of the alkyl chain of the surfactants used in the electroplating mixtures and by adding heptane as a co-solvent to swell the micellar surfactants in the lyotropic liquid crystalline phase. Moreover, Yamauchi et al. used the so-called contact plating method for the preparation of a highly ordered mesostructured platinum film on a gold surface by reduction of platinum ions in the presence of LLCs (Fig. 8) [103]. Mesoporous platinum microelectrodes were also fabricated by electrodepositing mesoporous platinum films from a hexagonal LLC plating mixture onto preformed microelectrodes, and the resulting materials possessed a high surface area with good mass transport characteristics.

The application of mesoporous platinum microelectrodes was studied by Watson et al. [104] in the electrochemical reduction of oxygen (Fig. 9). They found that the mass transport characteristics and reduction kinetics of the materials are superior to those of polished microelectrodes.

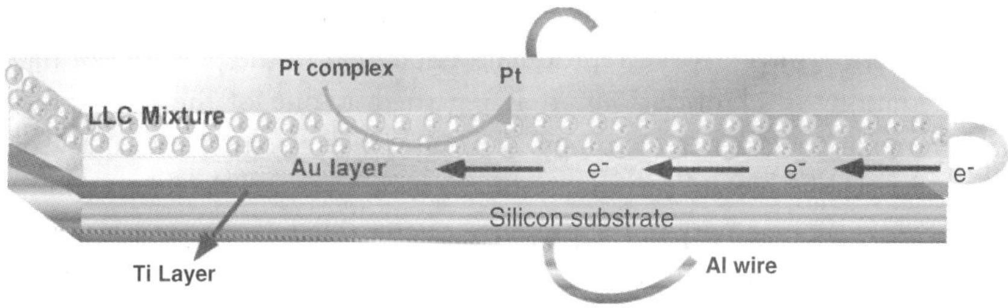

Figure 8 Schematic representation of the contact plating assembly process for the preparation of mesostructured platinum films templated by lyotropic liquid crystal.

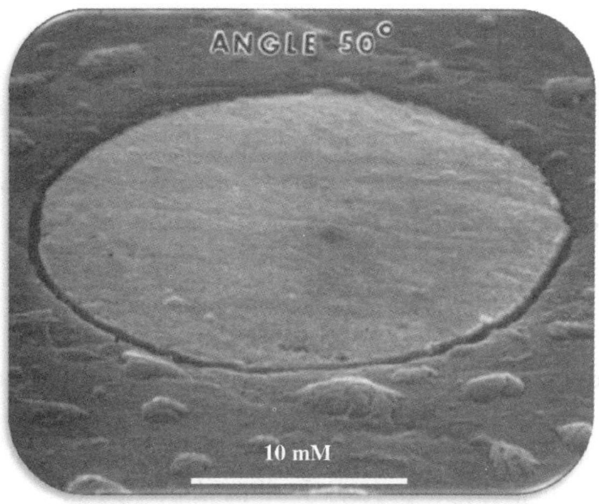

Figure 9 HR-SEM image of a nanostructured platinum micro-electrode.

These materials can also be used as amperometric sensors for the detection of hydrogen peroxide in aqueous solutions over a wide range of concentrations. Evans et al. showed that mesoporous platinum microelectrodes can be used as amperometric hydrogen peroxide sensors that show efficient mass transport properties with outstanding qualitative and quantitative results, good reproducibility, high precision, and accuracy of the measurements [105]. Mesoporous platinum can also be used as the electrode for a nonenzymatic glucose sensor. This was successfully demonstrated by Park et al. who reported that mesoporous platinum gives a stronger response to the glucose molecule than other interfering species such as l-ascorbic acid and 4-acetamidophenol [106]. This can be achieved because the high surface area of the mesoporous platinum electrode can selectively amplify the current due to the direct oxidation of glucose without an enzyme-like glucose oxidase. Noteworthy, sensitivities as high as 9.6 $\mu A\,cm^{-2}\,mM^{-1}$ were reproducibly obtained in the presence of high concentrations of chloride ions. It is regarded as one of the outstanding achievements in sensor technology as most of the electrochemical sensors based on noble metals for the nonenzymatic glucose detection almost entirely lose their activity in the presence of chloride ions due to poisoning. A significant enhancement in the catalytic activity of mesoporous platinum in the oxidation of methanol has also been demonstrated [107].

The electrochemical deposition technique has also been extended to the fabrication of mesoporous palladium films [108]. It should be pointed out that the electrochemistry of palladium differs significantly from that of platinum because it is able to absorb large quantities of hydrogen forming palladium hydride phases. Mesoporous palladium films with different structures can be successfully prepared using $C_{16}EO_8$- or Brig 56-type surfactants. Bartlett et al. also demonstrated that mesoporous palladium films can be converted to the b-hydride phase and back to palladium without a change in mesoporosity and textural parameters. Moreover, these materials posses a very high surface-to-volume ratio, which can help to readily distinguish the formation of adsorbed and absorbed hydrogen in the voltammetry of these films.

High surface area mesoporous rhodium films were successfully synthesized using the LLCT method through electrochemical deposition [109]. The obtained films exhibit a regular array of cylindrical pores with a high specific surface area. The reduction of nitrate in basic solution was also achieved using such mesoporous Rh films deposited on gold microelectrodes. The electrodes allowed for significantly larger currents as compared to those of the nonporous rhodium electrodes coated on the gold microelectrodes. Moreover, the LLCT method was also applied for the fabrication of various other metals such as Co [110], Sn [111], and Ni [112,113] using chemical or electrochemical reduction of the corresponding metal salts dissolved in aqueous domains of the LCT. From these results, one can conclude that the use of LLCT is a fascinating method for the preparation of a wide variety of mesoporous metal nanostructures, and the approach can be extended not only for metals but also to other mesoporous materials such as alloys and semiconductors. This is discussed in Chapter 4. There is no doubt that mesoporous metals with well-ordered structures and good textural parameters opened the window to a new field of exciting research and are going to make a great contribution in the fields of electrochemistry and sensor technology.

4. MESOPOROUS ALLOYS AND METAL–METAL OXIDE NANOCOMPOSITES

Alloying or the preparation of nanocomposites with metal oxides can significantly change the physical and chemical properties of mesoporous metals. In this section, the synthesis of mesoporous metal alloys and metal–metal oxide nanocomposites and their applications in sensors and in electrocatalysis will be described. For nonporous alloys, it is generally known that the catalytic activity of the alloy in a given reaction is often significantly higher than one of the respective pure metal. Thus, the synthesis methodology used for the preparation of mesoporous metals was extended to mesoporous alloys. Recently, several research groups have extensively studied the LLCT methodology in the synthesis of mesoporous alloys such as Pt–Pd [114], Te–Cd [115,116], Ni–Co [117,118], Pt–Rh [119], and Pt–Ni [120] by chemical or electrochemical co-reduction of two-metal species in the presence of LLCT. The composition of the metal framework and the wall mesostructure of the alloys can then be controlled by simply varying the concentration of the metals and the surfactant. It has been found that the structure, the composition, and the nature of the metallic sites of the synthesis products significantly change with the metal–metal combination. In the case of Pt–Ni alloys, their composition can be varied even though the standard potentials of Pt and Ni are completely different. The mesoporous Pt–Ni exhibits a very small particle size with diameters of less than 100 nm and their specific surface area amounts to $40\,m^2\,g^{-1}$. Moreover, the pore walls of the mesoporous Pt–Ni are in a binary intermetallic state and the metal atoms are uniformly distributed within the wall. On the contrary, a phase separation of the constituting metals on the nanoscale was observed for Pt–Pd, Te–Cd, and Ni–Co mesoporous alloys [114–120]. Moreover, it has also been demonstrated that the prepared mesoporous Pt–Rh alloys exhibit an excellent performance in the electrochemical oxidation of methanol or CO and showed better results as compared to that of the corresponding alloys made from nanoparticles. This improvement is mainly attributed to altered electronic and surface textural properties as compared to those of the pure elements. These interesting results prompted different groups to modify the surface characteristics of the materials by preparing nanocomposites from metal oxides. Recently, Saramat et al. realized the concept of preparing nanocomposites through this route, and they reported the preparation of mesoporous Pt–Al_2O_3 nanocomposite catalysts. Moreover, they also studied the activity of this material in the CO oxidation at low temperatures. It was further demonstrated that the catalyst appears to be less sensitive to CO poisoning than the nonporous metal–metal oxide nanocomposites [121].

Although LLCT can be used for the synthesis of several mesoporous metal or metal–metal oxide nanocomposites, it has certain disadvantages. For example, planar mesoporous films and particles cannot be prepared using LLCT as it cannot be uniformly introduced into the micrometer channels due to high viscosity. Nevertheless, the preparation of planar films in a reproducible manner is highly critical for the assembly of advanced functional devices. Kuroda et al. proposed a novel convenient pathway, namely, a so-called evaporation-mediated direct templating (EDIT), for the preparation of mesoporous metals in a confined area [122–126] (Fig. 10). The process involves two basic steps: The precursor solution composed of water, surfactant, platinum compounds, and ethanol diffuses into the microchannels and the solvent is subsequently evaporated from the microchannels. The LLC together with the platinum complexes is formed inside the microchannels after the evaporation of the solvent (Fig. 11). The EDIT method can be used to deposit mesoporous platinum inside the microchannels of an electrode through electrodeposition as this can help to avoid the formation of mesoporous platinum on the external surface of the microelectrodes. Edler and coworkers applied a similar method for preparing mesoporous silver films, in which a mixed non-ionic-anionic surfactant system is used as the template [127]. This method can also help to predetermine the final mesophase of the mesoporous metals since the phase of the LLCT can be controlled by solvent evaporation onto the electrodes.

The so-called nanocasting strategy, which is an exciting method for creating mesoporous materials that are difficult to synthesize by the

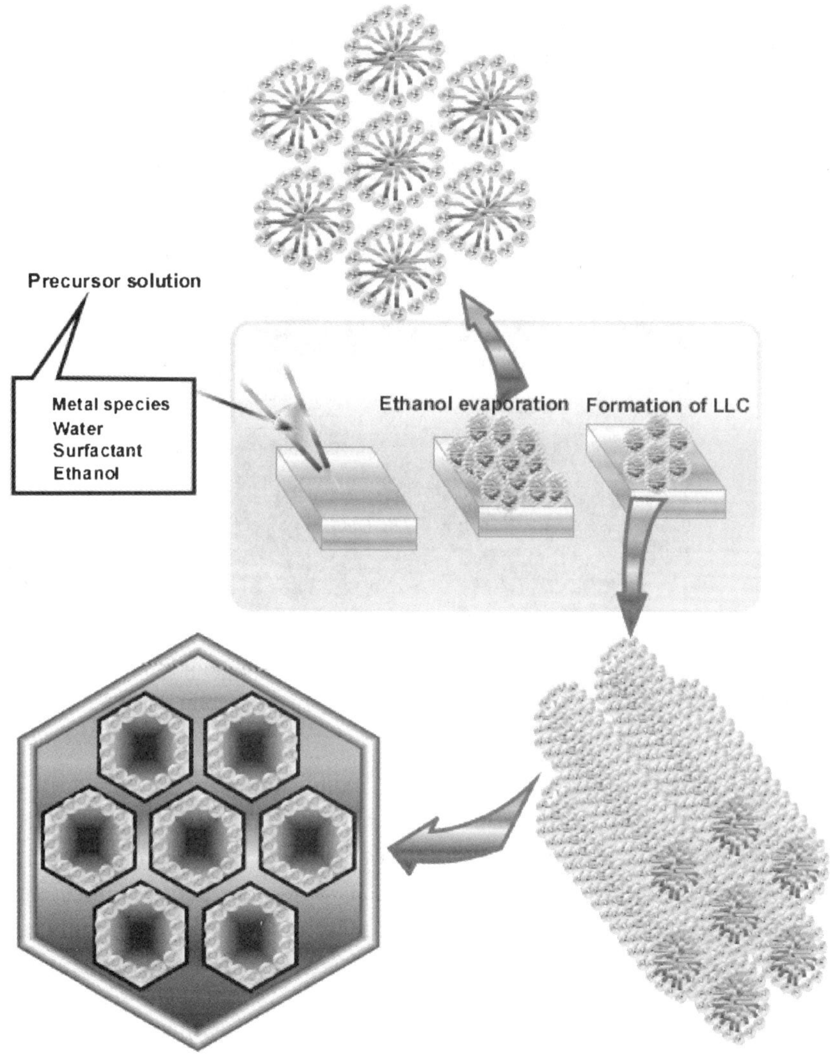

Figure 10 Schematic representation of the solvent-evaporation-mediated direct templating method for the preparation of mesoporous metals.

conventional cooperative self-assembly process, was also applied for the preparation of mesoporous metals with different structures and morphologies. In the early years, it was very difficult to synthesize mesoporous metals through the nanocasting strategy as the nature of the interaction between the metallic precursors and the respective template was not well established. It has been reported that only short metallic nanowires can be

Figure 11 HR-TEM image of mesoporous platinum prepared by the solvent-evaporation-mediated templating method. Ref. [103]. © 2008, John Wiley & Sons, Inc.

obtained if the connection between the metallic nanoparticles embedded in the channels of the mesoporous template is not properly linked [128]. Recently, Ryoo et al. successfully prepared mesoporous platinum with a 3D porous structure using MCM-48, which could offer better connectivity between the channels [129]. The procedure involves the simple impregnation of a metallic precursor solution into the mesochannels of the template over a longer period of time, followed by H_2 reduction. It is also demonstrated that a low heating temperature is preferable for the formation of well-ordered mesoporous metal nanostructures because higher heating temperature always generate large metal particles that may either destroy the connectivity between the particles inside the channels of the template or block the pore entrance of the template. A similar strategy has also been followed for the preparation of mesoporous osmium using organometallic complexes, for example, $Os_3(CO)_{12}$, as the osmium-containing precursors [130]. The obtained material showed a higher activity and excellent reusability in the oxidative cleavage of benzyl(5-hexynyl) ether in dimethylformamide at room temperature as compared to that of pure non-porous osmium.

5. MESOPOROUS SEMICONDUCTORS

Semiconducting materials are of tremendous importance in all areas of technological development and are required for the construction of modern electronic devices due to their electrical and band gap properties. The creation of nanoporosity in semiconducting materials such as CdS, ZnS, In_2S_3, CuS, Si, and Ge opens up a wide range of applications including biosensors, catalysis, and nanodevices. Among these semiconductors, elemental semiconductors are really attractive because they promise a wide range of application possibilities in optroelectronic and semiconducting devices. Recently, much progress has been made in the synthesis of mesoporous semiconductors using soft- and hard-templating techniques. The methodology used for the preparation of mesoporous semiconductors and their applications in nanotechnology are discussed in this section. Preference is given to the various methods available in the literature on the preparation of mesoporous semiconductors.

The soft-templating approach was initially used for the synthesis of mesoporous semiconductors by Armatas and Kanatzidis [131]. They described the synthesis of cubic mesostructured germanium with gyroidal channels separated by amorphous walls [131]. The symmetry of the obtained mesoporous germanium can be controlled by adopting different synthesis methodologies. 3D mesostructured germanium was prepared by employing Mg_2Ge, which is linked through a metathesis reaction with $GeCl_4$ in the presence of *N*-Eicosane-*N*-methyl-*N*-*N*-bis(2-hydroxyethyl) ammonium bromide surfactant under inert atmosphere. The material exhibits two independent 3D channels with a *Ia*3*d* space group symmetry. The structural order of the materials was confirmed by high-resolution transmission electron microscopy (HR-TEM) images. On the contrary, 2D hexagonally ordered mesoporous germanium can be prepared by linking Ge_9^{4-} clusters in formamide solutions with $GeCl_4$ in the presence of a structure-directing surfactant through a metathesis reaction [132]. The obtained material exhibits a BET surface area of $363\,m^2\,g^{-1}$ and a pore volume of $0.23\,cm^3\,g^{-1}$. As can be seen, the specific surface area is lower than for typical well-ordered mesoporous silica materials, which is mainly due to the heavy nature of germanium. HR-TEM images show a well-defined, long-range hexagonal arrangement of the mesopores (Fig. 12). It is also interesting to note that the energy bandgap of the mesoporous germanium is 1.70 eV, which is almost three times larger than that of bulk germanium, which corresponds to 0.66 eV.

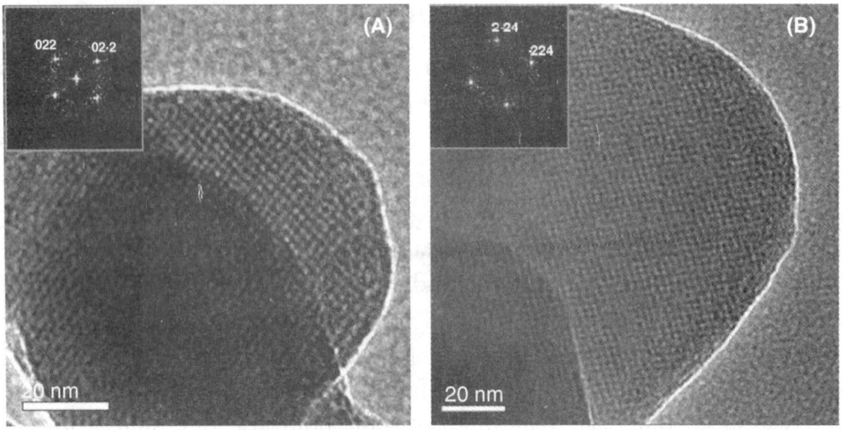

Figure 12 HR-TEM image of the mesostructured germanium semiconductor MSU-Ge-1. (a) Image taken along the [100] direction. (b) Image taken along the [110] direction. The pore-to-pore spacings are consistent with those deduced from X-ray diffraction. Insets in each panel show fast Fourier transform (FFT) images taken along an area of the HR-TEM images. Reproduced with permission from Ref. [132]. © 2006, American Association for the Advancement of Science.

Sun et al. also prepared hexagonally ordered mesostructured germanium by linking polymeric forms of the anion K_2Ge_9 with the cationic surfactants through electrostatic interactions followed by the oxidation with ferrocenium cations [133]. It should be mentioned that all the mesostructured Ge samples are highly sensitive toward air and can easily be converted into the respective oxides. Noteworthy, the materials prepared by Sun et al. [133] and Kanatzidis et al. are not composed of zero-valent Ge. Instead, they are composed of continuous porous framework Ge–Ge bonds and have a wall thickness of 1–2 nm [131,132]. Thus, care should be taken when a comparison of the physical and chemical properties of mesoporous Ge with bulk Ge nanocrystals is made. Moreover, the properties of the mesoporous Ge are completely different from those of disordered porous Ge prepared through chemical [134] or electrochemical deposition methods [135].

Metal sulfides are another class of interesting, mostly semiconducting materials, which offer potential applications in the electrical and electronic industries. Preparation of nonporous metal sulfides in the channels of mesoporous silica or carbon materials is quite easy. However, introducing well-ordered pores into metal sulfides is very difficult, mainly due to the difficult synthesis conditions. Mesoporous metal sulfides were first prepared by the group of Stupp through a soft-templating approach using surfactants

[136–138]. Mesostructured CdS and ZnS were prepared through combining the corresponding metal ions and H_2S in the presence of the surfactant molecules [136–139]. Mesostructured CdTe films were also prepared through a direct liquid crystal templating method using nonionic polymeric surfactants. Moreover, mesostructured tin sulfides and selenides, which resemble the structures of zeolites and their adsorption and sensing behavior, were also synthesized [139–142]. Unfortunately, the prepared materials were mesostructured individual nanocrystals but did not consist of continuous semiconductor frameworks. Gao et al. used the nanocasting method to introduce mesoporosity in CdS semiconductors [143]. It should be mentioned here that nanocasting techniques always require a high loading of the precursors in the mesochannels of the template. High loading with the Cd and S precursors is achieved by choosing suitable precursor species that should be highly soluble and form a proper link with the surface silanol groups of the template. Thus, $Cd_{10}S_{16}C_{32}H_{80}N_4O_{28}$ was used as the combined Cd and S precursor, which contains many OH groups and possesses high solubility and which can strongly bind with the surface silanol groups of the mesoporous silica template. An ordered mesostructured CdS was successfully prepared by impregnating a high loading of the above precursor into the mesopores of the template followed by the removal of the mesoporous silica template using either HF or NaOH. The resulting material possesses interconnected polycrystalline CdS nanowires with a well-ordered mesostructure. A specific surface area of almost 150 $m^2 g^{-1}$ was reported for the material. Mesoporous CdS can also be obtained by impregnating conventional Cd salts as precursor of the metal and thiourea as the sulfur source into the mesochannels of the template. After heat treatment at a temperature of 120 °C, followed by dissolution of the silica template by NaOH, the porous structure was obtained [144]. Similarly, several mesoporous metal sulfides such as ZnS, In_2S_3, and CuS can be prepared through the nanocasting technique [145]. The textural parameters of the above materials are excellent and are expected to offer various potential applications in the field of electronics, biosensors, and optronic devices.

6. MESOPOROUS POLYMERS

Another challenge in the development of non-siliceous mesoporous materials is the creation of a mesoporosity in conducting or crystalline polymers, which can offer a lot of applications including adsorption, separation/chromatography, and catalysis with organic substrates. However, the preparation of mesoporous polymers is not that easy and a lot of synthesis parameters have to be optimized to get well-ordered mesoporous polymers. Wu and Bein prepared conducting polymers inside the mesochannels of MCM-41 materials and found that the conductivity of the resulting solids was significantly enhanced as compared to either the pure mesoporous silica host or the pure conducting polyaniline molecules [146,147]. Ozin and his coworkers successfully prepared polymeric fibers through the nanocasting technique by polymerizing formaldehyde and phenol inside the channels of MCM-41 followed by the etching of the mesoporous silica template by HF or NaOH [148]. Similarly, polythiophene nanowires were also synthesized through a nanocasting pathway using a mesoporous silica-polythiophene nanocomposite as template, which was prepared by employing surfactants with polymerizable groups and $FeCl_3$ as catalyst to induce the polymerization inside the mesochannels. Furthermore, it was demonstrated by Li et al. that the obtained polymers can be used in photoelectric devices [149]. Although the mesoporous polymer materials derived from the nanocasting technique show some interesting properties, unfortunately, the structural order and the textural parameter of the materials were poor.

A self-assembly approach was also utilized for the synthesis of mesoporous polymers. Recently, mesoporous polyacrylonitrile with a disordered porous structure was synthesized by a self-assembly process using reverse pluronic nonionic surfactants and acrylonitrile, which generates well-dispersed micelles in a *N*-methyl-2-pyrrolidone solution. The polymers were generated by adding the radical initiator, azobisbutyronitrile, at low temperature. Finally, the disordered mesoporous polyacrylonitrile was obtained from the composite by removing the surfactant using solvent extraction. The materials possess a disordered pore structure with a broad pore size distribution, which is in the range from 6 to 10 nm [150].

Zhang et al. reported that the pore diameter of the mesoporous polymers can be controlled by the simple adjustment of the amount of hydrocarbons in the synthesis gel. However, in this case, only amorphous mesoporous polymers were obtained [151]. It was the brilliant work of Zhao et al. who successfully prepared highly ordered mesoporous polymeric resins through the EISA process [152–154]. The addition of amphiphilic triblock copolymers to low-molecular-weight polymers of phenol and formaldehyde, followed by a thermopolymerization process, yielded highly ordered mesoporous polymers. Polymeric precursors with varying amounts of hydroxyl groups were also used for the preparation of mesoporous polymers. It was found that precursors with a large number of hydroxyl groups are advantageous for obtaining well-ordered mesoporous polymers, as they can strongly interact with the non-ionic surfactants through hydrogen bonding interaction (Fig. 13). With this demonstration, the authors clearly pointed out that the proper choice of the polymeric precursor is the key to control the successful organization of the surfactant-polymeric mesostructures. It was further demonstrated that the obtained mesoporous polymers were stable up to 350 °C, which indeed helps to remove the structure-directing agents from the mesoporous polymers leaving behind the well-ordered mesochannels inside the polymeric network. The mesoporous polymers can also be converted into well-ordered mesoporous carbon molecules sieves through a carbonization process at high temperature without the use of the sacrificial mesoporous silica template. Zhao et al. also reported the techniques for controlling the structure and the textural parameters of the mesoporous polymers [152–154]. By varying the polymeric precursor-to-surfactant ratio, a whole family of mesoporous polymers with different structures, including mesostructures with lamellar, 2D hexagonal

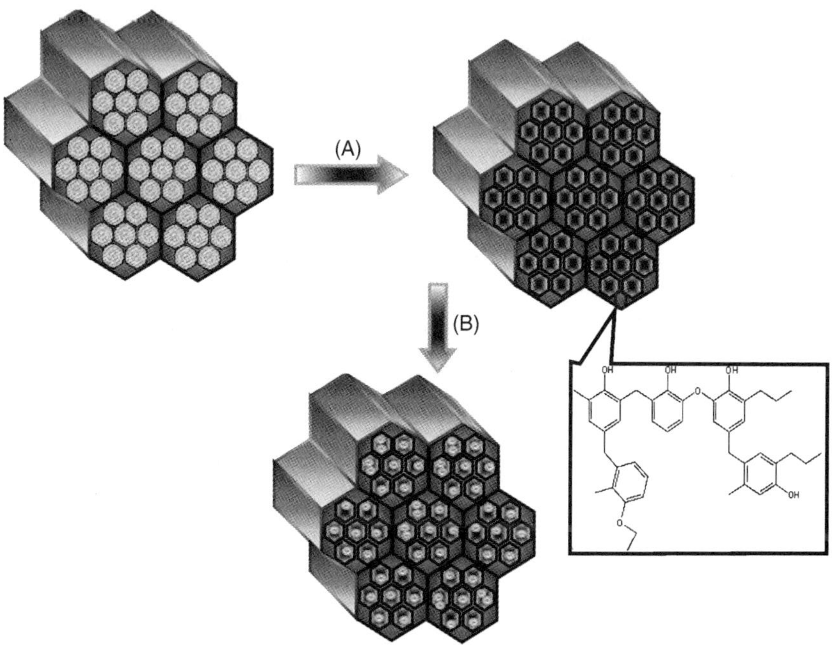

Figure 13 (a) Schematic representation of the direct templating method for the preparation of mesoporous polymer and (b) applying the evaporation-induced self-assembly process.

(*p6mm*), body-centered cubic, and bicontinuous 3D cubic symmetries (*Ia3d*), has been reported by Zhao et al. [152–155]. The idea of the interaction between the polymeric precursors and the surfactants has been followed by several researchers and generated several mesoporous polymer materials. Ikkala and his coworkers prepared mesoporous phenolic resins by crosslinking phenolic resin precursors with a new block copolymer (polystyrene-block-poly(4-vinylpyrinidine) (PS-block-P4VP), followed by thermo-polymerization [156]. After the thermopolymerization process at 100–190 °C and the subsequent removal of the surfactant by pyrolysis at 420 °C, mesoporous phenolic resins with hollow cylindrical cavities with the periodicity of 50 nm were successfully obtained. Recently, Jaroniec and his coworkers prepared mesoporous carbons with extremely large specific pore volumes of 6 $cm^3 g^{-1}$ by carbonizing phenolic resin-colloidal silica nanocomposites under inert atmosphere followed by the removal of the colloidal silica template with HF or NaOH [157].

Mallouk and coworkers used a different strategy for the preparation of ordered mesoporous polymers using colloidal crystals as template instead of polymeric surfactants (Fig. 14) [158]. The mesoporous polymers were prepared by replication of colloidal crystals made from silica spheres with diameters of 35 nm. Into the gaps of the sintered nanocrystals, monomers such as divinylbenzene, ethyleneglycol dimethacrylate, or a mixture of the two were impregnated. Polymerization and subsequent dissolution of the silica template resulted in a polycrystalline network of interconnected pores. When monomer mixtures are used, the pore diameter of the polymer replicas can be controlled between 15 and 35 nm. From these results, one can conclude that well-ordered mesoporous polymers can be easily prepared by optimizing the synthesis conditions and choosing the appropriate precursors and the templates. As a whole, it can be stated at present that the preparation of mesoporous polymer materials is still in its infancy. The progress in this field is giving us the impression that the above synthesis strategy can be applied for the synthesis of mesoporous polymers with different structure, frameworks, and electronic properties such as conducting and semiconducting, which could offer various potential applications in the field of sensors, nanodevices, and adsorption/separation.

7. MESOPOROUS CARBONS

Porous carbon materials with nanoscale pore sizes prepared from periodic inorganic silica templates have received much attention because of their

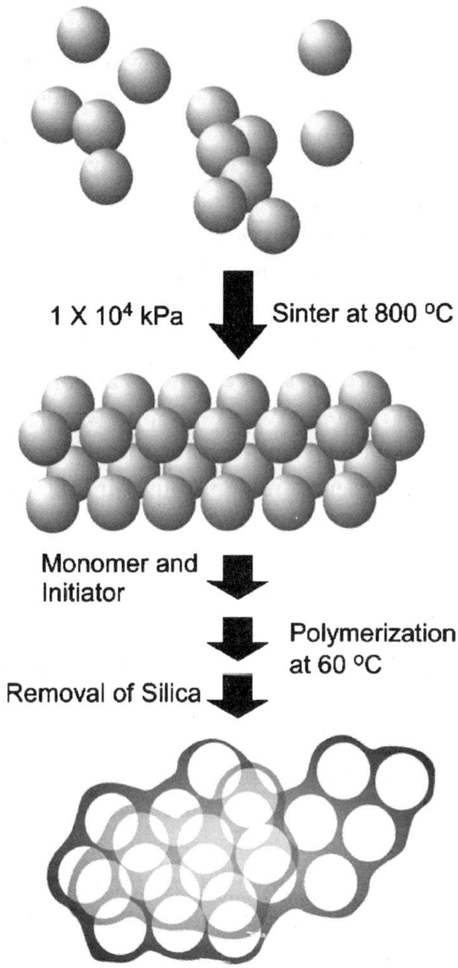

Figure 14 Schematic representation of the synthesis of ordered mesoporous polymers using colloidal crystals as template.

versatility and shape selectivity, which could be potentially useful for chromatographic separation systems, catalysts, nanoreactors, battery electrodes, capacitors, energy-storage devices, and biomedical devices. In this section, the synthesis of mesoporous carbons with different structures and their applications in various fields including adsorption, catalysis, and fuel cells will be discussed.

For the synthesis of porous carbon materials, the replica (or nanocasting) route is most often used. The replica method involves the preparation of new materials with the nanostructure being the replica to that of the template used in the process. Appropriate sources for the target materials are first impregnated into the template structures containing 3D networks of well-ordered mesoporous channels that are mutually intertwined, followed by solidification and template removal, thereby resulting in nanostructured porous carbon materials. This concept was widely used for the preparation of irregularly structured nanocarbons. During the late 1990s, ordered carbon materials have been synthesized using regularly structured templates such as zeolites and mesoporous silicas. Kyotani and coworkers successfully synthesized highly regular microporous carbon [159,160]. Carbon materials with mesoporous structures (CMK-*x*) were first prepared by Ryoo et al. using sucrose as the carbon source and mesoporous silicates such as MCM-48, SBA-1, and SBA-15 [161–165] as templates. Independently, and somewhat later, Hyeon et al. used a similar approach and reported the synthesis of well-ordered mesoporous carbons designated SNU-*x* [166–168].

The synthesis process of mesoporous carbons is schematically illustrated in Fig. 15. The selection of the mesoporous silica template is crucial for the successful preparation of mesoporous carbons with regular structures. MCM-41, one of the most frequently studied mesoporous silica, is not appropriate for the replica synthesis, because of the independent nature of the meso-channels in MCM-41, which are not interconnected by micropores. At the end, this results in non-connected rod-like carbon fibers. In contrast, the mesoporous silica MCM-48 with a cubic 3D structure is one of the most appropriate templates, because it has an enantiomeric pair of independently interpenetrating 3D networks of mesoporous channels that are mutually intertwined. A representative mesoporous carbon, CMK-1, is synthesized using mesoporous silica MCM-48 as template (Fig. 15a). Another mesoporous silica with large pores, SBA-15, has a 2D porous structure composed of hexagonal arrays of cylindrical pores that are finely interconnected by micropores through the walls. This material is widely used in the synthesis of mesoporous carbon. As the mesoporous channels in SBA-15-type materials are interconnected by micropores where the sucrose molecules can easily enter and undergo polymerization, an ordered mesoporous carbon structure is maintained even after silica removal. Using this concept, a mesoporous carbon with hexagonal symmetry, CMK-3, is obtained from a SBA-15 template (Fig. 15b). The pore size in mesoporous carbon materials can be tuned by controlling the structure of the silica replica. Recently, Hartmann and Vinu reported the successful systematic variation of the SBA-15 structure by controlling the synthesis temperature [3]. By using such pore size-tuned

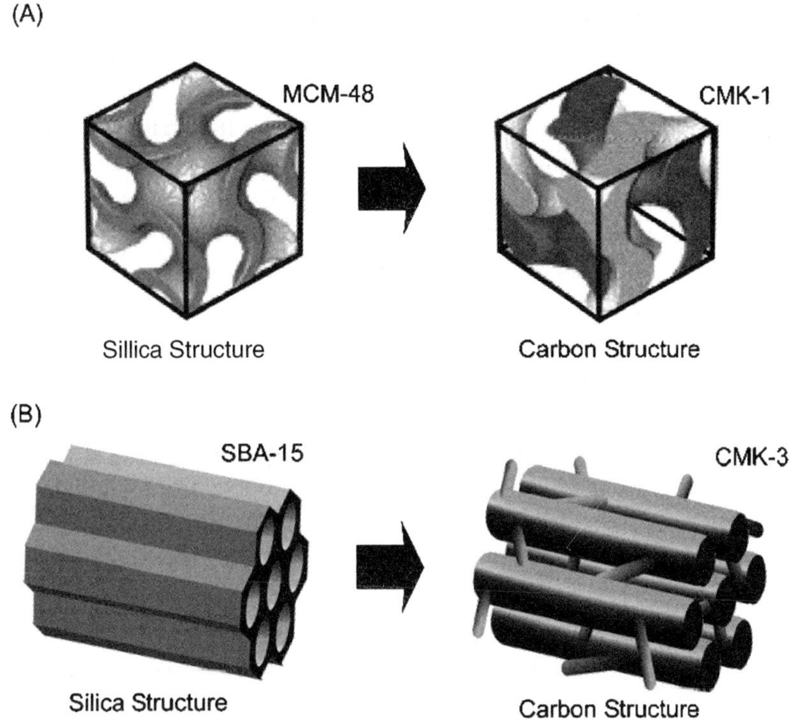

Figure 15 Schematic illustration of the synthesis of mesoporous carbon materials: (a) CMK-1 from MCM-48 and (b) CMK-3 from SBA-15.

SBA-15 replicas, CMK-3 mesoporous carbon materials with pore diameters ranging from 3 to 6.5 nm were successfully prepared [169].

Mesoporous carbon composed of carbon nanopipes, CMK-5, was also prepared by the group of Ryoo et al. using SBA-15 as the template [164]. The well-ordered mesoporous carbon tubes were obtained by controlling the filling of furfuryl alcohol, which was used as the carbon source, in the mesochannels of the template. It was demonstrated that the partial filling of the pores with the template is highly necessary to synthesize the mesoporous carbon with tubular structure. The ordered mesoporous carbon possesses hexagonal mesopores that are rigidly interconnected by small carbon nanorods that are formed inside the complementary channels between the adjacent mesopores. The resulting materials exhibit a bimodal pore size distribution, corresponding to the inner diameter of the carbon cylinders that corresponds to 5.9 nm and the pores formed between the adjacent cylinders with a diameter of 4.2 nm. Later, the same group optimized the synthesis conditions and obtained a high-quality CMK-5 material [170]. Mesoporous carbons, the structure of which was similar to that of CMK-5, were synthesized by the group of Schüth et al. who used mesoporous aluminosilicate as template [171]. The preparation involves the partial filling of the mesopores of AlSBA-15 materials with furfuryl alcohol and the subsequent polymerization using the acidic Al sites. After the carbonization at 900 °C followed by the removal of the template using HF or NaOH, mesoporous carbons composed of carbon tubes were obtained. Similar mesoporous carbon materials were also prepared through the catalytic chemical vapor deposition technique using transition metal–incorporated SBA-15 materials as the templates, which were prepared by dispersing ethylene-diamine functionalized SBA-15 in aqueous solutions containing transition metal ions. After a thermal treatment at high temperature, followed by the removal of the template [172], well-ordered mesoporous carbons can be obtained. It was also demonstrated that the deposition time has a strong influence in controlling the structural order of the materials, namely, the longer the deposition time, the higher was the structural order. The deposition technique has been further used to prepare mesoporous carbon hollow spheres, in which styrene was used as the carbon source and SBA-15 as the template.

Mesoporous carbon materials with different structures were also prepared by using templates with different structures. Che et al. prepared large-pore 3D-bicontinuous mesoporous carbons with cubic *Ia3d* symmetry using mesoporous monoliths with cubic structure as the template [173]. Kleitz et al. reported the synthesis of mesoporous carbon with *Ia3d* structure using mesoporous silica materials with 3D structures, prepared by different hydrothermal treatment using butanol as a structure modifier [174]. It is interesting to note that the same research group also reported that the pore size distribution of the obtained materials can be further tailored by controlled polymerization of a carbon source inside the mesochannels.

For mesoporous carbon materials, many potential applications have been proposed. As an example, their adsorption properties with respect to l-histidine and lysozyme are shown in Fig. 16. The adsorption of biomolecules from solution on solid surfaces has attracted much attention due to its scientific importance and potential applications in many areas such as biology, medicine, biotechnology, and food processing. In the medical and food industries, the specific removal of certain protein molecules is one of the critical processes. Fig. 16A shows the equilibrium adsorption isotherms of l-histidine on the mesoporous carbon CMK-3, activated carbon, and mesoporous silica SBA-15 [175,176]. In the range of equilibrium concentrations shown in Fig. 16A, the adsorption capacities

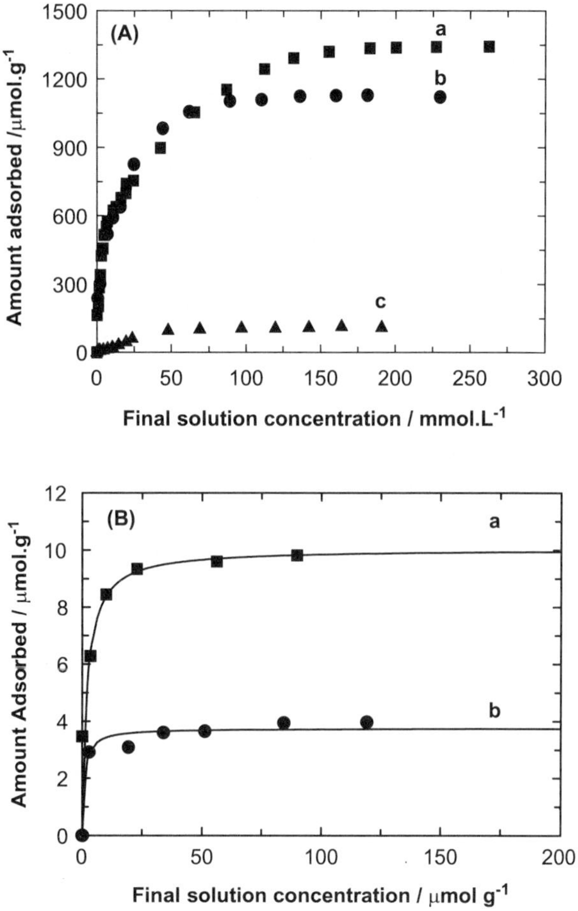

Figure 16 (A) l-histidine adsorption isotherms on mesoporous materials: (a) CMK-3, (b) activated carbon, and (c) SBA-15; (B) adsorption isotherms of lysozyme on mesoporous carbon materials: (a) CMK-3 and (b) CMK-1.

of these materials are tremendously different, as reported by Vinu et al.: Close to the isoelectric point of histidine, namely, at pH 7.5, CMK-3 showed a histidine adsorption capacity of approximately 1350 μmol g^{-1}, which is more than 12 times higher than the total adsorption capacity of SBA-15 (110 μmol g^{-1}). This indicates that in the case of SBA-15, the surface characteristic of the adsorbent plays an important role in determining the adsorption capacity. It has been previously reported that the number of surface hydroxyl groups of SBA-15 amounts to 4.1 OH groups per nm^2. These surface hydroxyl groups may prefer to form hydrated water layers rather than to adsorb the hydrophobic l-histidine.

In contrast, the surface of CMK-3 is highly hydrophobic in nature and thus favors accommodation of the l-histidine molecules. Similar results concerning the importance of hydrophobicity have been also recognized in other cases. For example, adsorption of proteins by the mesoporous carbon was found to occur most effectively at the isoelectric point of the protein, where the net charge of the protein is neutral. The efficient adsorption of lysozyme at its isoelectric point on the mesoporous carbon materials CMK-1 and CMK-3 is clearly demonstrated in Fig. 16B [177]. The CMK-3 material showed a greater capacity in the lysozyme adsorption than CMK-1 due to a larger pore size of the former material.

Recently, composites of mesoporous materials especially with carbon have received much attention owing to their excellent thermal, mechanical, and chemical properties. Mesoporous carbon has been used for the preparation of novel composites with organic substances. Choi and Ryoo reported the synthesis of polymer–carbon composite mesoporous materials (Fig. 17) [178], which exhibited the same chemical properties of the organic polymers and in addition showed a strong improvement of the stability of the mesopores against mechanical compression, thermal, and chemical treatments. As illustrated in Fig. 17, organic monomers were first impregnated into the pores of mesoporous carbon, followed by polymerization of the monolayers. Hydrophilic and hydrophobic organic monomers, such as styrene, methyl methacrylic acid, and poly(2-hydroxyethyl

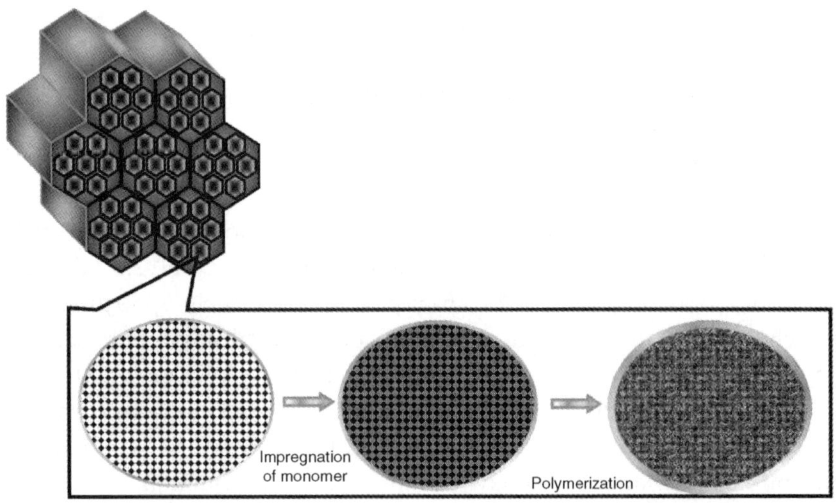

Figure 17 Synthesis of polymer–carbon composite mesoporous materials.

methylmethacrylate), were successfully used in this strategy. The mechanical stability increased upon the incorporation of polymers, showing a synergistic enhancement of the mechanical stability between the carbon framework and the polymer. The composite material from CMK-3 and polystyrene maintained the highly ordered nanostructure after being pressed for 10 min at 1500 MPa. In addition, the structure was also maintained even after heating to 150 °C. The presence of a thin layer of polymer does not cause a significant decrease in the electric conductivity of mesoporous carbon frameworks. The latter property may offer a remarkable advantage for applications in functional electrode materials such as amperometric biosensors.

The preparation of mesoporous carbons with reliable control of the textural parameters, such as specific pore volume, specific surface area, and pore diameter, is an exciting topic as these properties are critical in many industrial applications, particularly in separation and adsorption of large molecules. These carbon materials can provide a large density of sites on which the molecules can adsorb and a porous network to achieve a fast accessibility for adsorbate molecules to the porous matrix and the adsorption sites on their internal surface. For example, Mokaya and coworkers reported the use of chemical vapor deposition to supply the carbon source in a replica-type synthesis for the preparation of graphitic mesoporous carbons with controlled morphology and pore size [179]. Recently, Vinu et al. have reported the synthesis of mesoporous silica and aluminosilicates with variable pore diameters (SBA-15-*x* and AlSBA-15-*x*) by changing the synthesis temperature. They further applied these materials as replica for the synthesis of mesoporous carbon materials [25,169,180,181]. In the case of AlSBA-15, it acts as both an acid catalyst for the polymerization of the carbon source and as the template. Interestingly, the specific surface area, the specific pore volume, and the pore diameter of the obtained materials are significantly higher as compared to those of CMK-3.

Another strategy to control the pore diameter and the surface properties of the mesoporous carbon materials is tuning the wall thickness of the mesoporous template, and the walls of the mesoporous silica template is generally converted into pore after the dissolution with HF. Lee et al. followed this technique and employed mesoporous silica templates with different wall thicknesses as the templates to control the pore diameter of the resulting mesoporous carbon materials. By adjusting the thickness of the silica wall, the pore diameter of the mesoporous carbon materials can be successfully controlled [182].

Very recently, Vinu et al. reported a simple method for controlling the porous structure and the textural parameters of mesoporous carbons by a controlled pore filling technique. The controlled pore filling technique involves a simple variation of the concentration of sucrose molecules during their infiltration into the mesoporous matrix of the silica template [183]. The porous structure, morphology, and the textural parameters of the mesoporous carbons materials can be controlled by this method. The materials were denoted as CMK-3-*x*, where *x* represents the sucrose-to-silica weight ratio. It is noteworthy that the unit cell constant of the resulting mesoporous carbon materials increases from 9.80 to 10.62 nm with increasing the sucrose-to-silica weight ratio from 1.25 to 5. This is probably due to changes in the void spaces inside the channels upon filling with different amounts of sucrose molecule. There is significant strain in the mesopores as the amount of sucrose adsorbed at the pore walls is increased, and this strain is reduced by a structural relaxation of the carbon network following silica removal, which leads to a larger unit cell constant. However, the unit cell constant of CMK-3-0.8 is 11.25 nm, which is much higher than the unit cell constant of other materials prepared in this study. This is tentatively attributed to a large reduction in the structural shrinkage because of an incomplete filling of the channels in mesoporous silica by sucrose molecules, which may help to reduce the compressive forces inside the channels. This may also be attributed to the opening and merging of primary mesoporous channels due to the incomplete pore filling leading to a large unit cell constant.

The specific surface area of the mesoporous carbon materials systematically increases with decreasing the sucrose-to-silica weight ratio in the preparation step: The specific surface area increases from 724 to 1570 $m^2\,g^{-1}$ with decreasing the sucrose-to-template ratio from 5 to 0.8. The higher specific surface area of CMK-3-0.8 could be attributed to an increase in the surface heterogeneity and the presence of micropores in the material because of the incomplete filling of the pores with the sucrose molecules. The thickness of the pore wall of the mesoporous carbon materials increases from 5.62 to 6.74 nm with increasing the sucrose-to-silica weight ratio from 1.25 to 3.3 and then decreases to 6.42 nm with further increase of the amount of sucrose. On the contrary, the specific pore volume of the materials also increases from 0.57 to 1.31 $cm^3\,g^{-1}$ with decreasing the sucrose-to-silica weight ratio to 1.25 and then decreases to 1.23 $cm^3\,g^{-1}$ for the material prepared at a sucrose-to-silica ratio of 0.8.

The structural order and the microscopic features of the materials prepared at different sucrose-to-silica weight ratios were studied by

HR-TEM. The HR-TEM images of CMK-3-0.8, CMK-3-1.25, and CMK-3-5 are shown in Fig. 18. It is very clear from the HR-TEM images that the structural order of the materials prepared at different sucrose-to-silica weight ratios shows significant differences. It is also interesting to follow changes in the structure of carbon materials prepared at different sucrose contents. CMK-3-0.8 shows disordered mesoporous channels because of incomplete pore filling, whereas CMK-3-1.25 exhibited a highly ordered mesoporous structure with a linear array of mesopores, originating from the mesoporous silica framework that would convert into the mesopores of the resultant carbon after the HF treatment. The above results are consistent with the results and the conclusions derived from the XRD patterns and the N_2 adsorption–desorption isotherms of those samples. CMK-3-5 also exhibited an ordered mesoporous structure with a linear array of mesopores separated by carbon walls and some carbon deposits on the external surface, which is clearly confirmed by the presence of dark spots in the HR-TEM image. This interesting result implies that a higher amount of sucrose molecules could enhance the pore blocking in the mesoporous channels and carbon deposition on the external surface of the mesoporous carbon. Thus, it can be concluded that an optimized carbon-to-silica ratio is highly necessary to ascertain the formation of a mesoporous carbon structure with a very high degree of order and good textural parameters.

The surface morphology of the obtained carbon materials was characterized by high-resolution field emission scanning electron microscopy (HR-FESEM). Fig. 19 shows the HR-FESEM images of the carbon materials prepared at different sucrose-to-silica ratios and the original silica template. All the carbon samples except CMK-3-5 show rod-like morphology, which is almost similar to that of the original silica template. It is interesting to note that the morphological order of the mesoporous carbon materials decreases with increasing the sucrose-to-silica weight ratio. CMK-3-5 exhibits an irregular rod-like morphology with disordered particle clusters. This is mainly attributed to the carbon deposition on the external surface of the mesopores of CMK-3-5. This result is very much consistent with the results obtained from the HR-TEM and N_2 adsorption–desorption analysis.

Mesoporous carbons with ultra-large mesopores can be prepared by using mesocellular silica foams that have a large mesopore size. Lee et al. synthesized mesoporous carbon materials with pore diameters larger than 10 nm using mesocellular foam as the template [184]. Large-pore diameters can only be achieved through a partial filling of the pores of the

Figure 18 HR-TEM images of mesoporous carbons prepared at different sucrose-to-silica weight ratios: (a) CMK-3-0.8, (b) CMK-3-1.25, and (c) CMK-3-5.

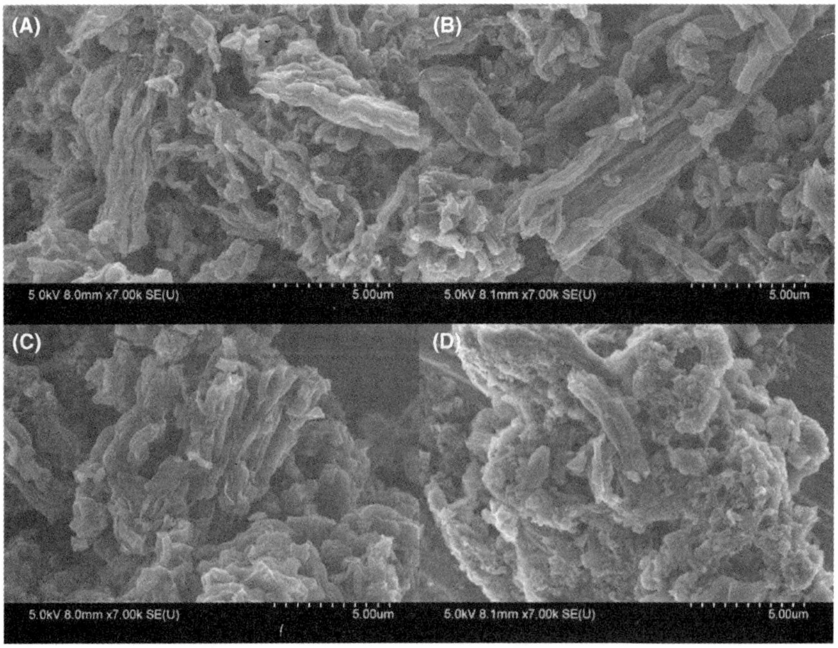

Figure 19 HR-FESEM images of mesoporous carbons prepared at different sucrose-to-silica weight ratios: (a) CMK-3-0.8, (b) CMK-3-1.25, (c) CMK-3.3, and (d) CMK-3-5.

mesocellular foams with phenol that is later polymerized inside the channels with formaldehyde. When mesocellular foam with a main cell diameter of 27 nm and a window size of 14 nm was used, mesoporous carbon materials with pore diameter larger than 27 nm and open cellular pores were obtained. Somewhat later, Oda et al. reported a mesocellular foam with a main cell size of 24 nm and a window size of 18 nm with closed hollow spherical pores using double impregnation of sucrose into the channels of mesoporous cellular silica foam materials [185]. From these results, it can be concluded that the pore opening of the mesoporous carbon cellular foam can be controlled by varying the simple impregnation procedure. It must be noted that this kind of mesoporous carbon materials with large pore diameters is extremely interesting for cultivating cells or for the immobilization of large enzyme molecules whose size is larger than 10 nm. Moreover, such materials also possess a lot of other potential applications, namely, as adsorbents for microorganisms or large toxic molecules.

The nanocasting technique generally results in mesoporous carbon materials with amorphous walls. This is a certain limit for pertinent applications, including electrodes for electrochemical double-layer capacitors, fuel cells, and biosensors. Introducing a graphitic structure into the walls of mesoporous carbons has been a great challenge in recent years because it is very difficult to maintain both the textural parameters of the mesoporous carbon materials and their graphitic behavior at the same time. Kim et al. tried to use aromatic carbon precursors instead of sucrose molecules to prepare graphitic mesoporous carbons [186]. However, they could only obtain mesoporous carbons with disordered graphitic sheets that are aligned perpendicular to the template walls. They further found that the mechanical strength of mesoporous carbons with disordered graphitic walls is much better than that of mesoporous carbons derived from sucrose or furfuryl alcohol. The group of Pinnavaia et al. also used aromatic carbon precursors such as naphthalene, anthracene, and pyrene with the aim of producing graphitic mesoporous carbons [187]. They showed that the electrical conductivity of materials obtained through this synthesis route was much higher than the one of amorphous mesoporous carbons. Polyvinyl chloride was used as the carbon precursor by Fuertes et al., who were able to prepare graphitic mesoporous carbons with an electrical conductivity of $0.3\,S\,cm^{-1}$ [188]. Later, catalytic graphitization was also used by the same group for synthesizing graphitic carbons using polypyrrole as the carbon precursor and $FeCl_3$ as the catalyst. The iron catalyst promoted the formation of a graphitic structure, and the resulting material showed a performance for electrochemical double-layer capacitors superior to the one of non-graphitic carbons [189].

Gierszal et al. reported the synthesis of mesoporous carbon using KIT-6 as the template [190]. KIT-6 is a well-ordered mesoporous silica consisting of an interpenetrating bicontinuous network of channels that are also interconnected through irregular micropores present in the mesopore walls. Taking advantage of this structure, the corresponding mesoporous carbon materials were obtained from various carbon precursors such as sucrose, furfuryl alcohol, and mesophase pitches. Especially, the pitch-based carbons with sufficiently thick pore walls are relatively easily graphitizable.

7.1 Mesoporous carbon nanocages

Very recently, we have extended the concept of the replica route for the synthesis of a novel type of nanocarbon, namely, carbon nanocages [21,191] using 3D large cage-type face-centered cubic mesoporous silica materials (KIT-5) as inorganic templates [192]. The obtained materials were named as CKT after carbon materials from KIT-5. It should be noted that the standard procedure for the preparation of the mesoporous carbon CMK-3, namely, using mesoporous silica SBA-15 as inorganic template, which has been reported by Ryoo et al. [161], is not suitable in our case to make a highly ordered CKT mesoporous carbon. This is mainly due to the higher bulk density and lower pore volume of KIT-5 as compared to the other mesoporous silicates such as MCM-48 and SBA-15. Therefore, we modified the synthesis procedure and fine-tuned many parameters. Especially, the appropriate weight ratio (ca. 2.5) of water-to-silica template (KIT-5) was found to be a crucial factor for a successful synthesis. The synthesis principle for the preparation of carbon nanocages is illustrated in Fig. 20. Note that this representation gives just a rough idea on the nature of the obtained materials and does not precisely display the actual structure of the obtained carbon material.

Carbon nanocage materials with different pore diameters were prepared using different KIT-5-*T* mesoporous silica samples (*T* denotes the synthesis temperature in °C of the mesoporous silica) as templates and sucrose as the carbon source. The synthesized materials were designated as CKT-1, CKT-2, and CKT-3, and they have been prepared from KIT-5-100, KIT-5-130, and KIT-5-150, using a sucrose-to-silica weight ratio of 0.75. Another set of samples was prepared at different sucrose-to-silica weight ratios, namely, 0.45, 1.2, and 2.0, and the resulting materials were labeled as CKT-3(A), CKT-3(B), and CKT-3(C), respectively.

Fig. 21A shows the powder XRD pattern of CKT-1, CKT-2, and CKT-3. All the samples show an intense (111) reflection and a broader

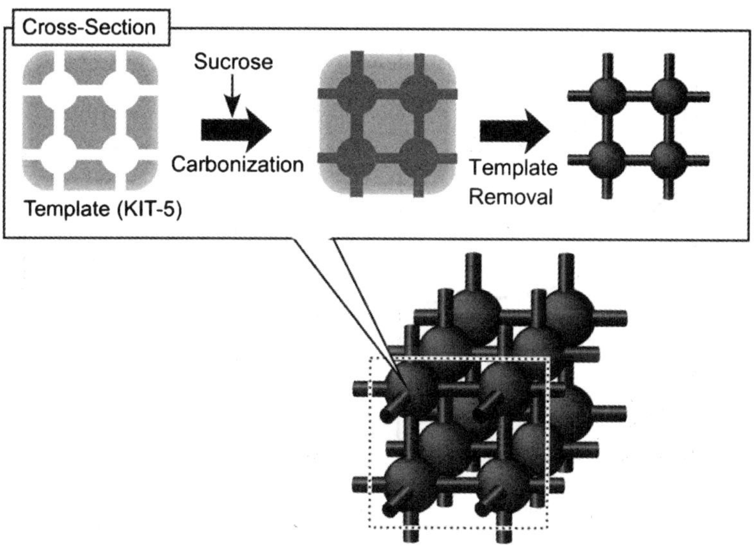

Figure 20 Synthesis principle for the preparation of carbon nanocage-type materials.

(200) reflection, demonstrating that the mesoporous structure was preserved even after the removal of the mesoporous silica by HF etching. Moreover, thermogravimetric analysis under an oxygen atmosphere revealed that the maximum silica residue is in the 1–1.5 wt.% range, confirming that the intense XRD peak does not result from the KIT-5 mesoporous silica template. Fig. 21B shows the XRD pattern of CKT-3(A), CKT-3(B), and CKT-3(C). The XRD pattern of the CKT-3(A) and CKT-3(B) also shows a main (110) reflection with a broader (200) reflection, while the XRD pattern of CKT-3(C) is completely different from its parent mesoporous silica and exhibits a broad peak at higher angle. This implies that some transformation into a new disordered structure has occurred during the removal of the silica framework in the case of CKT-3(C). HR-TEM images of the CKT-3(A) material are shown in Fig. 22. The images were recorded along two different crystallographic directions. Both micrographs confirm that the mesoporous carbon possesses a highly ordered structure with uniform pore size distribution. A regular arrangement of bright spots also reveals that the mesoporous material is of the 3D cage-type.

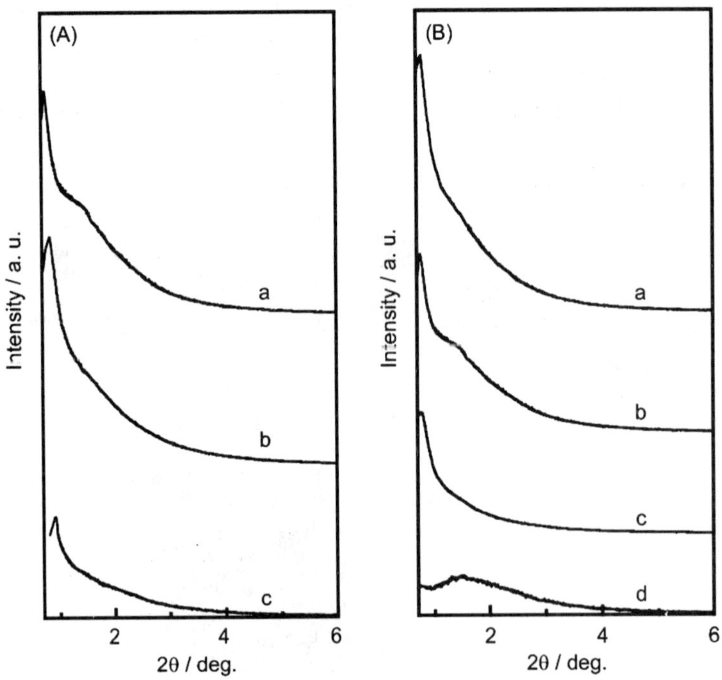

Figure 21 Powder XRD pattern of carbon nanocage materials: (A) (a) CKT-3, (b) CKT-2, and (c) CKT-1. (B) (a) CKT-3(A), (b) CKT-3, (c) CKT-3(B), and (d) CKT-3(C).

The nitrogen adsorption isotherms of CKT-1, CKT-2, and CKT-3 and their corresponding BJH desorption pore size distributions are shown in Fig. 23A and 23B, respectively. The isotherms of CKT-2 and CKT-3 are of type IV exhibiting an H2 hysteresis loop. The pore diameter of CKT-3 is larger than that of CKT-2, which reflects the influence of the pore diameter

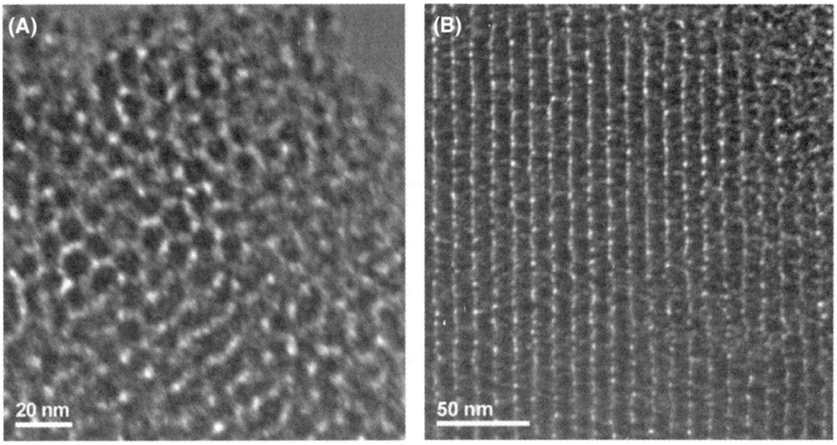

Figure 22 HR-TEM images of CKT-3(A): (a) cross-sectional projection, and (b) longitudinal projection. Reprinted with permission from Ref. [21]. © 2005, The Royal Society of Chemistry.

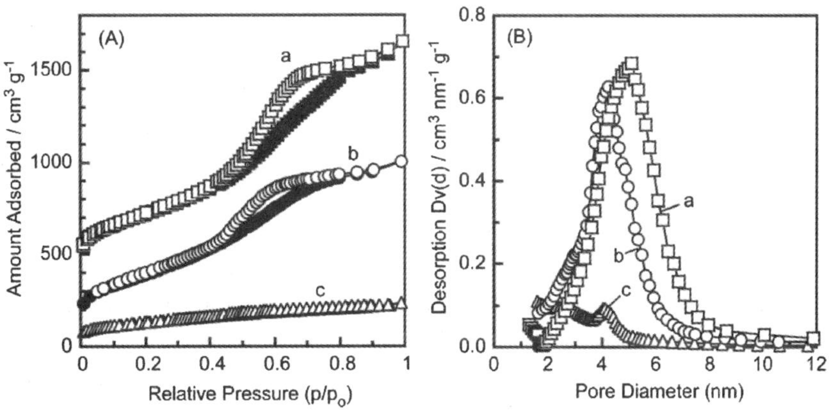

Figure 23 (A) Nitrogen adsorption–desorption isotherms and (B) BJH desorption pore size distribution: (a) CKT-3, (b) CKT-2, and (c) CKT-1. In diagram (A), closed symbols and open symbols represent adsorption and desorption isotherms, respectively. Reprinted with permission from Ref. [21]. © 2005, The Royal Society of Chemistry.

Table 1 Textural parameters of the mesoporous silica templates and the resulting mesoporous carbon nitrides with different pore diameters

Sample	a_o (nm)	A_{BET} ($m^2 g^{-1}$)	Pore volume ($cm^3 g^{-1}$)	Pore diameter (nm)
SBA-15-100	10.30	895	1.20	9.00
SBA-15-130	10.60	550	1.20	10.70
SBA-15-150	10.83	393	1.10	11.20
MCN-1-100	9.52	505	0.55	4.20
MCN-1-130	10.18	830	1.25	5.10
MCN-1-150	10.50	650	0.89	6.40

of their parent mesoporous silica. CKT-2 and CKT-3 possess pores with diameters of about 4.0 and 5.2 nm (Table 1), high BET surface areas of 1410 and 1515 $m^2 g^{-1}$ and large specific pore volumes of 1.46 and 2.0 $cm^3 g^{-1}$, respectively. The specific surface area and the specific pore volume of CKT-3 are significantly larger as compared to CMK-3 (1260 $m^2 g^{-1}$ and 1.1 $cm^3 g^{-1}$) and CMK-3-150 (1350 $m^2 g^{-1}$ and 1.6 $cm^3 g^{-1}$). In contrast, CKT-1 exhibited a type II isotherm with no well-pronounced capillary condensation step. The specific surface area and specific pore volume of CKT-1 are 475 $m^2 g^{-1}$ and 0.35 $cm^3 g^{-1}$, respectively, which suggests disordering in the structure. The nitrogen adsorption isotherms of CKT-3 mesoporous carbons prepared by using different sucrose-to-silica weight ratios are shown in Fig. 24A. The specific surface area (Fig. 24C) and specific pore volume (Fig. 24D) systematically increases with decreasing the sucrose-to-silica weight ratio, while the change in the pore diameter was not significant (Fig. 24B). The specific surface area and specific pore volume reached to 1600 $m^2 g^{-1}$ and 2.1 $cm^3 g^{-1}$, respectively, in the case of CKT-3(A). To calculate the cage diameter of the carbon nanocage, the theoretical model proposed by Ravikovitch et al. [193] was used. Thus, a cage diameter of 15 nm was obtained for CKT-3(A).

7.2 Application of mesoporous carbon nanocage materials

Because of the large-pore volume of the carbon nanocage materials, they are expected to possess superior capability in the adsorption of biomolecules. As a simple demonstration, the adsorption behavior of the

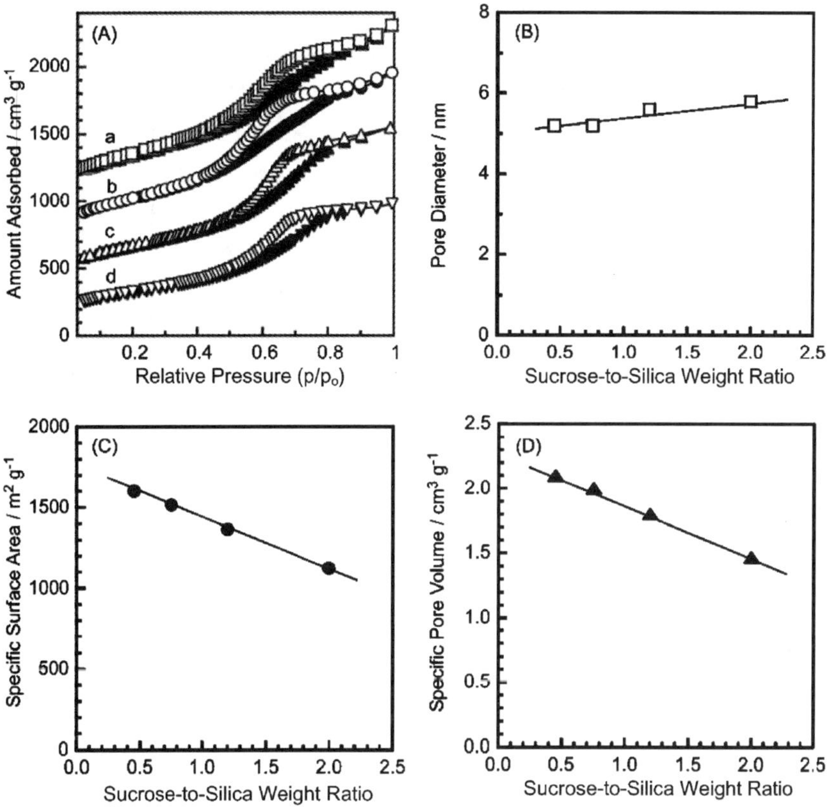

Figure 24 (A) Nitrogen adsorption–desorption isotherms: (a) CKT-3(A), (b) CKT-3, (c) CKT-3(B), and (d) CKT-3(C). In this diagram, closed symbols and open symbols represent adsorption and desorption isotherms, respectively. Effect of the sucrose-to-silica weight ratio on (B) pore diameter, (C) specific surface area, and (D) specific pore volume is summarized. Reprinted with permission from Ref. [21]. © 2005, The Royal Society of Chemistry.

carbon nanocage materials (CKT-3(A) and CKT-3(B)) with respect to lysozyme was investigated. The adsorption isotherms shown in Fig. 25 clearly indicate a higher capacity of the carbon nanocages for lysozyme adsorption as compared to that observed for CMK-3. The maximal monolayer adsorption capacities of CKT-3(A) and CMK-3(B) are 26.5 and 23.8 $\mu mol\,g^{-1}$, respectively, while that of CMK-3 amounts to only 9.8 $\mu mol\,g^{-1}$. Not limited to protein adsorption, carbon nanocage materials would have superior capability in molecular adsorption, extraction, and removal, which are currently investigated in our research group.

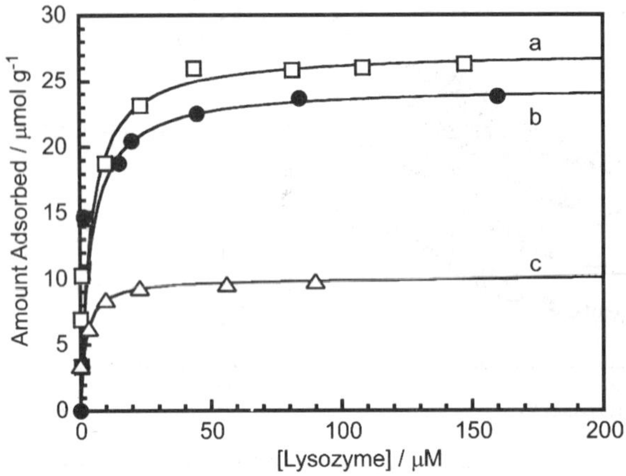

Figure 25 Adsorption isotherms for lysozyme on selected mesoporous materials: (a) CKT-3(A), (b) CKT-3(B), and (c) CMK-3. Reprinted with permission from Ref. [21]. © 2005, The Royal Society of Chemistry.

8. MESOPOROUS CARBON NITRIDES

The replica route can be utilized for the synthesis of novel mesoporous materials other than carbon, because its concept is general and its procedure is rather simple. In this section, the main focus is on the synthesis of mesoporous carbon nitride (CN), to demonstrate the wide diversity of the replica route. CN is a well-known and fascinating material that has attracted worldwide attention because the incorporation of nitrogen atoms in the carbon nanostructure can enhance the mechanical, conducting, field emission, and energy storage properties [194–205]. CN materials with five different structures have been predicted so far: one is 2D graphitic C_3N_4 and the other four are 3D CNs, namely, α-C_3N_4, β-C_3N_4, cubic-C_3N_4, and pseudocubic-C_3N_4. Among the CN materials, β-C_3N_4 and its allotropic cubic and pseudo-cubic phases are superhard materials, the structure and properties of which are expected to be similar to those of diamond and β-Si_3N_4. Owing to its unique properties such as semiconductivity, intercalation ability, and hardness, CN is regarded as a promising material that could find potential applications in many fields. Non-porous CN materials can be prepared from chemical precursors at very high temperatures. Very recently, Gao and Giu have reported the chemical synthesis of non-porous CN with turbostratic crystalline structure from polymerized ethylenediamine (EDA) and carbon tetrachloride (CTC) [206]. By constructing CN materials with porous structure, many novel applications could emerge: from catalysis to separation and adsorption of very bulky molecules to the preparation of devices with low dielectric constants. However, only little attention has been given to the synthesis of porous CN materials so far. Vinu et al. have combined the chemical synthesis and nano-templating routes for the preparation of CN with porous structure using mesoporous silica as template. This process generated a highly ordered mesoporous CN material [207], designated MCN-1 (Fig. 26), which has a uniform pore size distribution, high specific surface area, and a high specific pore volume. The material possesses a highly ordered hexagonal array of a 2D pore system with a lattice constant of 9.52 nm, and the structural symmetry of the well-known parent 2D mesoporous silica template, SBA-15, is retained. The details of the synthesis and the characterization of the novel material are described below.

In a typical synthesis, the calcined mesoporous silica SBA-15 used as the template was added to a mixture of ethylene diamine and CTC. The composites from the template silica and the CN polymer were then heat-treated in a nitrogen flow to carbonize the polymer. Mesoporous CN was recovered after dissolution of the silica framework in HF. Thermogravimetric analysis under an oxygen atmosphere revealed that the maximum silica residue is less than 1 wt.%.

The structure of mesoporous CN MCN-1 was first investigated by powder XRD. The XRD pattern of the MCN-1 material (Fig. 27) showed three clear peaks, which can be assigned to the (100), (110), and (200) diffractions of a 2D hexagonal lattice (space group *p6mm*) with a lattice constant $a_{100} = 9.52$ nm. The pattern is similar to the XRD pattern of the

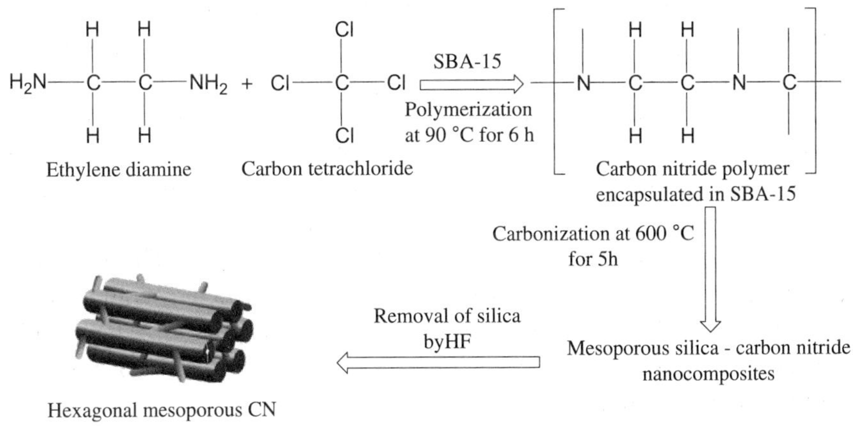

Figure 26 Preparation of hexagonally ordered mesoporous carbon nitride (MCN).

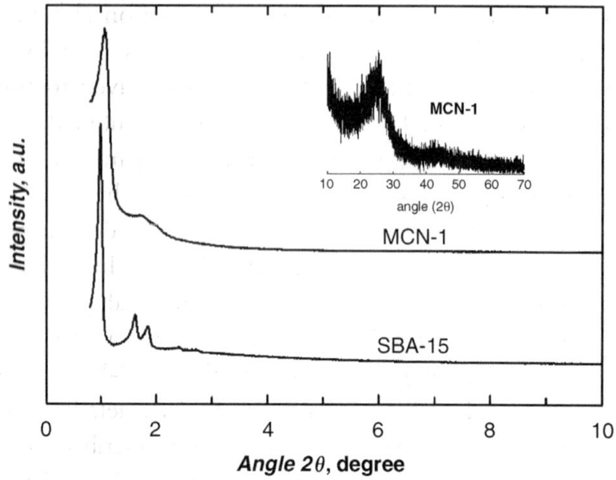

Figure 27 Powder XRD patterns of mesoporous carbon nitride (MCN-1) and mesoporous silica SBA-15. Inset: Powder XRD pattern of MCN-1 in the wider angle region.

parent mesoporous silica template SBA-15, although some peak broadening was observed as compared to the parent SBA-15 (Fig. 27). The powder diffraction pattern of MCN-1 also shows a single broad diffraction peak (Fig. 27 inset) close to 25.8°, corresponding to an interlayer d-spacing of 0.342 nm, which is similar to the d-spacing obtained in the non-porous CN spheres. This indicates turbostratic ordering of the carbon and nitrogen atoms in the graphene layers of MCN-1.

The pore structure of the mesoporous CN MCN-1 was investigated in more detail by nitrogen adsorption. The nitrogen adsorption isotherms of MCN-1 and parent mesoporous silica template SBA-15 are shown in Fig. 28A. The isotherm of MCN-1 is of type IV with a H1 hysteresis loop. The hysteresis loop in the isotherm of SBA-15 is larger than the hysteresis loop in the isotherm of MCN-1, suggesting a smaller amount of micropores. While the pore size distribution determined by nitrogen adsorption on SBA-15 is centered at 7.1 nm, the corresponding pore size distribution of MCN-1 is centered at 4.0 nm (Fig. 28B). The pore size distribution of the latter material is somewhat broadened as compared to that of SBA-15, but still exhibits narrow pore size distribution. The specific surface area and the specific pore volume were calculated to be 505 $m^2 g^{-1}$ and 0.55 $cm^3 g^{-1}$, respectively.

HR-TEM studies of MCN-1 were also carried out. When viewed down the [99] direction of MCN-1, only a stripe pattern could be detected

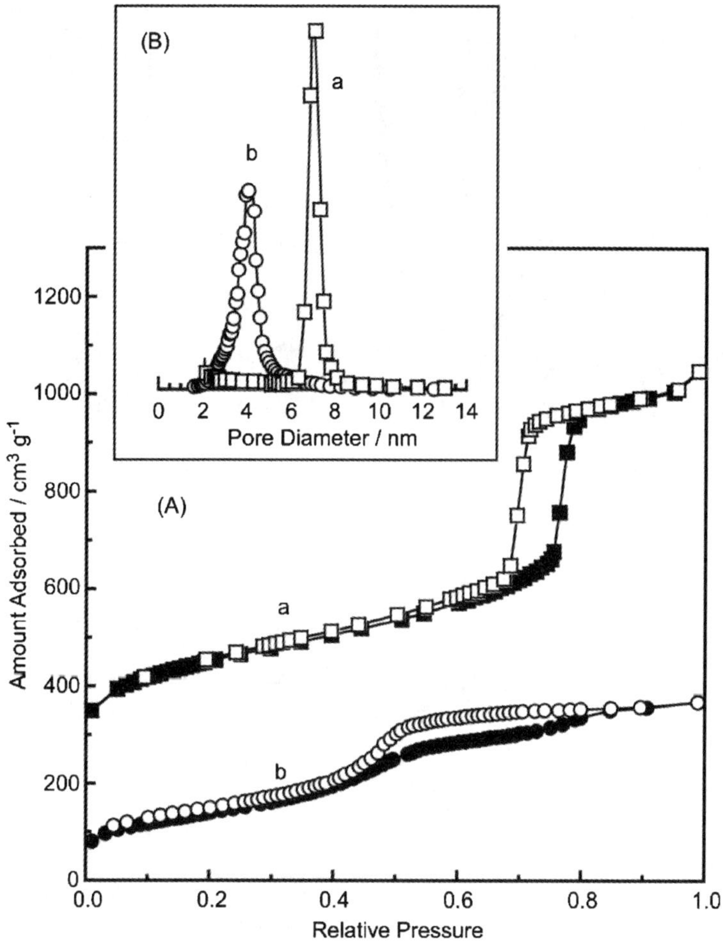

Figure 28 (A) Nitrogen adsorption–desorption isotherms and (B) BJH pore size distribution of: (a) SBA-15 and (b) MCN-1. In diagram (A), closed and open symbols represent adsorption and desorption isotherms, respectively.

(Fig. 29a). Bright contrast stripes on the under-focused image represent images of the pore walls, whereas dark contrast cores represent empty channels. The corresponding Fourier transform optical diffraction pattern derived from the image only shows a 1D array of spots along the [99] direction, indicating that there is no crystallographic ordering along the axis of the empty mesopores. The cross-sectional HRTEM image, Fig. 29b, clearly displays a hexagonal (honeycomb-like) arrangement of the mesopores. The inset shows a Fourier transform optical diffraction pattern. Elemental maps constructed revealed that the material is indeed composed

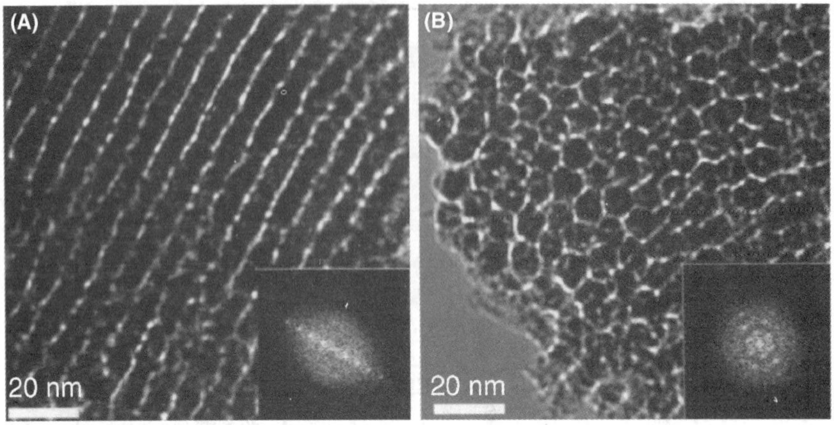

Figure 29 HR-TEM images of MCN-1: (a) longitudinal projection and (b) cross-sectional projection.

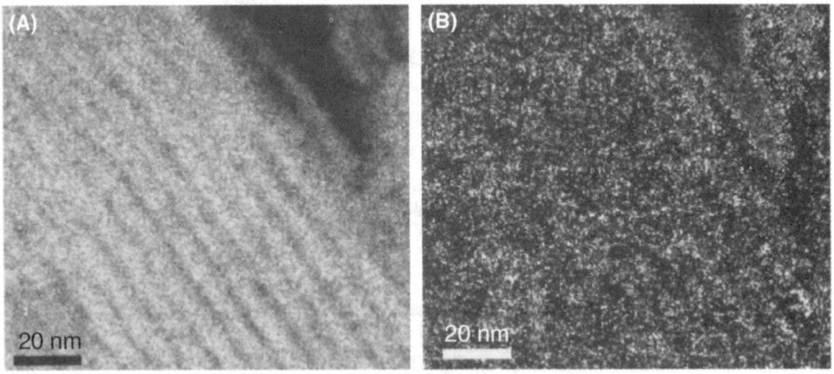

Figure 30 Elemental mapping of MCN-1: (a) carbon and (b) nitrogen.

of carbon (Fig. 30a) and nitrogen (Fig. 30b). Traces of other elements were not detected during mapping. An electron energy loss (EEL) spectrum of MCN-1 exhibited C and N K-edges located at 284 and 401 eV. The fine structure of the edges, in particular their left-hand shoulders revealing $1s\text{-}\pi^*$ electronic transitions, is a fingerprint of sp^2 hybridization.

In addition to the above-mentioned results, elemental analysis and the X-ray photoelectron spectroscopy (XPS) data after silica removal confirmed the absence of silicon in the MCN-1 materials. The XPS survey spectrum of MCN-1 after the silica removal shows a large signal for oxygen and does not show any peaks for elements other than carbon, nitrogen, and oxygen (Fig. 31A). The oxygen peak may come from moisture, ethanol,

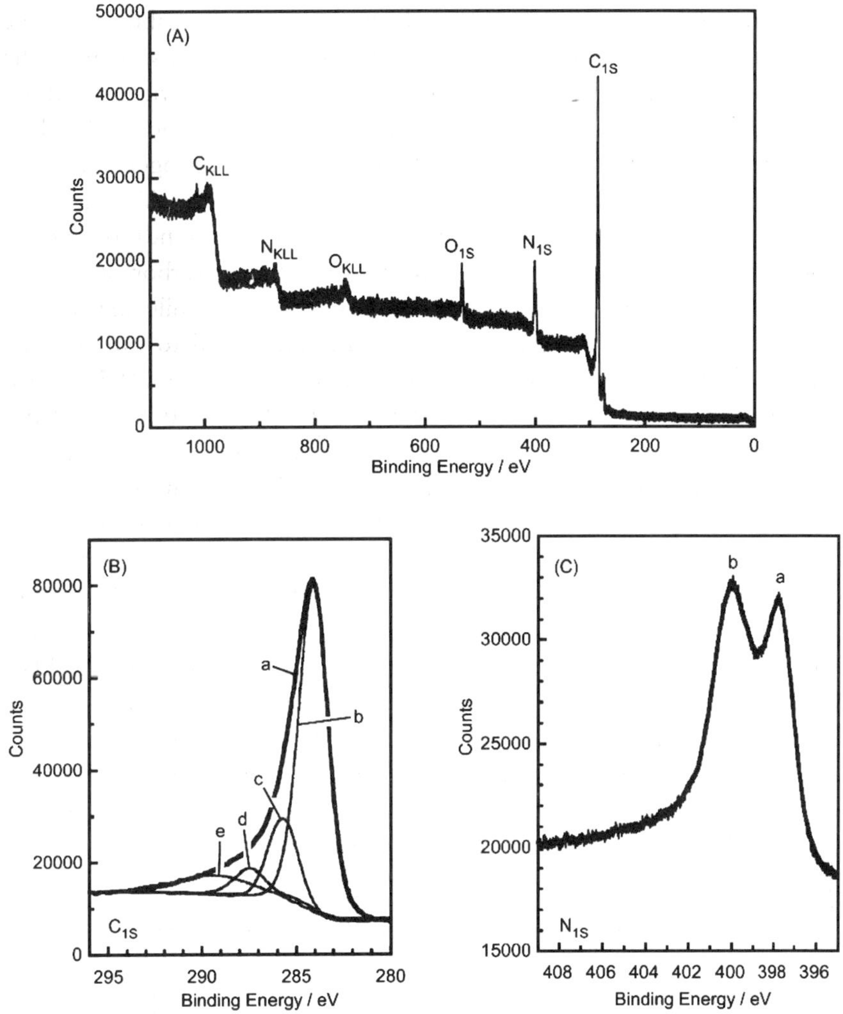

Figure 31 (A) XPS survey spectrum of MCN-1 and (B) the XPS C_{1s} spectrum (curve (a)) can be deconvoluted into four peaks with binding energies of (b) 284.1 eV, (c) 285.7 eV, (d) 287.5 eV, and (e) 289.3 eV. (C) The XPS nitrogen core-level 1s spectrum can be deconvoluted into two peaks centered at (a) 397.8 eV and (b) 400.2 eV.

atmospheric O_2, or CO_2 adsorbed on the surface of MCN-1. The nature of the chemical bonding in the MCN-1 CN walls was further investigated by XPS measurements, and the results are given in Fig. 31B. The XPS C_{1s} spectrum could be deconvoluted into four peaks with binding energies of 289.3, 287.5, 285.7, and 284.1 eV. The lowest energy contribution fitted for C_{1s} in MCN-1 (284.1) is assigned to pure graphitic sites in the amorphous CN matrix and the peak at 285.7 eV is attributed to the sp^2-hybridized carbon bound to nitrogen inside the aromatic structure. The highest energy contribution, 287.5 eV, is assigned to sp^2-hybridized carbon, while the contribution at 289.3 eV is assigned to the sp^2-hybridized carbon in the aromatic ring attached to NH_2 groups. The XPS nitrogen core-level 1s spectrum shows two peaks centered at 397.8 and 400.2 eV (Fig. 31C). The peak at the highest binding energy (400.2 eV) corresponds to nitrogen atoms trigonally bound to all sp^2 carbons or to two sp^2 carbon atoms and one sp^3 carbon atom in an amorphous CN network, while the peak at 397.8 eV is attributed to nitrogen sp^2-hybridized bound to carbon. The nature of the chemical bond of carbon and nitrogen in MCN-1 was also investigated by Fourier transform infrared (FT-IR) spectroscopy. Bands occurring at 1257.3 and 1570.7 cm^{-1} were assigned to aromatic CN stretching bonds and aromatic ring modes, respectively, while a broad peak centered at 3412 cm^{-1} was attributed to the stretching mode of NH groups in the aromatic ring. A similar result has also been observed in the non-porous CN samples. The absence of a peak around 2200 cm^{-1} confirms that there are no C≡N components in the MCN-1 material. These spectral features confirm the presence of a 1,3,5-triazine ring. In addition, the UV-Vis spectrum of MCN-1 consists of three components with maxima at 240, 255, and 275 nm. These bands can be unambiguously assigned to π-π^* electronic transitions in the aromatic 1,3,5-triazine compounds (Fig. 32). These spectroscopic data clearly indicate that the atomic environment of carbon atoms and nitrogen atoms in the walls of MCN-1 is similar to other

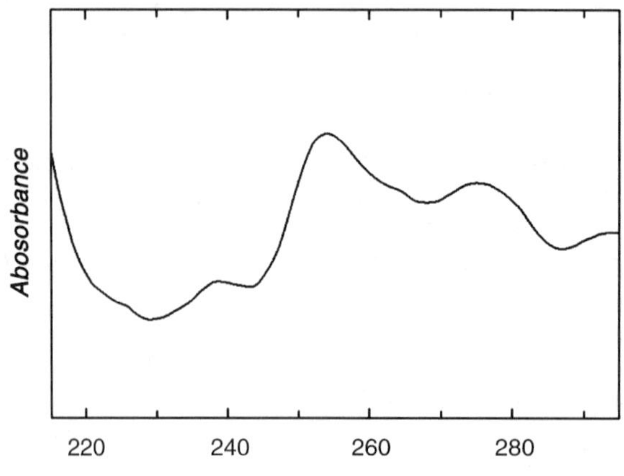

Figure 32 UV-Vis absorbance spectrum of MCN-1.

non-porous amorphous CN materials. Goettmann et al. reported that nanoparticles of graphitic C_3N_4 with different diameters and morphology could be immobilized inside the channels of mesoporous silica host matrices [208,209]. It has also been shown that graphitic C_3N_4 nanoparticles immobilized in mesoporous silica exhibit excellent photoluminescence properties.

8.1 Mesoporous carbon nitrides with varying pore diameters

Porous materials with uniform and tunable pore sizes are expected to play an important role in a number of applications that range from catalysis to molecular separations and sorption of very bulky molecules and to the preparation of semiconductors, semiconducting nanowires, and low dielectric constant devices. The pore structure and the textural parameters of mesoporous silica materials can be easily controlled by varying the chain length of the surfactants that are used as the structure directing agent and by adjusting the synthesis conditions such as reaction temperature, reaction time, and the solution pH. On the contrary, the pore structure of mesoporous carbons is entirely dependent on the pore size and structure of the silica template and the nature and the size of the carbon source used for their preparation. There has been a lot of reports on the pore size and textural parameter control of mesoporous silica and mesoporous carbon materials [161,167,177,183,210,211]. However, reports on the preparation of mesoporous CN materials with tunable pore diameters and textural parameters such as specific surface area and specific pore volume are very limited. In the following section, the preparation and the characterization of mesoporous CNs with different pore diameters, specific surface areas, pore volumes, and nitrogen contents will be described.

MCN-type materials with different pore diameters were prepared by Vinu et al. [212] using the same procedure as described in the previous section for the preparation of mesoporous CN except that the pore diameter of the SBA-15 templates was varied. The synthesized samples are denoted MCN-1-*T*, where *T* indicates the synthesis temperature of the mesoporous silica template. Another set of samples was prepared using different weight ratios of EDA-to-CTC and the samples were labeled as MCN-1-130-*x*, where *x* represents the weight ratio of EDA-to-CTC.

Fig. 33 shows the powder XRD diffraction pattern of the obtained MCN-1 materials. The materials presented were synthesized using SBA-15 with different pore diameters as template. Remarkably, the quality of the XRD pattern and the position of the main peak vary significantly with the pore diameter of the SBA-15 template used. MCN-1-100 and

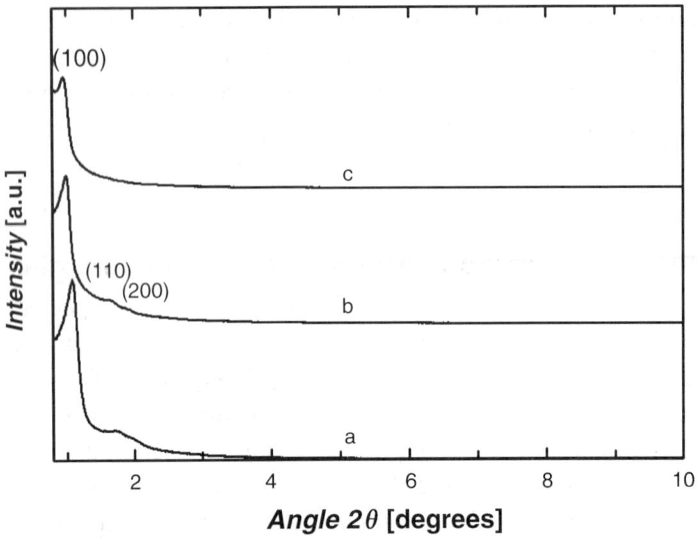

Figure 33 Powder XRD patterns of mesoporous carbon nitride with different pore diameters prepared using SBA-15-T templates: (a) MCN-1-100, (b) MCN-1-130, and (c) MCN-1-150.

MCN-1-130 exhibit three clear peaks, which can be indexed as 100, 110, and 200 reflections of a highly ordered 2D hexagonal mesostructure with the space group *p6mm*, similar to the XRD pattern of the parent mesoporous silica template. MCN-1-150 shows a sharp (100) peak together with only a weak (110) peak, indicating that the increase of the pore diameter of the template leads to some loss of higher order reflections in the diffraction pattern of the MCN product. The (110) peak intensity of MCN-1-150 is much lower as compared to that of MCN-1-100 and MCN-1-130. This could be attributed to empty CN mesopores, which are created by the incomplete filling of the ultra-large hexagonal mesoporous SBA-15 and weaken the peak intensity by interference of XRD between the inner and the outer carbon walls. It can also be observed that an increase in the pore diameter of the templates causes a shift of the peak toward the lower angle region, which provides the evidence of an increase of the d-spacing and the unit cell parameter. The unit cell constants of MCN-150, MCN-130, and MCN-1-100 are calculated to be 10.50, 10.18, and 9.52 nm, respectively. Remarkably, in addition to the peaks at low 2θ values, all the samples showed a single broad diffraction peak near 25.8° (not shown), corresponding to an interlayer d-spacing of 0.342 nm, which is almost similar to the d-spacing observed in the nonporous CN spheres.

This indicates the presence of a turbostratic ordering of the carbon and nitrogen atoms in the graphene layers of all the mesoporous CN samples with a uniform distribution of nitrogen atoms throughout the samples (Fig. 34) [207].

The textural parameters and the mesoscale ordering of the MCN materials prepared using SBA-15 with different pore diameters as template were analyzed by nitrogen adsorption–desorption measurements. Fig. 35a shows the nitrogen adsorption isotherms of MCN-1-100, MCN-1-130, and MCN-1-150. All isotherms are of type IV according to the IUPAC classification and feature capillary condensation in the mesopores. This is indicative of the presence of well-ordered mesopores in all the samples. The textural parameters such as the specific surface area, specific pore volume, and the pore diameter of the MCN samples are summarized in Table 1. As the unit cell constant of the materials increases from 9.52 to 10.50 nm, the position of the capillary condensation step shifts from lower relative pressure to higher relative pressure. The capillary condensation step for MCN-1-150 is significantly shifted to higher relative pressure, indicating a huge expansion in the pore size, which is quite consistent with the unit cell expansion observed from the XRD patterns in Fig. 33 and the

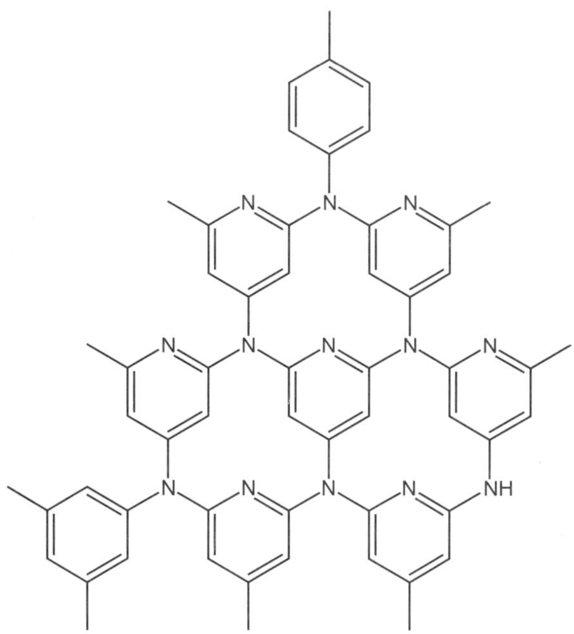

Figure 34 Schematic presentation of the uniform nitrogen distribution in the wall structure of mesoporous CN (MCN).

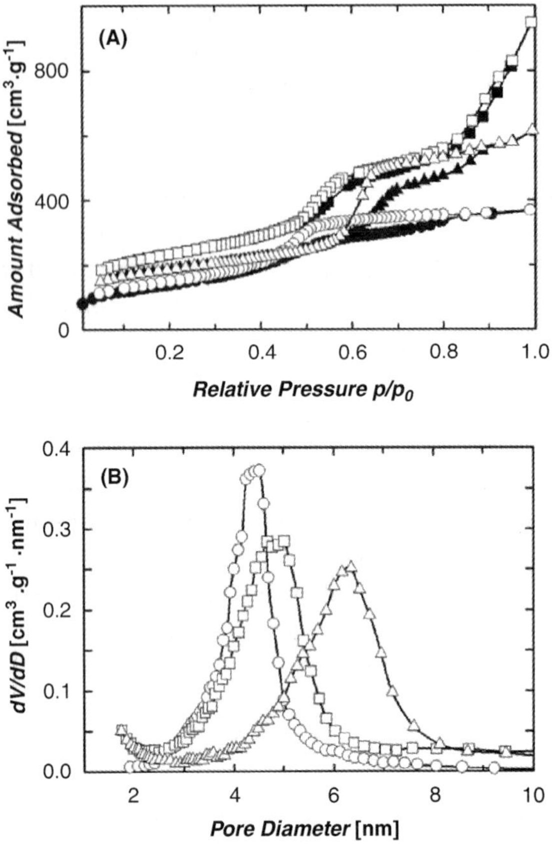

Figure 35 (A) Nitrogen adsorption–desorption isotherms and (b) BJH pore size distributions of mesoporous carbon nitride with various pore diameters (open symbols: desorption; closed symbols: adsorption): (●) MCN-1-100, (■) MCN-1-130, and (▲) MCN-1-150.

corresponding increase in the d-spacing. The change in the pore diameter of the MCN-type materials upon increasing the pore diameter of the template is clearly evident from the pore size analysis. Fig. 35b shows the pore size distribution of MCN-1-100, MCN-1-130, and MCN-1-150 as calculated from the nitrogen adsorption–desorption data using the BJH method. All the samples show a major peak, which mainly comes from the mesopores formed after dissolution of the silica matrix from the template. The pore diameter of the MCN materials increases with increasing the pore diameter of the silica template used. It should also be mentioned that the full width at half maximum of the BJH pore size distribution of MCN-1-150 is much larger than that of MCN-1-100 and MCN-1-130.

Among the MCN samples prepared by using SBA-15-T as template, MCN-1-150 exhibits a very large pore diameter, which is around 6.4 nm. This could be mainly due to an incomplete filling of CN into the polymer matrix in the ultra-large mesopores of SBA-15-150 as the same weight ratio of EDA-to-CTC is used for filling the mesopores of the templates with different pore diameter.

The shape of the isotherm for the sample synthesized using SBA-15-130 as template is markedly different from those for the MCN samples prepared from SBA-15-100 and SBA-15-150: A sharp rise in the isotherm for the MCN-1-130 occurs at relative pressures higher than 0.80, which indicates the presence of textural mesopores in the sample. These textural mesopores can be also observed on the corresponding high resolution-scanning electron microscope (HR-SEM) images, as presented in Fig. 36. It can be clearly seen that the void spaces in the MCN-1-130 sample are highly uniform and distributed throughout the samples, whereas the void spaces are hardly seen for the MCN-1-100 and the MCN-1-130 samples. We surmise that the significant characteristics of the textural mesoporosity in MCN-1-130 might be due to the difference in the pore connectivity and the size and surface

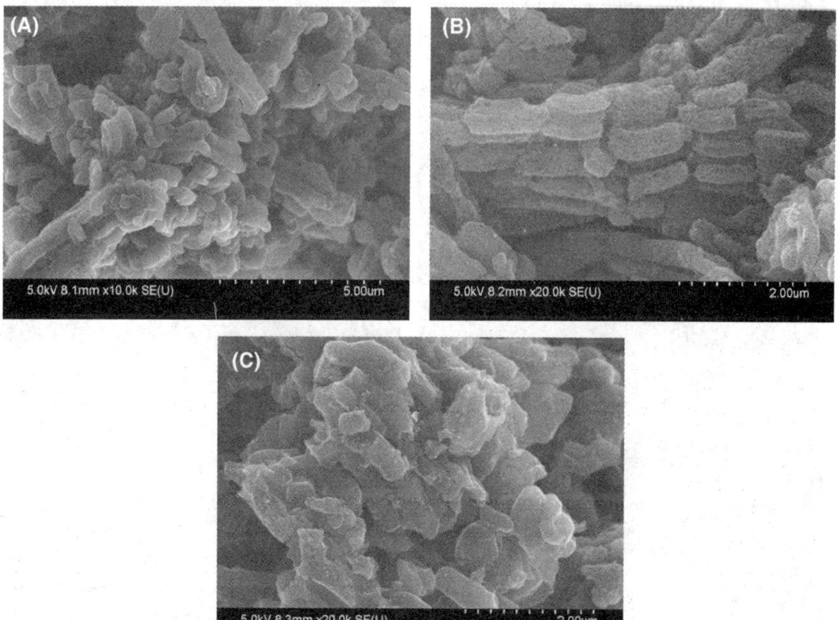

Figure 36 HR-SEM images of mesoporous carbon nitrides with different pore diameters: (a) MCN-1-100, (b) MCN-1-130, and (c) MCN-1-150.

texture of the silica particles in SBA-15-130. However, the exact reason for the formation of textural mesopores only in MCN-1-130 is currently not known and is under investigation. It is also important to note that the specific surface area and the pore volume of MCN-1-130 are much higher as compared to those for MCN-1-100 and MCN-1-150. The specific surface area and the specific pore volume of MCN-1-130 are $830\,m^2\,g^{-1}$ and $1.25\,cm^3\,g^{-1}$, respectively, whereas MCN-1-100 and MCN-1-150 possess specific surface areas of 505 and $650\,m^2\,g^{-1}$, respectively, and specific pore volumes of 0.89 and $0.55\,cm^3\,g^{-1}$.

HR-TEM was used to further examine the structural order and the morphology of the mesoporous CN materials with different pore diameters. The planar and the cross-sectional HR-TEM images of MCN-1-130 and MCN-1-150 are shown in Fig. 37a–37d. Bright contrast stripes on the under-focused image represent the pore walls, whereas dark contrast cores display empty channels. The contrast pattern may be converted by changing the lens defocus. The porous structure of both

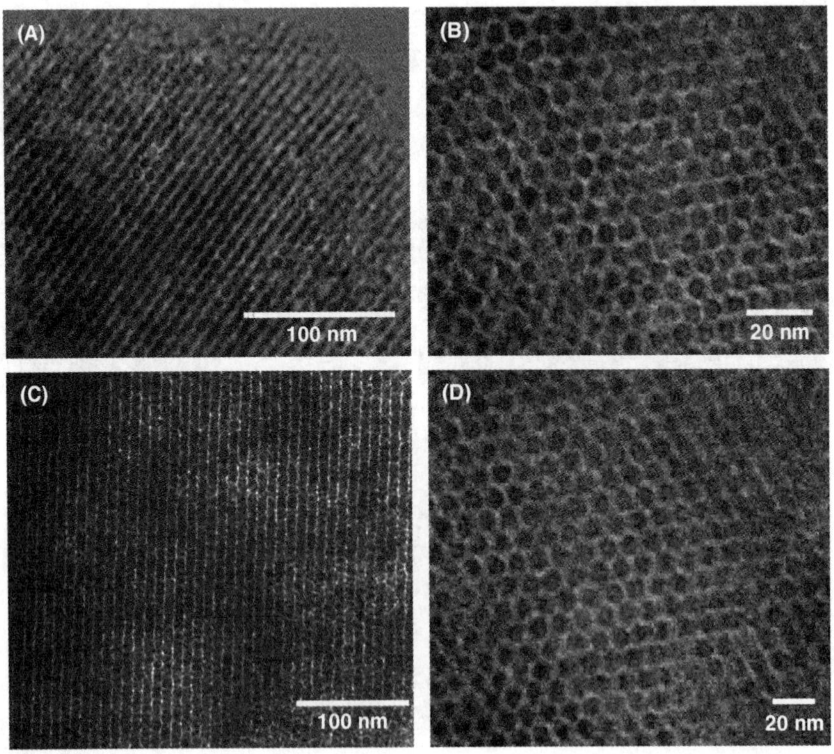

Figure 37 HR-TEM images of (a, b) MCN-1-130 and (c,d) MCN-1-150.

samples, when viewed in the direction perpendicular to their axis, shows a linear array of mesopores that are arranged in patches composed of regular tubular rows more than 500 nm long. HR-TEM images viewed down the pore axis reveal a hexagonally ordered honeycomb-like structure with uniform mesoporous channels. It is clearly seen from the HR-TEM images that the mesopore diameter of the MCN-1-150 is much larger than that of the MCN-1-130, which is in good agreement with the results obtained from XRD and nitrogen adsorption measurements. The carbon-to-nitrogen ratio calculated from the EEL spectrum and the CHN analysis for the samples is found to be 4.3 ± 0.05, which is in good agreement with the nonporous CN materials prepared through the chemical polymerization technique. The nature and the coordination of the carbon and nitrogen atoms in the materials were studied by electron energy loss spectroscopy (EELS), XPS, and FT-IR analysis. Almost similar results were obtained as those for mesoporous CNs prepared using SBA-15-100.

8.2 Mesoporous carbon nitrides with different textural parameters and nitrogen contents

The amount of the carbon and nitrogen source in the mesoporous silica template has a great impact on the textural parameters such as specific surface area, specific pore volume, and pore size distribution of the mesoporous CN replica materials. The MCN-1-130-type material has been chosen for studying the effect of the composition of the carbon and nitrogen source in the synthesis mixture, as this material shows an excellent structural order and interesting textural parameters. The MCN materials were prepared by varying the weight ratio of EDA-to-CTC in the synthesis gel and characterized by XRD, nitrogen adsorption, HR-SEM, EELS, and CHN analysis. The results of the structural characterization consistently show that the change in the weight ratio of the carbon-to-nitrogen source in the synthesis mixture greatly influences the textural parameters, morphology, and the nitrogen content in an unexpected way. Vinu also reported that the degree of polymerization in the pores of the silica template can easily be controlled by varying the weight ratio of EDA-to-CTC in the pore channels [212]. Fig. 38 shows the powder XRD pattern of MCN-1-130-*x* materials prepared from different weight ratios of EDA-to-CTC. All the samples exhibit a sharp diffraction peak at low angle and two higher order peaks that can be indexed as the (100), (110), and (200) reflections of the hexagonal space group *p6mm*. This indicates that all the materials possess a hexagonally ordered uniform mesoporous structure with long-range order and that the structural order of the materials is not

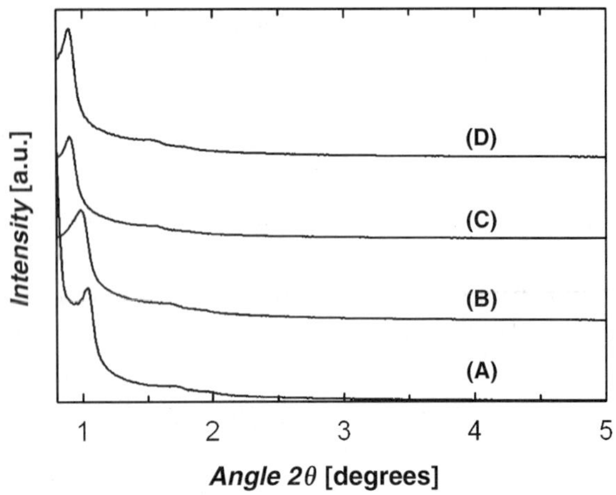

Figure 38 Powder XRD patterns of mesoporous carbon nitride samples prepared at different weight ratios of EDA and CTC: (a) MCN-1-130-0.30, (b) MCN-1-130-0.45, (c) MCN-1-130-0.60, and (d) MCN-1-130-0.90.

altered by varying the weight ratio of EDA-to-CTC. It can be also observed that the peaks are shifted toward lower angles upon increasing the weight ratio of EDA-to-CTC in the synthesis gel. This suggests that the d-spacing and the unit cell parameter of the sample vary significantly when the composition of the synthesis mixture is changed. The unit cell constant increases from 9.7 to 11.3 nm upon increasing the weight ratio of EDA-to-CTC from 0.3 to 0.9. This could mean that the increase of EDA in the synthesis mixture reduces the density of the reactant mixture, which makes the diffusion of the reactant molecules inside the porous matrix easier and improves the access to all the adsorption sites. This would facilitate the complete polymerization of EDA and CTC and support the filling of the mesoporous void space of the template with the CN polymer matrix. In addition, a significant compression of CN in the mesoporous matrix may develop during the carbonization process due to the shrinkage of the polymers in the pore matrix, leading to a larger unit cell constant. The significant raise of the unit cell constant and the d-spacing also suggests that the pore structure of the MCN materials greatly expands with increasing the weight ratio of EDA-to-CTC.

The nitrogen adsorption–desorption isotherms of the MCN-1-130-*x* materials prepared using different weight ratios of EDA-to-CTC are shown in Fig. 39. All isotherms are of type IV according to the IUPAC classification and exhibit a H2-type hysteresis loop and sharp capillary condensation

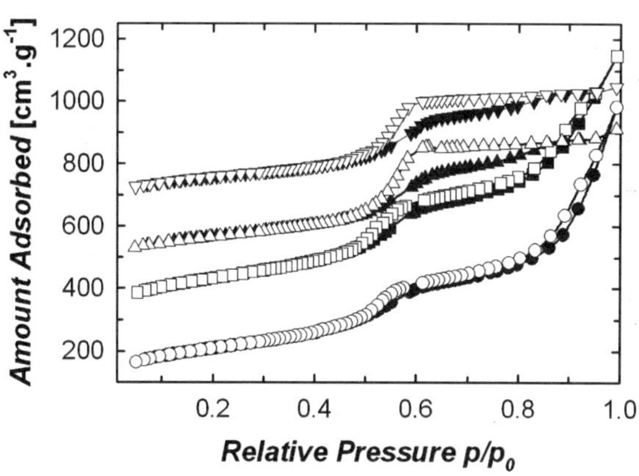

Figure 39 Nitrogen adsorption–desorption isotherms of mesoporous carbon nitrides prepared at different weight ratios of EDA and CTC (open symbols: desorption; closed symbols: adsorption): (●) MCN-1-130-0.30, (■) MCN-1-130-0.45, (▲) MCN-1-130-0.60, and (▼) MCN-1-130-0.90.

steps characteristic of nitrogen within the uniform mesopores. It is interesting to note that the capillary condensation step, which is generally correlated to the diameter of the mesopores, shifts toward a higher relative pressure with increasing the weight ratio of EDA-to-CTC. This indicates that the pore diameter of the material increases with increasing the weight ratio of EDA-to-CTC, which is quite consistent with the unit cell size obtained from the XRD analysis. On the contrary, the specific pore volume of the MCN materials systematically decreases with increasing the weight ratio of EDA-to-CTC while the specific surface area of the materials increases from 731 to 818 $cm^2 g^{-1}$ with increasing the weight ratio of EDA-to-CTC from 0.3 to 0.45 and then decreases to 552 $cm^2 g^{-1}$ for MCN-1-130-0.9 (Fig. 40). The shapes of the isotherms of MCN-130-0.3 and MCN-1-130-0.45 are completely different from that of the MCN material prepared using weight ratios of EDA-to-CTC above 0.45. It must be noted that a sharp rise in the isotherm at higher relative pressures is observed for MCN-1-130-0.30 and MCN-1-130-0.45. This is exclusively originated from textural mesopores generated by the void spaces between the particles that are formed due to the incomplete polymerization of the EDA and CTC, leading to a larger mesopores and macropore volume. The void spaces together with the rod-like morphology similar to the parent mesoporous silica template are confirmed by the HR-SEM images of the MCN samples prepared at different weight ratio of EDA-to-CTC (Fig. 41). It is interesting to see in the HR-SEM images that the size of the void space

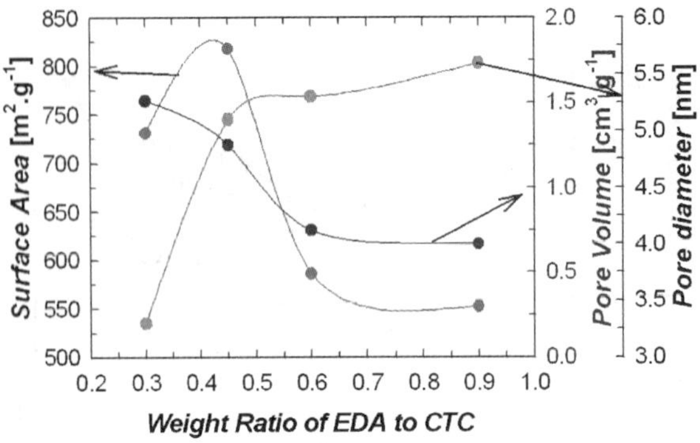

Figure 40 Effect of the EDA and CTC weight ratio on the textural parameters of MCN-1-130.

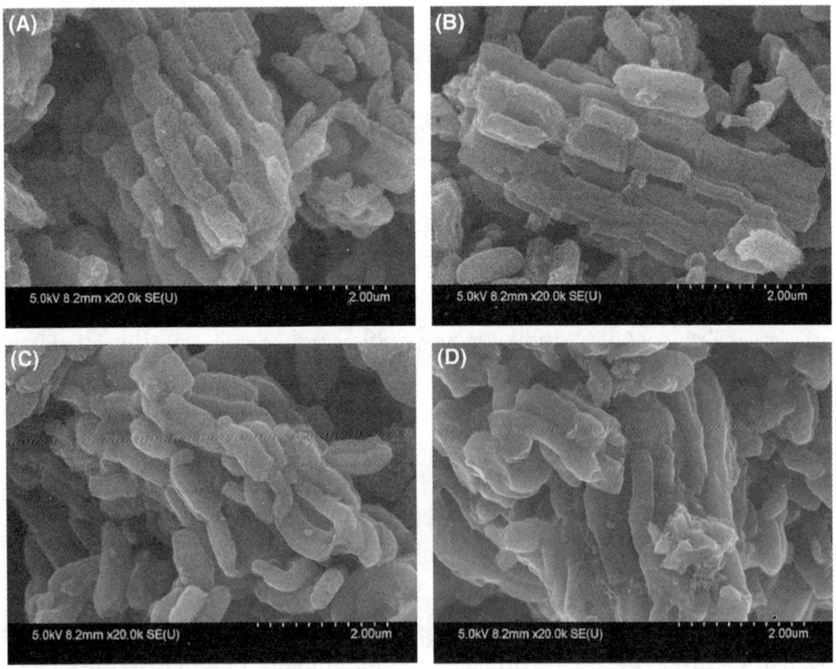

Figure 41 HR-SEM images of MCN-1-130 prepared at different weight ratios of EDA and CTC showing the control of textural mesopores: (a) MCN-1-130-0.30, (b) MCN-1-130-0.45, (c) MCN-1-130-0.60, and (d) MCN-1-130-0.90.

increases with decreasing the EDA-to-CTC ratio that undoubtedly supports the assumption that the low weight ratio of EDA-to-CTC does impede the complete polymerization of the reactant mixture. An incomplete polymerization promotes the decomposition of loosely bound organic species from the CN walls during the carbonization step, which leaves behind a lot of void space, namely, textural mesopores on the rod-like CN walls, leading to a large mesopore and macropore volume. It is worth mentioning that when the weight ratio of EDA-to-CTC was further increased to 1.20, only a thick gel kind of material was obtained, which exhibited a very low surface area and pore volume.

Controlling the nitrogen content in the MCN-type materials is critical as it dictates the basic strength of the materials. Vinu reported that the amount of nitrogen in the MCN-1-130 can be easily controlled by simply adjusting the weight ratio of EDA-to-CTC [212]. Fig. 42 shows the effect of the EDA-to-CTC weight ratio in the synthesis mixture on the amount of nitrogen content in the final materials. Surprisingly, the nitrogen content of the material significantly increases when the weight ratio of EDA-to-CTC is increased. The carbon-to-nitrogen ratio calculated from the CHN analysis decreases from 4.5 to 3.3 with increasing the weight ratio of EDA-to-CTC from 0.3 to 0.9 (Table 2). This is further confirmed by the EEL spectra of the samples prepared at weight ratios of EDA-to-CTC of 0.45 and 0.90, as shown in Fig. 43. The intensity of the N K-edge in the MCN-1-130-0.90 sample, which is located at 401 eV and typically assigned to the nitrogen atoms sp^2 hybridized with carbon, is significantly higher as

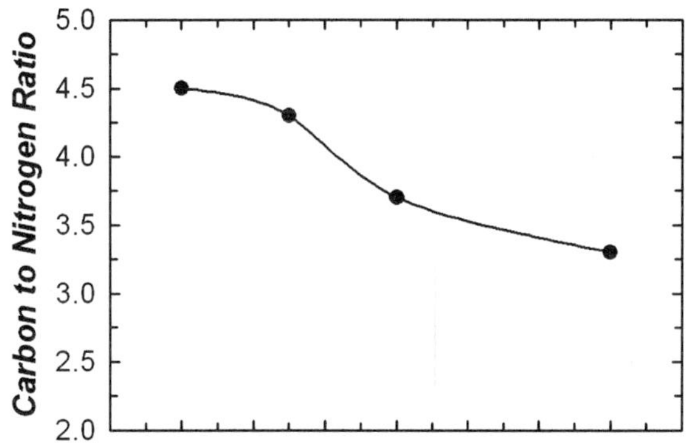

Figure 42 Effect of the EDA-to-CTC weight ratio on the final nitrogen content of MCN-1-130.

Table 2 Elemental composition of MCN-1 materials as determined from CHN analysis

Material	C (wt.%)	N (wt.%)	H (wt.%)	Other elements (Si, O, Cl, and F) (wt.%)	C/N molar ratio
MCN-1-100	74.1	17.1	2.2	6.6	4.3
MCN-1-130	73.7	17.3	2.5	6.5	4.3
MCN-1-150	74.4	17.4	2.0	6.2	4.3
MCN-1-130-0.3	74.0	16.5	2.3	7.2	4.5
MCN-1-130-0.45	73.7	17.3	2.5	6.5	4.3
MCN-1-130-0.6	71.0	19.2	3.9	5.9	3.7
MCN-1-130-0.9	68.0	20.5	5.9	5.6	3.3

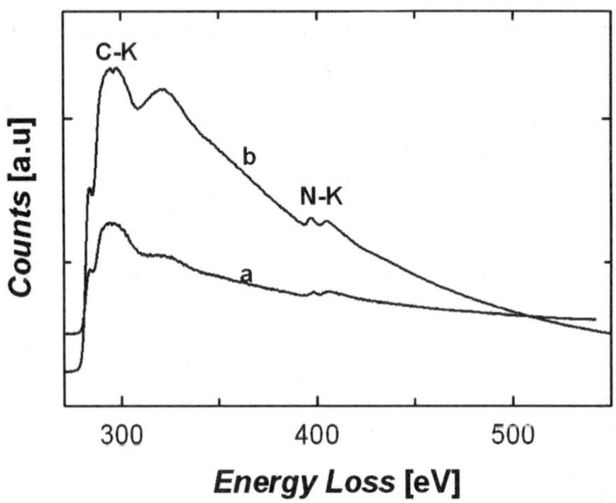

Figure 43 Electron energy loss spectra of (a) MCN-1-130-0.45 and (b) MCN-1-130-0.9.

compared to that of the MCN-1-130-0.45. Moreover, the carbon-to-nitrogen ratio as calculated from the EEL spectrum matches very well with the CHN analysis data. The huge rise in the nitrogen content in the sample with increasing the EDA content in the synthesis mixture could be mainly due to the fact that the degree of polymerization of the EDA and CTC is greatly enhanced when a large amount of EDA is used. This would hinder the release of the nitrogen atoms from the completely polymerized matrix during the carbonization process that is normally carried out at higher temperature.

From all the above results, it can be concluded that the pore diameter of MCN-type materials can be easily controlled by varying the pore diameter of the mesoporous silica template, but the weight ratio of EDA-to-CTC also plays an important role for controlling the textural parameters and most importantly the nitrogen content of the ordered MCN-type materials. It is also confirmed that the optimum weight ratio of EDA-to-CTC is around 0.45 for the synthesis of well-ordered MCN-type materials with appreciable textural parameters and high nitrogen content. Finally, the catalytic activity of the MCN-1-130 materials prepared at different weight ratios of EDA-to-CTC was tested in the Friedel–Crafts acylation of benzene using hexanoyl chloride as the acylation agent and *n*-heptane as the solvent. Among the catalysts studied, MCN-1-130-0.3 shows the highest conversion and 100% product selectivity to caprophenone. This could be mainly due to the fact that when the materials are prepared using a low amount of EDA, the formation of defect sites in the walls is high due to the incomplete polymerization of EDA and CTC molecules. These defect sites mainly consist of imine and amino groups that are responsible for the higher basic activity of MCN-1-130-0.3.

A novel highly ordered 3D cage-type mesoporous CN material (MCN-2) with very high surface area and pore volume has also been prepared using a 3D cage-type mesoporous silica, SBA-16, as a template through a simple polymerization reaction between ethylene diamine and CTC by Vinu et al. [213]. The material has been unambiguously characterized by various sophisticated techniques such as XRD, nitrogen adsorption, HR-TEM, EELS, XPS, ^{13}C CP MAS-NMR spectroscopy, and FT-IR spectroscopy. The XRD results reveal that MCN-2 possesses a 3D structure with an *Im3m* space group. The specific surface area and the pore volume of MCN-2 were determined to be $810\,m^2\,g^{-1}$ and $0.81\,cm^3\,g^{-1}$, respectively, which is significantly higher as compared to those of the template SBA-15 and MCN-1. A 3D cage-type structure of MCN-2 was confirmed by HR-TEM (Fig. 44).

8.3 Nitrogen-doped mesoporous carbon materials with different structures

Xia and Mokaya synthesized nitrogen-doped mesoporous carbon materials with graphitic pore walls through the chemical vapor deposition of acetonitrile on mesoporous silica template [214]. The pyrolysis temperature was varied between 950 and 1100 °C to obtain well-ordered N-doped

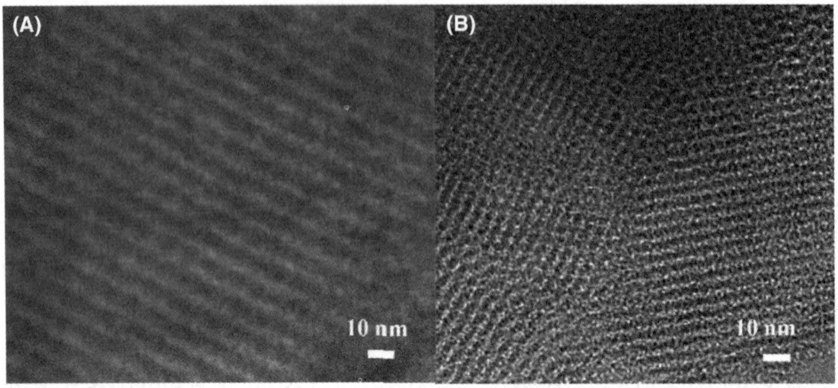

Figure 44 HR-TEM images of a 3D mesoporous carbon nitride (MCN-2): Images taken along (a) [200] and (b) [110] direction.

mesoporous carbons. The graphitization of the materials can be controlled by adjusting the carbonization temperature: It was found that the samples carbonized at 1100 °C possess a higher graphite content as compared to those prepared at temperatures lower than 1000 °C. Macroscopically, the materials have a spherical morphology, which is almost similar to that of the other mesoporous carbon materials prepared by chemical vapor deposition. The structure of the N-doped mesoporous carbon materials can also be varied by choosing the appropriate silica templates, including SBA-12, SBA-15, MCM-48, HMS, and MCM-41 [215]. The nitrogen content can also be controlled, namely, by varying the carbonization temperature. Hexagonally ordered N-doped mesoporous carbon was also synthesized using aniline as the source for both carbon and nitrogen and using SBA-15 as the template as reported by Vinu et al. [216]. The aniline molecule was successfully polymerized inside the channels of mesoporous silica with ammonium peroxydisulphate (APDS) (Fig. 45).

In a typical synthesis of N-doped mesoporous carbon, 0.5 g of calcined SBA-15 was thoroughly mixed with 2.5 g of aniline. The polymerization of aniline inside the nanochannels was done by adding 3.5 g of APDS, and the resultant greenish black mixture was heated at 100 °C overnight. Subsequently, 3.0 g of additional APDS was added to the above reaction mixture to achieve the complete polymerization, followed by drying in a vacuum oven at 40 °C for 24 h. The molar gel composition of the synthesis mixture was 0.008 SiO_2:0.027 $C_6H_5NH_2$:0.028 APDS. The resultant black colored SBA-15/polyaniline nanocomposite was ground into a fine powder and carbonized at 900 °C in a nitrogen flow of 100 mL min^{-1}

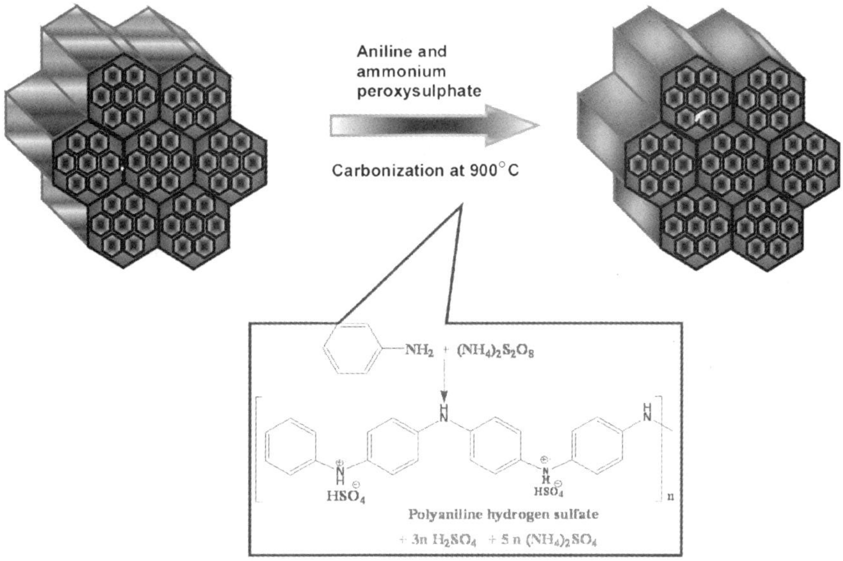

Figure 45 Preparation of nitrogen-doped mesoporous carbon by using a 3D mesoporous silica template KIT-6.

with a heating rate of 3.0 °C min^{-1} and kept at these conditions for 5 h. The N-doped mesoporous carbon was recovered after dissolution of the silica framework in 5 wt.% HF, followed by filtration, washing several times with ethanol, and drying at 100 °C in an air oven. With the help of CHN, XPS, and EEL analysis, the authors concluded that the nitrogen atoms have been successfully doped into the carbon walls and are trigonally bonded to graphitic carbon atoms. The material possesses a well-ordered pore structure, high specific surface area, uniform pore size distribution, and large specific pore volume.

Similarly, the method was extended to the preparation of N-doped mesoporous carbon materials with 3D large pore structure (N-MCK6) using KIT-6 as the template [217]. The obtained material possesses a well-ordered pore system with 3D cubic structure containing an enatiomeric pair of independent interpenetrating 3D continuous networks of mesoporous channels that are mutually intertwined and separated by carbon walls. The powder XRD patterns of N-MCK6 along with the parent mesoporous silica template KIT-6 are shown in Fig. 46. It can be clearly seen from Fig. 46 that the KIT-6 silica template exhibits a sharp, well-resolved (211) reflection and several higher order reflections at 2θ angles below 4°, indicating high structural ordering with the symmetry of the body-centered

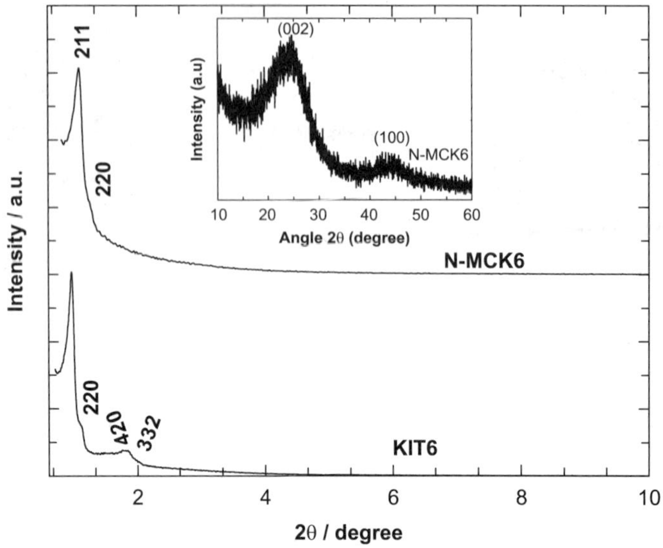

Figure 46 Powder XRD patterns of KIT-6 and N-MCK6 (the inset shows the wide-angle XRD pattern of N-MCK6).

cubic *Ia3d* space group. The unit cell constant, calculated from the (211) reflection of the cubic *Ia3d* space group using the equation $d_{211}\sqrt{6}$ is found to be 22.46 nm. The XRD pattern of the N-MCK6 shows two well-resolved peaks with d-spacings of 8.14 and 6.96 nm, respectively, and several weak higher order reflections. The first peaks can be indexed as (211) and (220) reflections of the body-centered cubic *Ia3d* type structure and the XRD pattern is almost similar to that of the parent mesoporous silica template KIT-6. This indicates that the prepared N-MCK6-type material possesses a 3D cubic structure with an enatiomeric pair of independent interpenetrating 3D continuous networks of mesoporous channels that are mutually intertwined and separated by carbon walls.

The unit cell parameter of N-MCK6, as calculated from the (211) reflection, amounts to 19.94 nm, which is much lower than that of the parent silica template. This could be ascribed either to a significant shrinkage of the mesoporous structure of the polymer–silica nanocomposite materials during the carbonization at very high temperature or to a silica removal process. It should be noted that the intensity of the (211) reflection of the XRD pattern of N-MCK6 is smaller than that of the parent silica template, suggesting the presence of defects in the carbon walls that separate the enantiomeric pair of mesoporous carbon channels created through the incorporation of nitrogen atoms in the carbon framework.

To check the microstructural and the graphitic character of N-MCK6, the material was characterized using a wide-angle powder XRD measurement. Fig. 46 (inset) shows the wide-angle powder XRD pattern of N-MCK6. The sample exhibits a sharp reflection at a 2θ value of 24.58° corresponding to an interlayer distance of 0.362 nm, which may be indexed to the (002) reflection of a pure graphitic lattice. The (002) peak of pristine graphitic carbon is around 26.5°, corresponding to the interlayer distance of 0.335 nm between the graphitic carbon sheets. The carbon interlayer distance of N-MCK6 is slightly higher than that of the pure graphitic carbon. This could be due to defects sites and curvature in the carbon walls and partial crystallization to a carbon framework. In addition, N-MCK6 also shows a broad reflection at 43.5°, which corresponds to a superposition of the (100) and (101) reflections of graphitic carbon, indicating a higher degree of graphitization in the material as compared to mesoporous carbon CMK-3. It is worthwhile to note that even though the material was carbonized even at a temperature of 900 °C, a carbon material with a partially graphitic structure was successfully obtained. It should be mentioned that the preparation of mesoporous carbons with graphitic walls from carbon sources such as sucrose and furfuryl alcohol is very difficult if the carbonization temperature is lower than 1100 °C. Thus, it is reasonable to assume that the formation of highly graphitic carbon in N-MCK6 could be due to the use of an aromatic carbon precursor aniline, consisting of a phenyl group attached to an amino group. Unlike other carbon sources, the phenyl group in the aniline molecules can provide the benzene ring for the formation of graphene sheets in the mesoporous carbon walls.

Representative HR-TEM images of N-MCK6 viewed along the two crystallographic directions are shown in Fig. 47. Fig. 47a clearly displays a highly ordered mesoporous structure with a linear array of mesopores, which are arranged in regular intervals. These well-ordered mesopores originate from the mesoporous silica framework that would convert into the mesopores of the resultant carbon after the HF treatment. The HR-TEM image also reveals that many graphene sheets are oriented along the axis parallel to the carbon rods of the 3D body-centered cubic structure of N-MCK6 (Fig. 47a inset). The cross-sectional HR-TEM image of N-MCK6 clearly shows the presence of a well-ordered 3D body-centered cubic mesostructure, which is very similar to that reported for the 3D mesoporous silica template KIT-6 (Fig. 47b). This clearly demonstrates that 3D N-doped mesoporous carbon with a body-centered cubic *Ia*3*d* structure has been truly replicated from the KIT-6 mesoporous silica template using polyaniline as a carbon precursor.

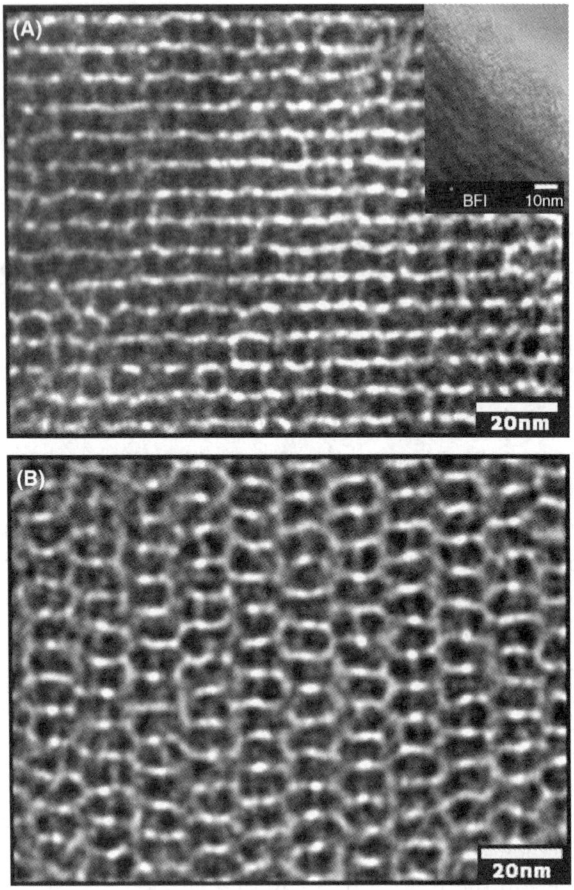

Figure 47 HR-TEM images of N-MCK6: (a) horizontal view and (b) cross-sectional view (inset: the magnified HR-TEM image of N-MCK6).

To explore the nature, the coordination and the extent of nitrogen doping, N-MCK6 was characterized by EELS. The carbon-to-nitrogen ratio calculated from the EELS analysis was found to be 13.0. The sample exhibits a sharp C K-edge located at 284 eV, which corresponds to transitions occurring from the orbital 1s-π^* states that are mainly caused by the higher electronegativity of nitrogen that decreases the electron density on the carbon atoms. This can be attributed to the trigonal sp^2-hybridized carbon bound to nitrogen. The EEL spectrum also displays the N K-edge positioned at 405 eV, which reveals that the nitrogen atoms in N-MCK6 are mostly sp^2-hybridized.

The parent KIT-6 mesoporous silica possesses a specific surface area of 728 $m^2 g^{-1}$ and a specific pore volume of 0.99 $cm^3 g^{-1}$, while the specific surface area and the specific pore volume of N-MCK6 amounts to 726 $m^2 g^{-1}$ and 0.82 $cm^3 g^{-1}$, respectively. Even though the specific surface area of the N-MCK6 is almost similar to that of the parent silica template, the specific pore volume is relatively small. This can be due to the fact that the carbonization of the carbon-polyaniline nanocomposite with a large amount of aromatic rings does not create much microporosity as it is very difficult to break the aromatic ring at the relatively low carbonization temperature of 900 °C in an inert atmosphere. Hence, the low specific pore volume in the resulting material.

The pore size distribution of KIT-6 appears narrow and is centered at about 8.0 nm, whereas N-KIT6 exhibits a broad peak centered around 4.6 nm, which is much larger than the wall thickness of the KIT-6 mesoporous silica (ca. 3.2 nm). It is a well-known fact that the mesopores are generated by the dissolution of the silica walls of the template by an HF treatment during the replication process. Interestingly, the size of the mesopores of N-MCK6 is almost 1.4 nm larger than the wall thickness of the template. It is therefore quite reasonable to assume that a structural shrinkage occurred in the polyaniline composites in the mesoporous matrix of the template, and/or the incomplete filling of the mesopores of the template with the carbon precursor generates some empty space in the mesoporous channels. Thus, it is quite likely that a part of the empty mesopores in the template generated by the above structural changes is responsible for the enlargement of the pore diameter of N-MCK6. It was also reported that the sample mainly possesses graphitic carbon atoms that are bound to nitrogen in the form of imine and amine groups [217]. This simple method for the preparation of mesoporous carbon can therefore deliver products with a considerable amount of nitrogen in the carbon lattice. Such materials possess considerable potential for applications in electrical and electronic devices and for energy storage.

9. MESOPOROUS BORON NITRIDES AND MESOPOROUS BORON CARBON NITRIDES

Gas-phase reactions are frequently used for the post-synthetic modification of mesoporous materials. Ozin and coworkers, for example, reported the preparation of periodic mesoporous aminosilicates that contain functional amine groups in the framework of a mesoporous network [218]. These materials were synthesized under a flow of ammonia gas through thermal ammonolysis of periodic mesoporous organosilica. Wan et al. demonstrated the preparation of mesoporous silicon oxynitrides by heating fresh SBA-15 precursors in ammonia in a flow-through quartz tube reactor [219].

9.1 The elemental substitution method

A dedicated extension of these methods could potentially lead to a new route to mesoporous materials with various framework components. Such a strategy was already realized in nanotube chemistry where nanotubes composed of boron nitride and boron CN are easily prepared through a substitution reaction starting from carbon nanotubes as templates. Very recently, Vinu et al. have applied this synthesis technique to mesoporous materials and realized the preparation of mesoporous boron nitride (MBN) and mesoporous boron CN (MBCN) from mesoporous carbon [220]. This strategy can be regarded as the third-generation approach for the synthesis of mesoporous materials (Fig. 48). We called it "elemental substitution method" in the field of mesoporous materials. In this section, the details on the preparation of MBN and MBCN are described as pioneering examples of the elemental substitution method.

Hexagonal boron nitride, which structurally resembles graphite, is an insulator exhibiting a band gap of approximately 5.5 eV, chemically inert, and thermally stable up to 1600 °C. By constructing boron nitride and boron CN structures, exhibiting ordered arrays of channels, novel applications could emerge: from catalysis to molecular separations and sorption of very bulky molecules to the preparation of semiconductors, semiconducting nanowires, and low dielectric constant devices. Vinu et al. have recently reported for the first time the preparation of MBN and MBCN with very high specific surface areas and specific pore volumes [220]. The materials were prepared through substitution reactions at high temperatures, using a well-ordered hexagonal mesoporous carbon (CMK-3) as the template and boron trioxide as boron source under nitrogen gas flow as nitrogen source. The MBNs were usually synthesized at higher temperature such as 1750 °C, while the MBCN can be obtained through the substitution reaction at considerably lower temperatures (1450–1550 °C) with short reaction times (Fig. 49).

The textural parameters of these mesoporous materials were obtained through nitrogen adsorption–desorption isotherm measurements, which allowed the determination of the specific surface area, specific pore volume, and the mesopore size distribution. Fig. 50A shows the nitrogen adsorption–desorption isotherms of the MBN and the MBCN. Both isotherms can be classified as type IV with a H1-type broad hysteresis loop,

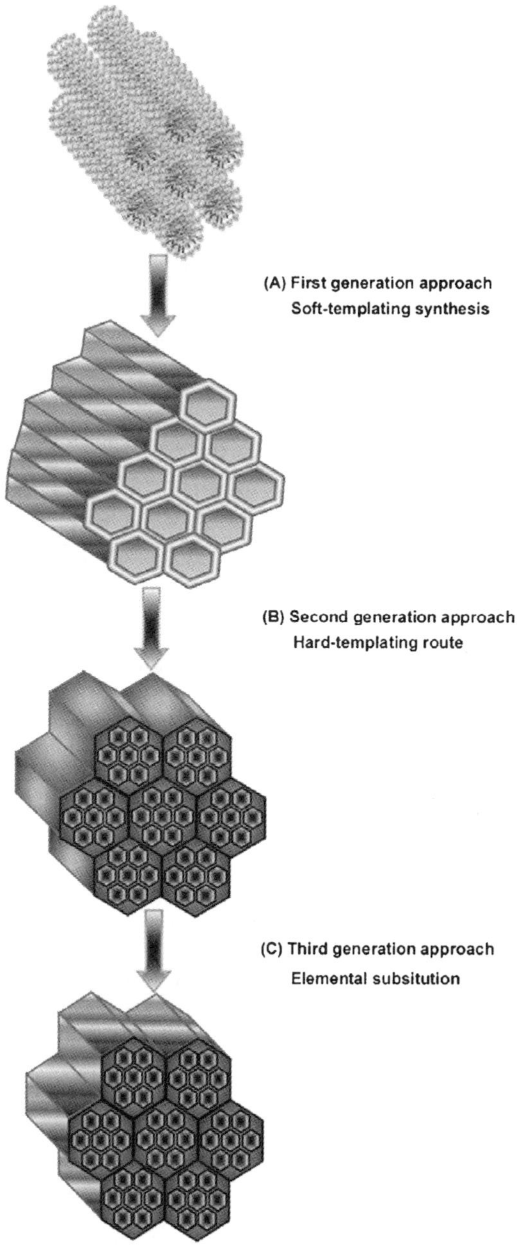

Figure 48 Historical flow in synthesis strategies of mesoporous materials: (a) template synthesis, (b) replica route, and (c) elemental substitution.

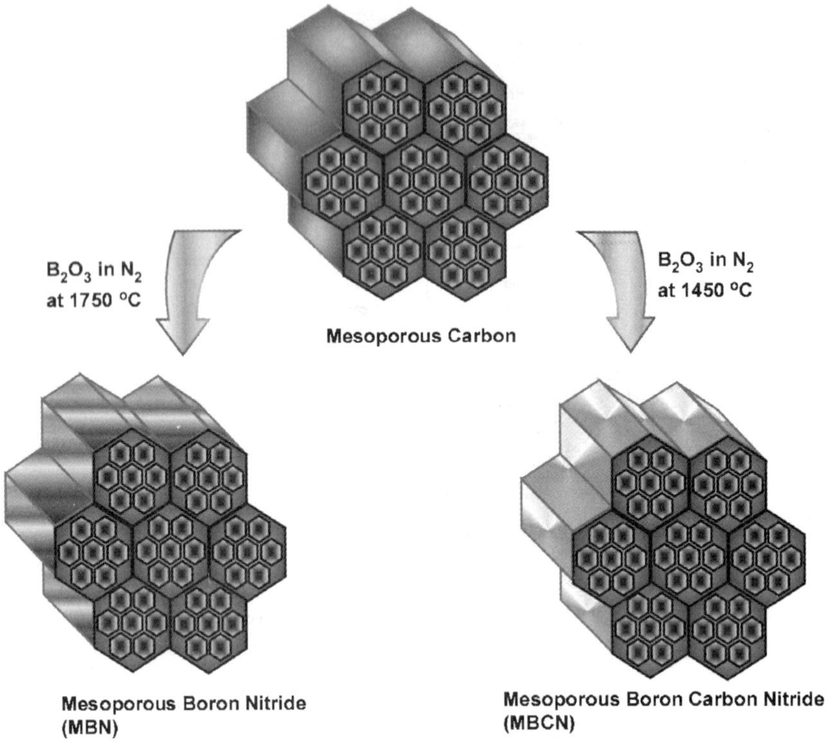

Figure 49 Synthesis principle of mesoporous boron nitride (MBN) and mesoporous boron carbon nitride (MBCN).

which is typical of mesoporous materials. The pore size distribution curves of these materials shown in Fig. 50B are somewhat broad as compared to those of the parent mesoporous carbon but are still sharp enough to suggest a narrow distribution of the pore size. However, they indicate that the pore structure of the MBN and the MBCN are not perfectly homogeneous and that they may contain an appreciable amount of mesopores of irregular shape. The BET surface area, mesopore volume, and pore diameter decrease with increasing synthesis temperature, that is, decreasing the carbon content in the mesoporous materials (see below). The MBCN synthesized at 1450 °C and the MBCN synthesized at 1550 °C possess relatively high specific surface areas of 740 and 650 $m^2\,g^{-1}$ and large specific pore volumes of 0.69 and 0.60 $cm^3\,g^{-1}$, respectively. On the contrary, the specific surface area and the specific pore volume of the MBN remained relatively low at 565 $m^2\,g^{-1}$ and 0.53 $cm^3\,g^{-1}$, respectively.

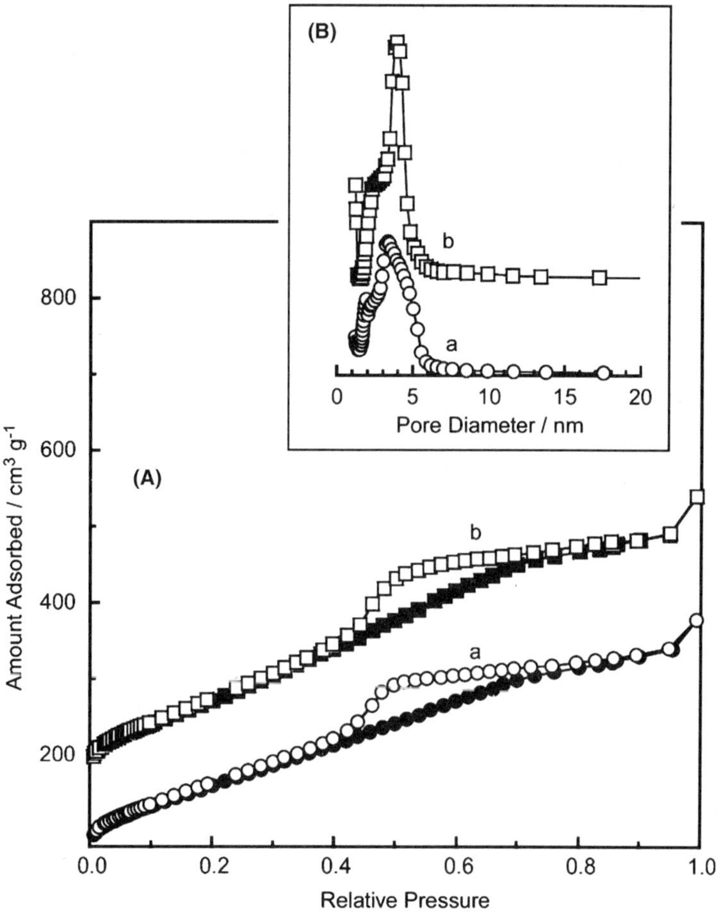

Figure 50 (A) Nitrogen adsorption–desorption isotherms and (B) BJH pore size distributions: (a) mesoporous boron nitride and (b) mesoporous boron carbon nitride synthesized at 1750 and 1450 °C, respectively. In diagram (A), closed symbols and open symbols represent adsorption and desorption isotherms, respectively. Reprinted with permission from Ref. [220]. © 2005, American Chemical Society.

Fig. 51 shows HR-TEM images of the MBN where the inset in Fig. 51A represents the corresponding diffraction pattern. The images in low and high magnification indicate that the prepared boron nitrides possess a mesoporous structure with a disordered boron nitride framework with highly crystalline layers. In contrast, a higher magnification of the HR-TEM image of the MBCN material (Fig. 52A) clearly displays well-ordered mesoporous structures with a local interlinking of crystalline boron CN layers. The mesoporous carbon walls, which were previously amorphous,

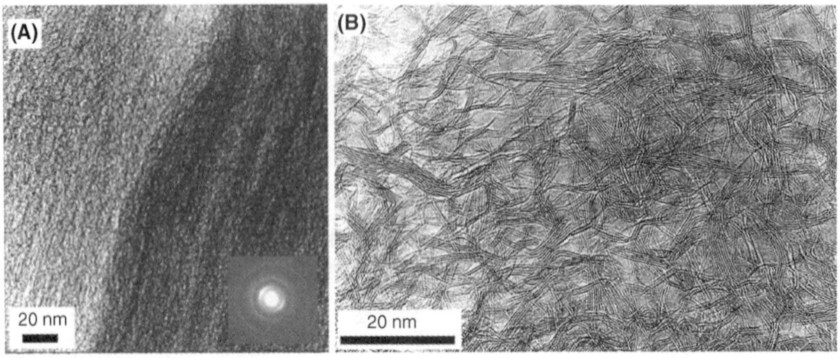

Figure 51 HR-TEM images of mesoporous boron nitride materials: (a) low magnification (the inset shows the diffractions pattern) and (b) high magnification. Reprinted with permission from Ref. [220]. © 2005, American Chemical Society.

could be completely transformed into crystalline boron CN walls during the substitution reaction. Elemental mapping studies of the MBCN sample revealed that the material is indeed composed of boron (Fig. 52B), carbon (Fig. 52B(c)), and nitrogen (Fig. 52B(d)). Traces of other elements were not detected using this technique. Judging from the comparison of these "partial" images with the whole image (Fig. 52B(a)), these elements can be regarded as homogeneously distributed throughout the sample.

The elemental composition and the structure of the MBN and the MBCN were also analyzed by EEL spectroscopy. The boron-to-nitrogen ratio calculated from the EEL spectrum of the MBN was 1.0, which is in good agreement with the boron nitride, and a carbon signal was virtually absent. The XPS survey spectrum of the MBN showed sharp signals for boron and nitrogen and did not show any peaks for other elements except oxygen as contaminant. The overall boron-to-nitrogen ratio of the MBN obtained from the XPS analysis is 0.94. In the case of the MBCN, the EEL spectrum showed an increase of the carbon content (8.0 and 20.1% for the corresponding materials synthesized at 1550 and 1450 °C, respectively) as the synthesis temperature decreased. The elemental composition of the MBCN synthesized at 1450 °C was determined by XPS to 21.1% of boron, 37.7% of carbon, and 23.0% of nitrogen with 18.2% of oxygen as unavoidable contaminant.

9.2 Synthesis of mesoporous boron nitrides by a nanocasting method

The nanocasting technique was also employed for the preparation of MBN materials with different structures. Bois and coworkers reported the

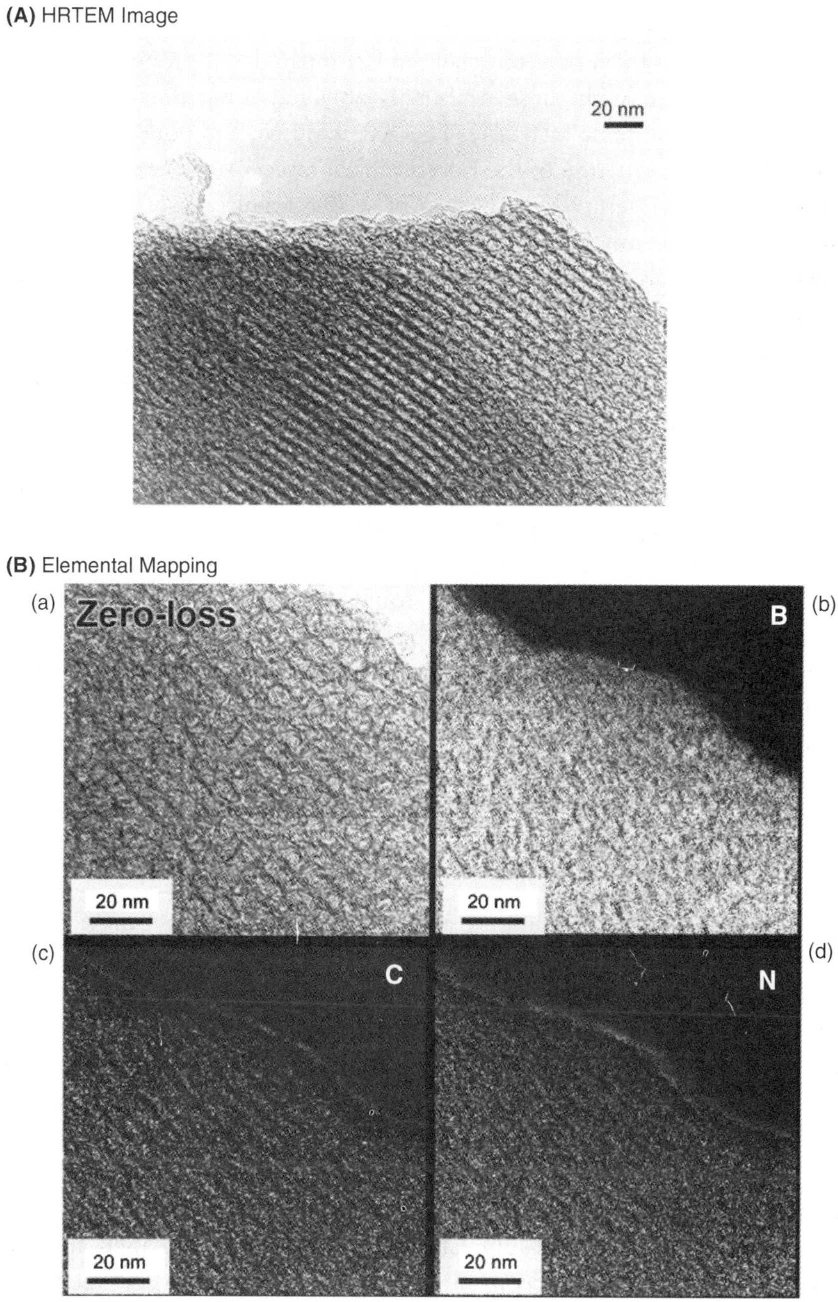

Figure 52 (A) HR-TEM image and (B) elemental mapping ((a) whole image, (b) boron, (c) carbon, and (d) nitrogen) of mesoporous boron carbon nitride materials. Reprinted with permission from Ref. [220]. © 2005, American Chemical Society.

preparation of a MBN using an aminoborazine as the boron nitride source with silica (SBA-15) as the template and compared the properties of the resulting materials with those of MBNs obtained using the mesoporous carbon CMK-3 as template [221]. The XRD, TEM, and pore size analysis showed that the structure of the boron nitride molecular sieves synthesized from the carbon template consisted of a 2D regular array of uniform mesopores of 3.4 nm in diameter. However, it is not clear whether the obtained materials contain other impurities such as carbon, oxygen, and hydrogen, as the synthesis of non-porous nitrides always require a high temperature treatment to get a high purity. The authors also tried to impregnate aminoborazine as boron and nitrogen precursors into the mesoporous silica channels. However, this procedure was not successful, most probably due to the hydrophilic nature of the silica template.

As demonstrated above, MBN and MBCN can be successfully synthesized through elemental substitution in mesoporous carbon. Such materials could be useful as catalyst supports operating at high temperature in an oxidative atmosphere, and they could also find a potential use in the preparation of batteries and fuel cells because they are chemically inert and quite resistant to oxidation. The synthesis of mesoporous materials using the elemental substitution technique is now ready for extension to the synthesis of other metal nitrides such as GaN, AlN, and Si_3N_4.

10. SUMMARY AND FUTURE PERSPECTIVES

Research on the synthesis and the application of mesoporous non-siliceous materials is an emerging area attracting a lot of attention by researchers in the field. It is rapidly growing because of the existence of a lot of opportunities in the design, the synthesis, and potential applications of novel non-siliceous materials. During the past few years, considerable progress has been made in the preparation of these exciting materials. This book chapter has presented an overview over the most significant advances in the synthesis of mesoporous non-siliceous materials such as metal oxides, metals, semiconductors, polymers, carbons, and nitrides.

Mostly, soft- and hard-templating methods have been used for the preparation of the non-siliceous mesoporous materials. However, depending on the nature of the mesoporous materials, the appropriate methodology should be used for obtaining well-ordered products with excellent textural parameters. Using the soft-templating approach and varying the synthesis conditions, surfactant concentration, solution pH, and the temperature, the structure and the textural parameters of the mesoporous non-siliceous materials can be controlled. Mesoporous metals are a fascinating group of materials derived through the soft-templating approach, and they are promising for the production of sensors, fuel cell supports, electronics, optics, and catalysts, since they possess a high specific surface area, conductivity, and a well-ordered porous structure. Some of the possible applications of these materials have been addressed in this chapter. Mesoporous metal alloys and metal–metal oxide nanocomposites are also interesting materials and showed a superior performance in the electrochemical oxidation of methanol or CO as compared to that of the corresponding mesoporous metals and non-porous metals and alloys.

Mesoporous carbons, CNs, and nitrogen-doped mesoporous carbon materials having variable pore sizes, structures, and specific surface areas are fascinating materials and can only be obtained by nanocasting techniques using different kinds of tailor-made inorganic mesoporous metal oxides with different structures as the template. A controlled pore filling of the precursors (namely, of the mesoporous template) can be used for tuning the textural parameters of the mesoporous carbon and nitride materials. This is also critical for controlling the properties of these materials for various industrial applications. It is important to note that filling of the channels of the parent mesoporous material with the desired precursor must be strictly controlled to prepare high-quality materials. Thus, the conversion of the laboratory-made products into industrial scale is going to be a major challenge to allow for the bulk production of such materials.

The direct substitution of constituting elements in mesoporous materials is an interesting approach and it provides new materials with maintained structural regularity. According to the third-generation method (the "elemental substitution method"), MBN and MBCN have been successfully synthesized for the first time by Vinu et al. [220]. These materials possess high specific surface area, pore volume, and uniform mesopores, which could be helpful in many applications including electrodes for batteries and fuel cells, catalytic supports, and field emission devices.

Although a huge amount of ordered mesoporous materials has been synthesized during the past few years, research in the field is still expanding. However, a lot of emphasis has still to be put on the transfer of a laboratory-scale approach to bulk-scale production with the aim of commercializing such materials. There are various other approaches that can be used for the design and the synthesis of diverse non-siliceous mesoporous nanostructures with different elements and functions. Moreover, the research on mesoporous non-siliceous materials can be further extended to the preparation of mesoporous proteins and fullerenes using the above approaches, which can have potential applications in the field of enzymatic catalysis, biosensors, and fuel cells. This would create a new field of research in the coming years and provide novel materials with a widespread potential applicability.

ACKNOWLEDGMENTS

The studies described in this contribution were partially supported by Ministry of Culture, Sports, Science, and Technology under the Strategic Program for Building an Asian Science and Technology Community Scheme. In addition to this funding institution, the author also thanks his students (Miss. J. Josena and Miss. Kalyani) and postdoc fellows (Dr. V. V. Balasubramanian, Dr. Logudurai, and Dr. P. Srinivasu) for assisting him in the preparation of this contribution.

GLOSSARY

Carbon nanocage: A cage-type mesoporous carbon with three-dimensional structure, a very high specific surface area and a large specific pore volume.

Elemental substitution in mesoporous materials: A method for the synthesis of mesoporous materials where elements in the parent starting mesoporous solid are substituted by other elements and at the same time keeping the ordered structure.

Mesoporous boron carbon nitride: A boron carbon nitride material with regularly arranged mesoscopic pores prepared through elemental substitution in mesoporous carbon using boron trioxide, N_2 gas, and mesoporous carbon as boron, nitrogen, and carbon sources, respectively.

Mesoporous boron nitride: A boron nitride material with regularly arranged mesoscopic pores prepared through elemental substitution in mesoporous carbon using appropriate boron sources under a flow of nitrogen gas.

Mesoporous carbon: A carbon material with regularly arranged mesoscopic pores prepared through a replica-type or a direct synthesis route using a mesoporous silica template or organic surfactants as template and appropriate carbon sources.

Mesoporous carbon nitride: A carbon nitride material with regularly arranged mesoscopic pores prepared through the replica route using mesoporous silica as the template and appropriate carbon and nitrogen sources.

Mesoporous materials: A family of materials with regularly arranged pores in the mesoscopic (mesoporous) range (2.0–50 nm).

Mesoporous silica: A silica material with regularly arranged mesoscopic pores through template synthesis mainly by using micelle-type assemblies.

Replica route: A method for the synthesis of mesoporous materials where regular structures of mesoporous materials are replicated into chemically different mesoporous materials with the structural replicate of the template used.

Template synthesis: A method for the synthesis of mesoporous materials where regular assemblies of micelles are structurally transcribed into a silica structure and so on.

REFERENCES

[1] C.T. Kresge, M.E. Leonowicz, W.J. Roth, J.C. Vartuli, J.S. Beck, Nature 359 (1992) 710.
[2] A. Corma, Chem. Rev. 97 (1997) 2373.
[3] M. Hartmann, A. Vinu, Langmuir 18 (2002) 8010.
[4] A. Vinu, V. Murugesan, M. Hartmann, J. Phys. Chem. B 108 (2004) 7323.
[5] D. Zhao, Q. Huo, J. Feng, B.F. Chmelka, G.D. Stucky, J. Am. Chem. Soc. 120 (1998) 6024.
[6] Q. Zhang, K. Ariga, A. Okabe, T. Aida, J. Am. Chem. Soc. 126 (2004) 988.
[7] A. Vinu, V. Murugesan, O. Tangermann, M. Hartmann, Chem. Mater. 16 (2004) 3056.
[8] S. Che, A.E.G. Bennet, Y. Yokoi, K. Sakamoto, H. Kunieda, O. Terasaki, T. Tatsumi, Nat. Mater. 2 (2003) 801.
[9] Q. Huo, R. Leon, P.M. Petroff, G.D. Stucky, Science 268 (1995) 1324.
[10] A. Vinu, V. Murugesan, M. Hartmann, Chem. Mater. 15 (2003) 1385.
[11] Q. Huo, D.I. Margolese, U. Ciesla, D.G. Demuth, P. Feng, T. Gier, P. Sieger, A. Firouzi, B.F. Chmelka, F. Schüth, G.D. Stucky, Chem. Mater. 6 (1994) 1176.
[12] D. Zhao, J. Feng, Q. Huo, N. Melosh, G.H. Fredickson, B.F. Chmelka, G.D. Stucky, Science 279 (1998) 548.
[13] A. Vinu, T. Krithiga, V. Murugesan, M. Hartmann, Adv. Mater. 16 (2004) 1817.
[14] A. Vinu, V. Murugesan, W. Bohlmann, M. Hartmann, J. Phys. Chem. B 108 (2004) 11496.
[15] A. Vinu, D.P. Sawant, K. Ariga, K.Z. Hossain, S.B. Halligudi, M. Hartmann, M. Nomura, Chem. Mater. 17 (2005) 5339.
[16] A. Vinu, P. Srinivasu, M. Miyahara, K. Ariga, J. Phys. Chem. B 110 (2006) 801.
[17] P.T. Tanev, T.J. Pinnavaia, Science 267 (1995) 865.
[18] S.A. Bagshaw, E. Prouzet, T.J. Pinnavaia, Science 269 (1995) 1242.
[19] F. Kleitz, D. Liu, G.M. Anilkumar, I.-S. Park, L.A. Solovyov, A.N. Shmakov, R. Ryoo, J. Phys. Chem. B 107 (2003) 14296.
[20] A. Vinu, M. Miyahara, V. Sivamurugan, T. Mori, K. Ariga, J. Mater. Chem. 15 (2005) 5122.
[21] A. Vinu, M. Miyahara, T. Mori, K. Ariga, J. Porous Mater. 13 (2006) 379.

[22] K. Ariga, A. Vinu, M. Miyahara, J. Hill, T. Mori, J. Am. Chem. Soc. 129 (2007) 11022.
[23] A. Vinu, M. Miyahara, K. Ariga, J. Nanosci. Nanotechnol. 6 (2006) 1510.
[24] P. Srinivasu, V.V. Balasubramanian, L. Kumaresan, D.P. Sawant, X. Jin, S. Alam, K. Ariga, T. Mori, A. Vinu, J. Nanosci. Nanotechnol. 7 (2007) 3250.
[25] A. Vinu, T. Mori, K. Ariga, Sci. Technol. Adv. Mater. 7 (2006) 753.
[26] A. Vinu, J. Dedecek, V. Murugesan, M. Hartmann, Chem. Mater. 14 (2002) 2433.
[27] M. Hartmann, A. Vinu, S.P. Elangovan, V. Murugesan, W. Böhlmann, Chem. Commun. 11 (2002) 1238.
[28] (a) J.M. Sun, D. Ma, H. Zhang, C.L. Wang, X.H. Bao, D.S. Su, A. Klein-Hoffmann, G. Weinberg, S. Mann, J. Mater. Chem. 16 (2006) 1507. (b) L.N. Wang, T. Qi, Y. Zhang, J.L. Chu, Micropor. Mesopor. Mater. 91 (2006) 156. (c) Y.J. Wang, F. Caruso, Adv. Mater. 18 (2006) 795.
[29] (a) S.M. Yang, I. Sokolov, N. Coombs, C.T. Kresge, G.A. Ozin, Adv. Mater. 11 (1999) 1427. (b) H.Y. Jin, Z. Liu, T. Ohsuna, O. Terasaki, Y. Inoue, K. Sakamoto, T. Nakanishi, K. Ariga, S. Che, Adv. Mater. 18 (2006) 593.
[30] (a) Q.S. Huo, D.Y. Zhao, J.L. Feng, K. Weston, S.K. Buratto, G.D. Stucky, S. Schacht, F. Schüth, Adv. Mater. 9 (1997) 974. (b) J.H. Jung, Y. Ono, K. Hanabusa, S. Shinkai, J. Am. Chem. Soc. 122 (2000) 5008. (c) P.J. Bruinsma, A.Y. Kim, J. Liu, S. Baskaran, Chem. Mater. 9 (1997) 2507. (d) J. Krishna Murthy, U. Gross, S. Rudiger, E. Kemnitz, J.M. Winfield, J. Solid State Chem. 179 (2006) 739. (e) S.C. Zhang, C. Zhang, H.P. Yang, Y.F. Zhu, J. Solid State Chem. 179 (2006) 873. (f) J.G. Yu, L. Zhao, B. Cheng, J. Solid State Chem. 179 (2006) 226. (g) U.G. Singh, R.T. Williams, K.R. Hallam, G.C. Allen, J. Solid State Chem. 178 (2005) 3405. (h) F. Marlow, M.D. McGehee, D.Y. Zhao, B.F. Chmelka, G.D. Stucky, Adv. Mater. 11 (1999) 632.
[31] (a) H. Yang, A. Kuperman, N. Coombs, A.S. Mamiche, G.A. Ozin, Nature 379 (1996) 703. (b) T. Asefa, M. Kruk, M.J. MacLachlan, N. Coombs, H. Grondey, M. Jaroniec, G.A. Ozin, J. Am. Chem. Soc. 123 (2001) 8520. (c) D. Grosso, A.R. Balkenende, P.A. Albouy, A. Ayral, H. Amenitsch, F. Babonneau, Chem. Mater. 13 (2001) 1848. (d) D.Y. Zhao, P.D. Yang, D.I. Margolese, B.F. Chmelka, G.D. Stucky, Chem. Commun. 22 (1998) 2499.
[32] A. Vinu, K. Ariga, S. Saravanamurugan, M. Hartmann, V. Murugesan, Micropor. Mesopor. Mater. 76 (2004) 91.
[33] A. Vinu, K.U. Nandhini, V. Murugesan, W. Böhlmann, V. Umamaheswari, A. Pöppl, M. Hartmann, Appl. Catal. A: Gen. 265 (2004) 1.
[34] M. Kathik, A.K. Tripathi, N.M. Gupta, A. Vinu, M. Hartmann, M. Palanichamy, V. Murugesan, Appl. Catal. A: Gen. 268 (2004) 139.
[35] A. Vinu, M. Hartmann, Chem. Lett. 33 (2004) 588.
[36] D.P. Swant, A. Vinu, N.E. Jacob, F. Lefebvre, S.B. Halligudi, J. Catal. 235 (2005) 341.
[37] A. Vinu, M. Karthik, M. Miyahara, V. Murugesan, K. Ariga, J. Mol. Catal. A: Chem. 230 (2005) 151.
[38] A. Vinu, G.S. Kumar, K. Ariga, V. Murugesan, J. Mol. Catal. A: Chem. 235 (2005) 57.
[39] T. Krithiga, A. Vinu, K. Ariga, B. Arabindoo, M. Palanichamy, V. Murugesan, J. Mol. Catal. A: Chem. 237 (2005) 238.
[40] A. Vinu, K. Shanmugapriya, G. Changrasekar, V. Murugesan, K. Ariga, J. Nanosci. Nanotechnol. 5 (2005) 542.
[41] V. Umamaheswari, W. Böhlmann, A. Pöppl, A. Vinu, M. Hartmann, Micropor. Mesopor. Mater. 89 (2006) 47.
[42] A. Vinu, D.P. Sawant, K. Ariga, M. Hartmann, S.B. Halligudi, Micropor. Mesopor. Mater. 80 (2005) 195.

[43] A. Vinu, B.M. Devassy, S.B. Halligudi, W. Böhlmann, M. Hartmann, Appl. Catal. A: Gen. 281 (2005) 207.
[44] D.M. Antonelli, J.Y. Ying, Angew. Chem. Int. Ed. Engl. 34 (1995) 2014.
[45] R. Takahashi, S. Takenaka, S. Sato, T. Sodesawa, K. Ogura, K. Nakanishi, J. Chem. Soc., Faraday Trans. 94 (1998) 3161.
[46] U. Bach, D. Lupo, P. Comte, J.E. Moser, F. Weissörtle, J. Salbeck, H. Spreitzer, M. Grätzel, Nature 395 (1998) 583.
[47] P. Yang, D. Zhao, D.I. Margolese, B.F. Chmelka, G.D. Stucky, Chem. Mater. 11 (1999) 2813.
[48] D. Grosso, G.J. de, A.A. Soler-Illia, F. Babonneau, C. Sanchez, P.-A. Albouy, A. Brunet-Bruneau, A.R. Balkenende, Adv. Mater. 13 (2001) 1085.
[49] P.C.A. Alberius, K.L. Frindell, R.C. Hayward, E.J. Kramer, G.D. Stucky, B.F. Chmelka, Chem. Mater. 14 (2002) 3284.
[50] F.-S. Xiao, Y. Han, Y. Yu, X. Meng, M. Yang, S. Wu, J. Am. Chem. Soc. 124 (2002) 888.
[51] M. Vettraino, M. Trudeau, A.Y.H. Lo, R.W. Schurko, D. Antonelli, J. Am. Chem. Soc. 124 (2002) 9567.
[52] R.C. Hayward, B.F. Chmelka, E.J. Kramer, Adv. Mater. 17 (2005) 2591.
[53] T. Streethawong, Y. Suzuki, S. Yoshikawa, J. Solid State Chem. 178 (2005) 329.
[54] D.M. Antonelli, J.Y. Ying, Chem. Mater. 8 (1996) 874.
[55] H. Kosslick, G. Lischke, G. Walther, W. Storek, A. Martin, R. Fricke, Micropor. Mater. 9 (1997) 13.
[56] J.N. Kondo, L. Lu, Y. Takahara, K. Maruya, K. Domen, N. Igarashi, T. Tatsumi, Bull. Chem. Soc. Jpn. 73 (2000) 1123.
[57] D. Antonelli, J.Y. Ying, Angew. Chem. Int. Ed. Engl. 35 (1996) 426.
[58] D.M. Antonelli, A. Nakahira, J.Y. Ying, Inorg. Chem. 35 (1996) 3126.
[59] F. Schüth, U. Ciesla, S. Schacht, M. Thieme, Q. Huo, G.D. Stucky, Mater. Res. Bull. 34 (1999) 483.
[60] B.T. Holland, C.F. Blanford, T. Do, A. Stein, Chem. Mater. 11 (1999) 795.
[61] M.S. Wong, J.Y. Ying, Chem. Mater. 10 (1998) 2067.
[62] S.A. Bagshaw, T.J. Pinnavaia, Angew. Chem. Int. Ed. Engl. 35 (1996) 1102.
[63] F. Vaudry, S. Khodabandeh, M.E. Davis, Chem. Mater. 8 (1996) 1451.
[64] P. Liu, I.L. Moudrakovski, J. Liu, A. Sayari, Chem. Mater. 9 (1997) 2513.
[65] S. Inagaki, S. Guan, Y. Fukushima, T. Ohsuna, O. Terasaki, J. Am. Chem. Soc. 121 (1999) 9611.
[66] T. Asefa, M.J. MacLachlan, N. Coombs, G.A. Ozin, Nature 402 (1999) 867.
[67] B. Chakraborty, A.C. Pulikottil, S. Das, B. Viswanathan, Chem. Commun. 911 (1997).
[68] D. Zhao, Z. Luan, L. Kevan, Chem. Commun. 1009 (1997).
[69] Y.Z. Khimyak, I. Klinowski, J. Phys. Chem. Chem. Phys. 2 (2000) 5275.
[70] T. Kimura, Y. Sugahara, K. Kuroda, Chem. Mater. 11 (1999) 508.
[71] B.T. Holland, P.K. Isbester, C.F. Blanford, E.J. Munson, A. Stein, J. Am. Chem. Soc. 119 (1997) 6796.
[72] G.S. Attard, P.N. Bartlett, N.R.B. Coleman, J.M. Elliott, J.R. Owen, J.H. Wang, Science 278 (1997) 838.
[73] J.L. Mohanan, I.U. Arachchige, S.L. Brock, Science 307 (2005) 397.
[74] D. Grosso, C. Boissière, B. Smarsly, T. Brezesinski, N. Pinna, P.A. Albouy, H. Amenitsch, M. Antonietti, C. Sanchez, Nature Mater. 3 (2004) 787.
[75] P. Yang, D. Zhao, D.I. Margolesse, B.F. Chmelka, G.D. Stucky, Nature 396 (1998) 152.
[76] A.L. Pruden, D.F. Ollis, J. Catal. 82 (1983) 404.
[77] R.W. Matthews, J. Phys. Chem. 91 (1987) 3328.
[78] S. Chappel, S.-G. Chen, A. Zaban, Langmuir 18 (2002) 3336.

[79] S. Ito, T. Takeuchi, T. Katayama, M. Sugiyama, M. Matsuda, T. Kitamura, Y. Wada, S. Yanagida, Chem. Mater. 15 (2003) 2824.
[80] H. Shibata, T. Ogura, T. Mukai, T. Ohkubo, H. Sakai, M. Abe, J. Am. Chem. Soc. 127 (2005) 16396.
[81] E.L. Crepaldi, G.J. de, A. Soler-Illia, D. Grosso, F. Cagnol, F. Ribot, C. Sanchez, J. Am. Chem. Soc. 125 (2003) 9770.
[82] B. Smarsly, D. Grosso, T. Brezesinski, N. Pinna, C. Boissière, M. Antonietti, C. Sanchez, Chem. Mater. 16 (2004) 2948.
[83] W. Yan, S.M. Mahurin, S.H. Overbury, S. Dai, Chem. Mater. 17 (2005) 1923.
[84] Y.-Y. Lyu, S.H. Yi, J.K. Shon, S. Chang, L.S. Pu, S.-Y. Lee, J.E. Yie, K. Char, G.D. Stucky, J.M. Kim, J. Am. Chem. Soc. 126 (2004) 2310.
[85] Z. Zhang, T.J. Pinnavaia, J. Am. Chem. Soc. 124 (2002) 12294.
[86] J.-J. Shyue, M.R. De Guire, J. Am. Chem. Soc. 127 (2005) 12736.
[87] B. Lee, D. Lu, J.N. Kondo, K. Domen, J. Am. Chem. Soc. 124 (2002) 11256.
[88] T.T. Emos, J. Li, L.F. Nazar, J. Am. Chem. Soc. 124 (2002) 8516.
[89] A. Bhaumik, S. Inagaki, J. Am. Chem. Soc. 123 (2001) 691.
[90] B. Tian, X. Liu, B. Tu, G. Yu, J. Fan, L. Wang, S. Xie, G.D. Stucky, D. Zhao, Nat. Mater. 2 (2003) 159.
[91] B. Tian, X. Liu, H. Yang, S. Xie, C. Yu, B. Tu, D. Zhao, Adv. Mater. 15 (2003) 1370.
[92] B. Tian, X. Liu, L.A. Solovyov, Z. Liu, H. Yang, Z. Zhang, S. Xie, F. Zhang, B. Tu, C. Yu, O. Terasaki, D. Zhao, J. Am. Chem. Soc. 126 (2004) 865.
[93] J. Roggenbuck, M. Tiemann, J. Am. Chem. Soc. 127 (2005) 1096.
[94] K. Zhu, B. Yue, W. Zhou, H. He, Chem. Commun. (2003) 98.
[95] K. Zhu, H. He, S. Xie, X. Zhang, W. Zhou, S. Jin, B. Yue, Chem. Phys. Lett. 377 (2003) 317.
[96] F. Jiao, B. Yue, K. Zhu, D. Zhao, H. He, Chem. Lett. 32 (2003) 770.
[97] K. Zhu, Y. Bin, S. Xie, S. Zhang, Z. Biao, S. Jin, H. He, Chin. J. Chem. 22 (2004) 33.
[98] G.S. Attard, J.C. Glydeand, C.G. Goltner, Nature 378 (1995) 366.
[99] J. Tanori, N. Duxin, C. Petit, I. Lisiecki, P. Veillet, M.P. Pileni, Colloid Polym. Sci. 886 (1995) 273.
[100] G.S. Attard, C.G. Goltner, J.M. Corker, S. Henke, R.H. Templer, Angew. Chem. Int. Ed. Engl. 36 (1997) 1315.
[101] J. Zhao, X. Chen, L.Y. Jiao, Y.C. Chai, G.D. Zhang, J. Liu, Chem. Lett. 33 (2004) 842.
[102] H. Boo, S. Park, B. Ku, Y. Kim, J.H. Park, H.C. Kim, T.D. Chung, J. Am. Chem. Soc. 126 (2004) 4524.
[103] (a) Y. Yamauchi, T. Yokoshima, T. Momma, T. Osaka, K. Kuroda, Chem. Lett. 33 (2004) 1576. (b) Y. Yamauchi, K. Kuroda, Chem. Asian J. 3 (2008) 664.
[104] P.R. Birkin, J.M. Elliott, Y.E. Watson, Chem. Commun. (2000) 1693.
[105] S.A.G. Evans, J.M. Elliott, L.M. Andrews, P.N. Bartlett, P.J. Doyle, G. Denuault, Anal. Chem. 74 (2002) 1322.
[106] S. Park, T.D. Chung, H.C. Kim, Anal. Chem. 75 (2003) 3046.
[107] A. Kucernak, J. Jian, Chem. Eng. J. 81 (2003) 93.
[108] P.N. Bartlett, B. Gollas, S. Guerin, J. Marwan, Phys. Chem. Chem. Phys. 4 (2002) 3835.
[109] (a) P.N. Bartlett, J. Marwan, Micropor. Mesopor. Mater. 62 (2003) 73. (b) T. Gabriel, I.S. Nandhakumar, G.S. Attard, Electrochem. Commun. 4 (2002) 610.
[110] P.N. Bartlett, P.N. Birkin, M.A. Ghanem, P. de Groot, M. Sawick, J. Electrochem. Soc. 148 (2001) 119.
[111] A.H. Whitehead, J.M. Elliott, J.R. Owen, G.S. Attard, Chem. Commun. (1999) 331.
[112] P.A. Nelson, J.M. Elliott, G.A. Attard, J.R. Owen, Chem. Mater. 14 (2002) 524.

[113] T. Gabriel, I.S. Nandhakumar, G.S. Attard, Electrochem. Commun. 4 (2002) 610.
[114] S. Guerin, G.S. Attard, Electrochem. Commun. 3 (2001) 544.
[115] I.S. Nandhakumar, T. Gabriel, X. Li, G.S. Attard, M. Markham, D.C. Smith, J.J. Baumberg, Chem. Commun. (2004) 1374.
[116] M.L. Markham, J.J. Baumberg, D.C. Smith, X. Li, T. Gabriel, G.S. Attard, I.S. Nandhakumar, Appl. Phys. Lett. 86 (2005) 11912.
[117] Y. Yamauchi, T. Yokoshima, T. Momma, T. Osaka, K. Kuroda, J. Mater. Chem. 14 (2004) 2935.
[118] Y. Yamauchi, S.S. Nair, T. Yokoshima, T. Momma, T. Osaka, K. Kuroda, Stud. Surf. Sci. Catal. 156 (2005) 457.
[119] G.S. Attard, S.A.A. Leclerc, S. Maniguet, A.E. Russell, I. Nandhakumar, P.N. Bartlett, Chem. Mater. 13 (2001) 1444.
[120] Y. Yamauchi, S.S. Nair, T. Momma, T. Ohsuna, K. Kuroda, J. Mater. Chem. 16 (2006) 2229.
[121] A. Saramat, M. Andersson, S. Hant, P. Thormahlen, M. Skoglundh, G.S. Attard, A.E.C. Palmqvist, Eur. Phys. J. D. 43 (2007) 209.
[122] Y. Yamauchi, T. Momma, T. Osaka, K. Kuroda (Waseda University), Japanese Patent 233272, 2006.
[123] Y. Yamauchi, T. Momma, H. Kitoh, T. Osaka, K. Kuroda, Electrochem. Commun. 7 (2005) 1364.
[124] Y. Yamauchi, K. Kuroda, Electrochem. Commun. 8 (2006) 1677.
[125] Y. Yamauchi, H. Kitoh, T. Momma, T. Osaka, K. Kuroda, Sci. Technol. Adv. Mater. 7 (2006) 438.
[126] Y. Yamauchi, M. Komatsu, A. Takai, R. Sebata, M. Sawada, T. Momma, M. Fuziwara, T. Osaka, K. Kuroda, Electrochim. Acta 53 (2007) 604.
[127] K. Luo, C.T. Walker, K.J. Edler, Adv. Mater. 19 (2007) 1506.
[128] Y.J. Han, J.M. Kim, G.D. Stucky, Chem. Mater. 12 (2000) 2068.
[129] H.J. Shin, R. Ryoo, Z. Liu, O. Terasaki, J. Am. Chem. Soc. 123 (2001) 1246.
[130] K. Lee, Y.H. Kim, S.B. Han, H.K. Kang, S. Park, W.S. Seo, J.T. Park, B. Kim, S.B. Chang, J. Am. Chem. Soc. 125 (2003) 6844.
[131] G. Armatas, M.G. Kanatzidis, Nature 441 (2006) 1122.
[132] G. Armatas, M.G. Kanatzidis, Science 313 (2006) 817.
[133] D. Sun, A.E. Riley, A.J. Cadby, E.K. Richman, S.D. Korlann, S.H. Tolbert, Nature 441 (2006) 1126.
[134] K. Busch, S. John, Phys. Rev. E 58 (1998) 3896.
[135] L.K. Van Vugt, A.F. Van Driel, R.W. Tjerkstra, L. Bechger, W.L. Vos, D. Vanmaekelbergh, J.J. Kelly, Chem. Commun. (2002) 2054.
[136] P. Osenar, P.V. Braun, S.I. Stupp, Nature, Adv. Mater. 8 (1996) 1022.
[137] P.V. Braun, P. Osenar, S.I. Stupp, Nature 380 (1996) 325.
[138] S.I. Stupp, P.V. Braun, Science 277 (1997) 242.
[139] J. Li, H. Kessler, M. Soulard, L. Khouchaf, M.-H. Tuiler, Adv. Mater. 10 (1998) 946.
[140] T. Jiang, G.A. Ozin, J. Mater. Chem. 8 (1998) 1099.
[141] M. Fröba, N. Oberender, Chem. Commun. (1997) 1729.
[142] T. Jiang, G.A. Ozin, R.L. Bedard, Adv. Mater. 6 (1994) 860.
[143] F. Gao, Q. Lu, D. Zhao, Adv. Mater. 15 (2003) 739.
[144] X. Liu, B. Tian, C. Yu, B. Tu, Z. Liu, O. Terasaki, D. Zhao, Chem. Lett. 32 (2003) 824.
[145] H. Yang, D. Zhao, J. Mater. Chem. 15 (2005) 1217.
[146] C.G. Wu, T. Bein, Science 264 (1994) 1757.
[147] C.G. Wu, T. Bein, Science 266 (1994) 1013.
[148] S.A. Johnson, D. Khushalani, N. Coombs, T.E. Mallouk, G.A. Ozin, J. Mater. Chem. 8 (1998) 13.
[149] G. Li, S. Bhosale, T. Wang, Y. Zhang, H. Zhu, J.H. Fuhrhop, Angew. Chem. Int. Ed. Engl. 42 (2003) 3818.

[150] J. Jang, J. Bae, Chem. Commun. (2005) 1200.
[151] F.Q. Zhang, Y. Meng, D. Gu, Y. Yan, C.Z. Yu, B. Tu, D.Y. Zhao, J. Am. Chem. Soc. 127 (2005) 13508.
[152] Y.H. Deng, T. Yu, Y. Wan, Y.F. Shi, Y. Meng, D. Gu, L.J. Zhang, Y. Huang, C. Liu, X.J. Wu, D.Y. Zhao, J. Am. Chem. Soc. 129 (2007) 1690.
[153] Y. Meng, D. Gu, F.Q. Zhang, Y.F. Shi, L. Cheng, D. Feng, Z.X. Wu, Z.X. Chen, Y. Wan, A. Stein, D.Y. Zhao, Chem. Mater. 18 (2006) 4447.
[154] Y. Meng, D. Gu, F.Q. Zhang, Y.F. Shi, H. Yang, Z. Li, C. Yu, B. Tu, D.Y. Zhao, Angew. Chem. Int. Ed. Engl. 44 (2005) 7053.
[155] Y. Meng, D. Gu, F.Q. Zhang, Y.F. Shi, L. Cheng, D. Feng, Z.X. Wu, Z.X. Chen, Y. Wan, A. Stein, D.Y. Zhao, Chem. Mater. 18 (2006) 4447.
[156] H. Kosonen, S. Valkama, A. Nykanen, M. Toivanen, G. Ten Brinke, J. Ruokolainen, O. Ikkala, Adv. Mater. 18 (2006) 201.
[157] K.P. Gierszal, M. Jaroniec, J. Am. Chem. Soc. 128 (2006) 10026.
[158] S.A. Johnson, P.J. Ollivier, T.E. Mallouk, Science 283 (1999) 963.
[159] T. Kyotani, T. Nagai, S. Inoue, A. Tomita, Chem. Mater. 9 (1997) 609.
[160] T. Kyotani, Carbon 38 (2000) 269.
[161] R. Ryoo, S.H. Joo, S. Jun, J. Phys. Chem. B 103 (1999) 7743.
[162] S. Jun, S.H. Joo, R. Ryoo, M. Kruk, M. Jaroniec, Z. Liu, T. Ohsuna, O. Terasaki, J. Am. Chem. Soc. 122 (2000) 10712.
[163] R. Ryoo, S.H. Joo, M. Kruk, M. Jaroniec, Adv. Mater. 13 (2001) 677.
[164] S.H. Joo, S.J. Choi, I. Oh, J. Kwak, Z. Liu, O. Terasaki, R. Ryoo, Nature 412 (2001) 169.
[165] L.A. Solovyov, V.I. Zaikovskii, A.N. Shmakov, O.V. Belousov, R. Ryoo, J. Phys. Chem. B 106 (2002) 12198.
[166] J. Lee, S. Yoon, T. Hyeon, S.M. Oh, K.B. Kim, Chem. Commun. 2177 (1999).
[167] J. Lee, S. Yoon, S.M. Oh, C.-H. Shin T. Hyeon, Adv. Mater. 12 (2000) 359.
[168] S. Han, S. Kim, H. Lim, W. Choi, H. Park, J. Yoon, T. Hyeon, Micropor. Mesopor. Mater. 58 (2003) 131.
[169] A. Vinu, C. Streb, V. Murugesan, M. Hartmann, J. Phys. Chem. B 107 (2003) 8297.
[170] M. Kruk, M. Jaroniec, T.-W. Kim, R. Ryoo, Chem. Mater. 15 (2003) 2815.
[171] A.H. Lu, W. Schmidt, B. Spliethoff, F. Schüth, Adv. Mater. 15 (2003) 1602.
[172] W.-H. Zhang, C. Liang, H. Sun, Z. Shen, Y. Guan, P. Ying, C. Li, Adv. Mater. 14 (2002) 1776.
[173] S. Che, A.E. Garcia-Bennett, X. Liu, R.P. Hodgkins, P.A. Wright, D. Zhao, O. Terasaki, T. Tatsumi, Angew. Chem. Int. Ed. Engl. 42 (2003) 3930.
[174] F. Kleitz, S.H. Choi, R. Ryoo, Chem. Commun. (2003) 2136.
[175] A. Vinu, K.Z. Hossain, G.S. Kumar V. Sivamurugan, K. Ariga, Stud. Surf. Sci. Catal. 156 (2005) 631.
[176] A. Vinu, K.Z. Hossain, G.S. Kumar, K. Ariga, Carbon 44 (2006) 530.
[177] A. Vinu, M. Miyahara, K. Ariga, J. Phys. Chem. B 109 (2005) 6436.
[178] M. Choi, R. Ryoo, Nature Mater. 2 (2003) 473.
[179] Y. Xia, Z. Yang, R. Mokaya, Chem. Mater. 18 (2006) 140.
[180] A. Vinu, K. Ariga, Chem. Lett. 34 (2005) 674.
[181] (a) A. Vinu, M. Hartmann, Catal. Today 102 (2005) 189. (b) A. Vinu, K. Ariga, Chem. Lett. 34 (2005) 674.
[182] J.S. Lee, S.H. Joo, R. Ryoo, J. Am. Chem. Soc. 124 (2002) 1156.
[183] A. Vinu, P. Srinivasu, M. Takahashi, T. Mori, V.V. Balasubramanian, K. Ariga, Micropor. Mesopor. Mater. 200 (2007) 20.
[184] J. Lee, K. Sohn, T. Hyeon, J. Am. Chem. Soc. 123 (2001) 5146.
[185] Y. Oda, K. Fukuyama, K. Nishikawa, S. Namba, H. Yoshitake, T. Tatsumi, Chem. Mater. 16 (2004) 3860.
[186] T.-W. Kim, I.-S. Park, R. Ryoo, Angew. Chem. Int. Ed. Engl. 42 (2003) 4375.

[187] C.H. Kim, D.-K. Lee, T.J. Pinnavaia, Langmuir 20 (2004) 5157.
[188] A.B. Fuertes, S. Alvarez, Carbon 42 (2004) 3409.
[189] A.B. Fuertes, T.A. Centeno, J. Mater. Chem. 15 (2005) 1079.
[190] K.P. Gierszal, T.-W. Kim, R. Ryoo, M. Jaroniec, J. Phys. Chem. B 109 (2005) 23263.
[191] A. Vinu, M. Miyahara, K. Ariga, Stud. Surf. Sci. Catal. 158B (2005) 971.
[192] F. Kleitz, D. Liu, G.M. Anilkumar, I.-S. Park, L.A. Solovyov, A.N. Shmakov, R. Ryoo, J. Phys. Chem. B 107 (2003) 14296.
[193] P.I. Ravikovitch, A.V. Neimark, Langmuir 18 (2002) 1550.
[194] A.Y. Liu, M. L Cohen, Science 245 (1989) 841.
[195] Y. Qiu, L. Gao, Chem. Commun. (2003) 2378.
[196] Q. Guo, Q. Yang, L. Zhu, C. Yi, S. Zhang, Y. Xie, Solid State Commun. 132 (2004) 369.
[197] E. Kroke, M. Schwarz, Coord. Chem. Rev. 248 (2004) 493.
[198] Y.-J. Bai, B. Lu, Z.-G. Liu, L. Li, D.-L. Cui, X.-G. Xu, Q.-L. Wang, J. Crystal Growth 247 (2003) 505.
[199] J.L. Zimmerman, R. Williams, V.N. Khabashesku, J.L. Margrave, Nano. Lett. 12 (2001) 731.
[200] V.N. Khabashesku, J.L. Zimmerman, J.L. Margrave, Chem. Mater. 12 (2000) 3264.
[201] J. Wang, D.R. Miller, E.G. Gillan, Carbon 41 (2003) 2031.
[202] E.G. Gillan, Chem. Mater. 12 (2000) 3906.
[203] D.R. Miller, J. Wang, E.G. Gillan, J. Mater. Chem. 12 (2002) 2463.
[204] M. Kim, S. Hwang, J.-S. Yu, J. Mater. Chem. 17 (2007) 1656.
[205] J. Kouvetakis, A. Bandari, M. Todd, B. Wilkens, N. Cave, Chem. Mater. 6 (1994) 811.
[206] Y. Qiu, L. Gao, Chem. Commun. (2003) 2378.
[207] A. Vinu, K. Ariga, T. Mori, S. Hishita, T. Nakanishi, D. Golberg, Y. Bando, Adv. Mater. 17 (2005) 1648.
[208] F. Goettmann, A. Fischer, M. Antonietti, A. Thomas, Angew. Chem. Int. Ed. Engl. 45 (2006) 4467.
[209] M. Groenewolt, M. Antonietti, Adv. Mater. 17 (2005) 1789.
[210] J.Y. Kim, S.B. Yoon, J.-S. Yu, Chem. Mater. 15 (2003) 1932.
[211] S.B. Yoon, J.Y. Kim, J.-S. Yu, Chem. Commun. (2001) 559.
[212] A. Vinu, Adv. Funct. Mater. 18 (2008) 816.
[213] A. Vinu, P. Srinivasu, D.P. Sawant, T. Mori, K. Ariga, J.-S. Chang, S.-H. Jhung, V.V. Balasubramanian, Y.K. Hwang, Chem. Mater. 19 (2007) 4367.
[214] Y. Xia, R. Mokaya, Adv. Mater. 16 (2004) 1553.
[215] Y. Xia, R. Mokaya, Chem. Mater. 17 (2005) 1553.
[216] A. Vinu, P. Srinivasu, T. Mori, T. Sasaki, A. Asthana, K. Ariga, S. Hishita, Chem. Lett. 36 (2007) 770.
[217] A. Vinu, S. Anandan, C. Anand, P. Srinivasu, K. Ariga, T. Mori, Micropor. Mesopor. Mater. 108 (2008) 398.
[218] T. Asefa, M. Kruk, N. Coombs, H. Grondey, M.J. MacLachlan, M. Jaroniec, G.A. Ozin, J. Am. Chem. Soc. 125 (2003) 11662.
[219] K. Wan, Q. Liu, C. Zhang, J. Wang, Bull. Chem. Soc. Jpn. 77 (2004) 1409.
[220] A. Vinu, M. Terrones, D. Golberg, S. Hishita, K. Ariga, T. Mori, Chem. Mater. 17 (2005) 5887.
[221] P. Dibandjo, F. Chassagneux, L. Bois, C. Sigala, P. Miele, J. Mater. Chem. 15 (2005) 1917.

CHAPTER 4

Catalysis with Microporous Aluminophosphates and Silicoaluminophosphates Containing Transition Metals

Martin Hartmann[1,*] *and* **S.P. Elangovan**[2]

[1]Advanced Materials Science, University of Augsburg, Universitätsstr. 1, D-86159 Augsburg, Germany and Erlangen Catalysis Resource Center, Friedrich-Alexander-Universität Erlangen-Nürnberg, 91058 Erlangen, Germany

[2]Nippon Chemical Industrial Co. Ltd., 9-11-1 Kameido, Koto-ku, Tokyo 136-8515, Japan

Contents

Abstract

The literature related to the catalytic activity of aluminophosphate ($AlPO_4$) and silicoaluminophosphate (SAPO) molecular sieves containing transition metal ions and/or transition metal clusters is reviewed. Microporous crystalline metal aluminophosphates (MeAPOs) and metal silicoaluminophosphates (MeAPSOs) represent an important group of inorganic materials because of their large potential as adsorbents and catalysts. This review focuses mainly on the catalytic activity of MeAPOs and MeAPSOs in

*Corresponding author. Tel.: +49 9131 85 28 792; Fax: +49 9131 85 27 421
E-mail address: martin.hartmann@ecrc.uni-erlangen.de

Advances in Nanoporous Materials, Volume 1 ISSN 1878-7959, DOI 10.1016/S1878-7959(09)00104-2

comparison with the corresponding MeAPSOs covering the literature through 2008. The catalysis over these materials is summarized and discussed with respect to the catalytic performance, possible reaction mechanisms, and potential applications.

1. INTRODUCTION

Aluminophosphate molecular sieves ($AlPO_4$-*n*), where *n* denotes a particular structure type, form an interesting class of microporous crystalline materials comparable to the well-known zeolites, which are aluminosilicate molecular sieves. Zeolites have pores or channels formed by alumina and silica tetrahedra linked through oxygen bridges. Substitution of other elements for Al and/or Si in the molecular sieve framework can yield various kinds of new materials. Wilson et al. [1] reported in 1982 the synthesis of microporous aluminophosphate molecular sieves ($AlPO_4$-*n*). $AlPO_4$ frameworks are constructed from an alternating set of AlO_2^- and PO_2^+ tetrahedra, and thus, the resulting frameworks are neutral. $AlPO_4$ molecular sieves cover a wide range of different structure types; some are analogous to certain zeolites such as $AlPO_4$-42 [zeolite A structure, Linde type A (LTA) – framework structure code of the International Zeolite Association (IZA)], $AlPO_4$-34 (chabazite structure, CHA), or $AlPO_4$-37 (faujasite structure, FAU). But there is also a considerable number of $AlPO_4$s such as $AlPO_4$-11, $AlPO_4$-41, or VPI-5, which are unique structures with no zeolite analog. At the time of writing this contribution, more than 50 $AlPO_4$s with different structures are known with 19 being analogous to natural or synthetic zeolites. Most of the structures have been assigned a three-letter framework structure code (Table 1). The majority of the studies deal with materials having the framework structure $AlPO_4$-5 (AFI), $AlPO_4$-11 (AEL), and CHA, but numerous other materials [$AlPO_4$-36 (ATS), $AlPO_4$-41 (AFO), $AlPO_4$-31 (ATO)] have also been investigated (Fig. 1).

Transition metal ions (TMIs) can be introduced into $AlPO_4$s and SAPOs by three different methods: impregnation, ion exchange, or isomorphous substitution. In the latter method, the TMI salt is directly introduced into the synthesis mixture. An exciting property of the $AlPO_4$-*n* materials is that Al and/or P can be partially replaced by silicon (SAPO-*n*) and/or other elements (EAPO-*n*, EAPSO-*n*) [2–8]. Initially, it has been claimed that 12 different elements (E = Li, Be, Mg, B, Ti, Mn, Fe, Co, Zn, Ga, Ge, and As) can be incorporated into the $AlPO_4$ structure besides Al, P, and Si [9,10]. Nowadays, the number of elements possibly incorporated into the framework has increased to 20 with the addition of V, Cr, Ni, Cu, Zr, Nb, Mo, and Cd. However, the evidence for true lattice incorporation of these

Table 1 Typical structures in $AlPO_4$-*n*-based molecular sieves

Species	Structure type	Pore size (nm)	Incorporation reported
Very large pore			
VPI-5	VFI	1.27	Yes
8	AET	0.79 × 0.87	Yes
Large pore			
5	AFI	0.73	Yes
36	ATS	0.62 × 0.75	Yes
37	FAU	0.74	Yes
40	AFR	0.67 × 0.69	Yes
46	AFS	0.7	Yes
DAF-1	DFO	0.73	Yes
EMM-3	EZT	0.61 × 0.65	No
Medium pore			
11	AEL	0.4 × 0.65	
31	ATO	0.54	
41	AFO	0.43 × 0.7	
Small pore			
14	AFN	0.31 × 0.43	Yes
17	ERI	0.36 × 0.51	Yes
18	AEI	0.38	Yes
25	ATV	0.3 × 0.49	Yes
33	ATT	0.42 × 0.46	No
34, 44, 47	CHA	0.38	Yes
35	LEV	0.38 × 0.48	Yes
39	ATN	0.4 × 0.4	Yes
42	LTA	0.41	Yes

Table 1 *(Continued)*

Species	Structure type	Pore size (nm)	Incorporation reported
43	GIS	0.31 × 0.45	No
52	AFT	0.43	No?
56	AFX	0.34 × 0.36	No?
STA-6	SAS	0.42	Yes
STA-7	SAV	0.39	Yes
Very small pore			
16	AST	0.3	Yes
20	SOD	0.3	Yes
28	Novel	0.3	No

elements is less persuasive, the number of structure types is smaller and the levels of incorporation are significantly lower (Fig. 2).

Incorporation of silicon or other elements creates negative framework charges, which are balanced by exchangeable extra-framework cations. The introduction of silicon or other elements including TMIs is used to generate Brønsted-acid sites. The origin of Brønsted acidity in silicoaluminophosphates (SAPOs) is the insertion of Si at the phosphorus site, which leads to the formation of a negatively charged framework that is balanced by protons attached to the Si–O–Al bridges. It is obvious that the local structure of Brønsted-acid sites in zeolites and SAPOs is similar. However, the acidity of these materials may vary greatly even in isostructural systems. The isomorphous substitution of TMIs for aluminum, phosphorus, or silicon in $AlPO_4$ or SAPO materials may also generate Brønsted-acid sites unless elements with equal oxidation states are replaced, for example, Cr(III) for Al(III) or Ti(IV) for Si(IV).

The incorporation of TMIs into framework sites (isomorphous substitution) of AlPO and SAPO molecular sieves yielding MeAPO and MeAPSO-*n* materials is of particular interest for the design of novel catalysts. Isomorphous substitution is typically defined as the replacement of an element in the crystalline framework by another element with similar cation radius and coordination requirements. However, many elements claimed to be incorporated into the $AlPO_4$ framework have radius ratios

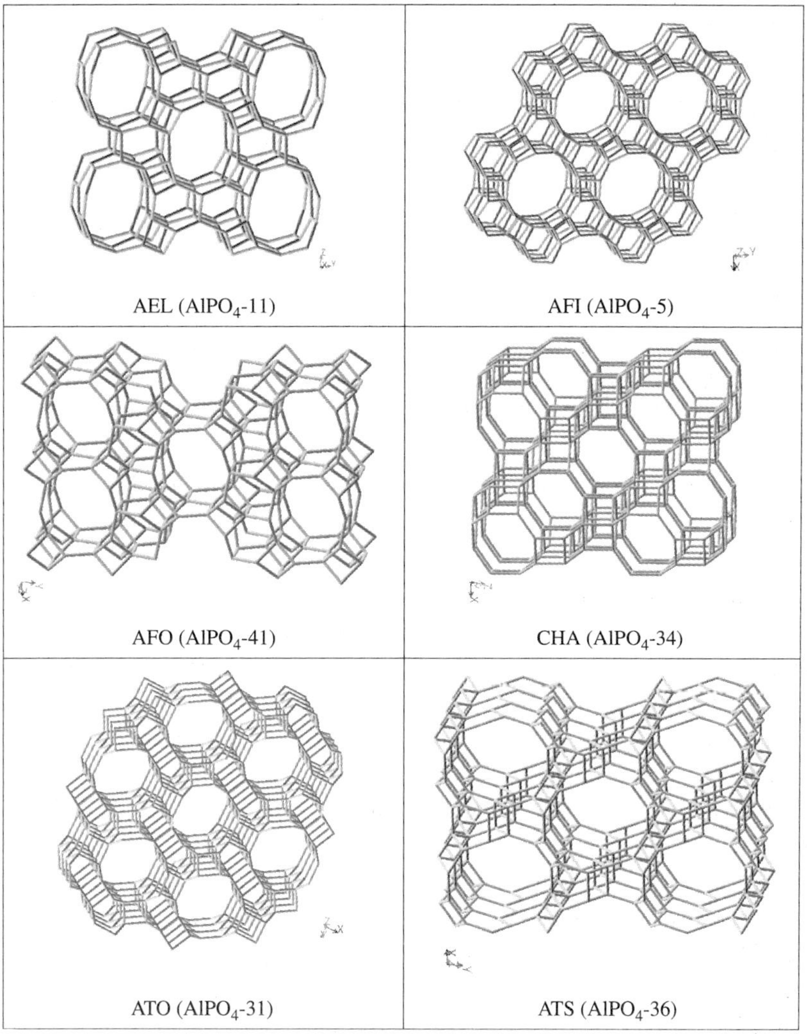

Figure 1 Structure of selected aluminophosphate (taken from the IZA database, namely www.iza-online.org).

and T–O distances that are inconsistent with the accepted crystal chemistry concept for tetrahedral coordination. It is anticipated that their successful incorporation is due to the flexibility of the microporous $AlPO_4$ framework, which is believed to be larger than that of the corresponding aluminosilicates, and due to specific interactions with the organic template coupled with the mildly acidic gel chemistry used in their synthesis. Despite the use of a large number of different characterization methods, information

+1	+2		+4	+4 +5	+3	+2 +3	+2 +3	+2 +3	+2	+2	+2	+3	+4	+5
Li	Be											B		
Na	Mg											Al	Si	P
K	Ca	Sc	Ti	V	Cr	Mn	Fe	Co	Ni	Cu	Zn	Ga	Ge	As
Rb	Sr	Y	Zr	Nb	Mo	Tc	Ru	Rh	Pd	Ag	Cd	In	Sn	Sb

Figure 2 Abbreviated Periodic Table showing elements isomorphously substituted into $AlPO_4$-based molecular sieves and their typical oxidation states (shaded = good evidence; underlined = weak evidence).

on the isomorphous substitution of TMIs into $AlPO_4$ frameworks is typically hard to obtain. Major problems are the low metal concentration and the presence of different framework and non-framework sites.

Isomophous substitution in $AlPO_4$ and SAPOs is typically carried out to design novel materials with unique acidic or redox properties. In particular, tuning the acid site strength by isomorphous substitution remains a significant issue in heterogeneous catalysis since microporous molecular sieves are extensively tested as solid acid catalysts. The acid site strength and, thus, also the catalytic activity depend on the type of metal substituted. In addition to a large number of experimental studies [11–13], which have investigated the correlation between acidity and local structure of the active site, recently computational studies have been undertaken to rationalize the effect of isomorphous substitution on the Brønsted and Lewis acidity of MeAPOs [14–16]. Periodic *ab initio* quantum mechanical calculations have been employed to study the structure and acidity of Mg, Ca, Sr, Cr, Mn, Fe, Co, Ni, and Zn isomorphously substituted into AlPOs with CHA framework structure. The generated charge is compensated by an acid proton on a neighboring oxygen. Saadoune et al. [15] showed that the local environment of the divalent cations is a distorted tetrahedron in which the Me–OH bond of the metal ion to the protonated oxygen is 15 pm longer than the other three M–O bonds. The calculations reveal that the nature of the bonding is ionic, which explains the Lewis acidity of the Me^{2+} ions. Moreover, it is concluded that the acid site strength is attributable to a complex combination of the structural and electronic features of the substituted metal ion and does not have any considerable contribution from the local environment or electronic distribution of the Me^{2+} ion in the framework. The relative strength of Brønsted-acid sites in isomorphously substituted CHA has been studied using density functional theory (DFT) with periodic boundary conditions by Elanany et al. [17]. It is deduced from calculating the adsorption energy of ammonia that the Brønsted-acid site strength follows the order MnAPO-34 > CoAPO-34 > FeAPO-34 > SAPO-34 > CrAPO-34 and TAPO-34. The authors, however, point out that prediction of the acid site strength in MeAPOs is difficult due to the ionic bonding and because many parameters contribute to the acidity at the same time. A semi-quantitative scale to correlate the Brønsted and Lewis acidity to the type of metal ion isomorphously substituted into MeAPOs is suggested by Chatterjee [16] from DFT calculations. It is observed that for bivalent metal cations Lewis acidity linearly increases with the ionic radius, whereas the Brønsted acidity is solely dependent on the nearest oxygen environment. The relative strength of Brønsted-acid sites in MeAPO-5 molecular sieves was studied by Elanany et al. using periodic DFT [18]. It is revealed that MeAPO-5 (Me = Si, Ti or Zr) has weaker acid sites compared to MeAPO-34, which is supported by experimental evidence from temperature-programmed desorption (TPD) of ammonia [19].

Apart from their activity in acid-catalyzed reactions, MeAPOs play a prominent role in selective oxidation of hydrocarbons. Redox activity in MeAPOs is achieved when the TMIs in the framework, for example, Co, Cr, Fe or Mn, can reversibly change their oxidation state [20]. The calculated redox energies of the Me^{2+}/Me^{3+} couples indicate that among the transition metals investigated, Fe is the most stable in the 3^+ oxidation state, whereas Mn is the most stable as 2^+ ion. Cr and Co have intermediate behavior and can switch more easily between the two oxidation states. These few examples show that efforts are undertaken to correlate the observed catalytic activity in acid-catalyzed and oxidation reactions with the structural and electronic properties of the catalysts. However, in the studies described here, the occurrence of metal ions trapped in the cavities as charge-balancing cation was not investigated.

In this chapter, we review the utilization of transition-metal-containing $AlPO_4$ and SAPO molecular sieves in catalysis through 2008. We believe that this field has reached a certain maturity and calls for a critical summary of the achievements, and the questions still remaining open. Previous reviews [6,7,21] mainly focus on the synthesis and the physicochemical characterization of these materials, which is certainly a prerequisite for understanding and tuning the catalytic properties. The recent review by Pastore et al. [22] covers the synthesis and characterization of AlPOs and SAPOs in particular with respect to the evolution of synthesis procedures as well as experimental and theoretical approaches to the description of Si insertion into the neutral AlPO framework. Mesoporous $AlPO_4$s, which have received some attention recently, are not covered. A recent review by Kimura [23] covers the relevant literature in this area.

2. ACID CATALYSIS

2.1 Isomerization and cracking of olefins

The skeletal isomerization of the much more abundant *n*-butenes has been recognized as an attractive alternative for increasing the availability of the more demanded isobutene, which is required as the feedstock in the synthesis of various monomers (metacrylic acid, methacrolein), polymers (e.g., poly-isobuten) and chemicals including methyl-*tertiary*-butylether (MTBE) and ethyl-*tertiary*-butylether (ETBE). Oxygenates, alcohols, and especially *tertiary* ethers have become important components in gasoline due to tightening of legislation concerning fuels. In the case of gasoline, the Clean Air Act (CAA) amendments in USA established, among others, the reduction of volatility and aromatics as well as the introduction of oxygenated compounds. Oxygenates improve the combustion of fuels and thus reduce the exhaust emissions and simultaneously boost the gasoline research octane number (RON). The most frequently used oxygenates are MTBE and ETBE, which are prepared from isobutene and methanol or ethanol, respectively, by acid catalysis. Among the higher alkenes, *n*-pentenes are the most interesting feedstock. They are used to manufacture the octane booster *tert*-amyl-methylether (TAME), which exhibits a much better biodegradability than MTBE. However, the availability of isoalkenes is a limiting factor in expanding MTBE and TAME production.

A few years ago, a definite shortage of MTBE was predicted, and hence, several processes for the production of MTBE have been developed. Industrial processes for *n*-butene isomerization based on acidic zeolites such as ferrierite and ZSM-35, which possess a pore system of interlinked 10-membered and 8-membered ring channels, have been developed and operated in the United States for a limited period of time [24]. The skeletal isomerization of butenes has received much interest not only from industry but also from academia, as it is a reaction which is difficult to catalyze. For thermodynamic reasons, low temperatures are desired, while the main non-selective side reaction is the dimerization/oligomerization of butene. The patent and the open literature suggest that the undesired side reactions can be suppressed by choosing appropriate reaction conditions [25]. The first option, which is frequently used with amorphous catalyst, for example based on alumina, is to operate well above 400 °C. At this temperature,

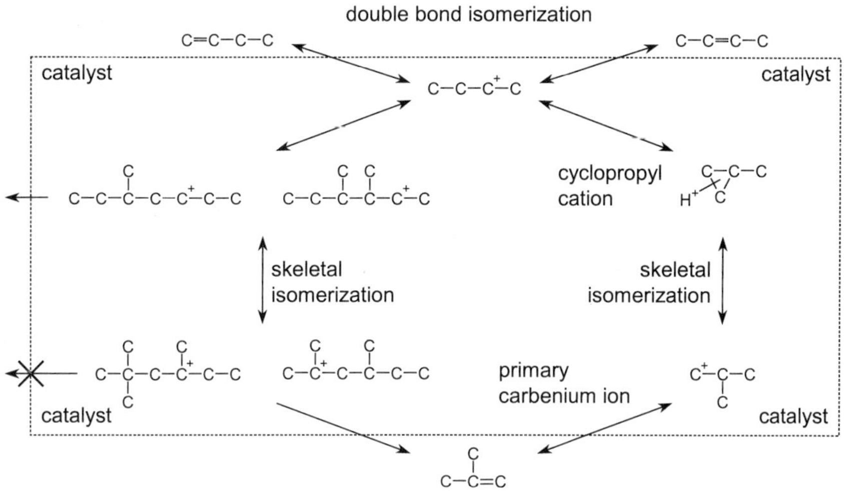

Figure 3 Proposed mechanism for the skeletal isomerization of butenes (after Ref. [25]).

dimerization and oligomerization are thermodynamically unfavorable. However, at this high temperature, the isobutene yield is reduced due to the thermodynamic equilibrium distribution of the butenes. Even more serious is the rapid catalyst deactivation by coke formation resulting in a short catalyst life. Another option is to operate at reduced butene partial pressure ($p_{\text{butene}} < 0.2$ bar) by dilution of the butane feed with inerts. Both options are cost intensive, and thus, a more appealing alternative is the application of a highly selective isomerization catalyst that strongly reduces the formation of oligomers and dimers (Fig. 3) [25].

The isomorphous substitution of TMIs (e.g., Fe and Mn) into the framework of medium-pore $AlPO_4$ such as $AlPO_4$-11 has been shown to enhance the selectivity of this catalyst for skeletal isomerization of olefins even beyond that observed with SAPOs [26]. Over SAPO-11, the yield of skeletal isomers from the conversion of hexene amounts to 42%, whereas the yield of cracked products is only 3%. Over MnAPO-11, an isomer yield of 64% is found, and the yield of cracking products amounts to 2% [27]. FAPO-11 is even more selective, giving yields of 71% and 1.5% for isomerization and cracking products, respectively. Similar trends are observed for MnAPO-31 and FAPO-31 [27]. MnAPO-31 and MnAPSO-31 have also been tested for catalytic isomerization of 1-butene, which proceeds by either double bond or skeletal isomerization. MnAPSO-31 ($MnO/P_2O_5 = 0.01$) yields the highest percentage of isobutene, whereas the parent material $AlPO_4$-31 promotes the double bond shift with only minor skeletal isomerization [28]. However, compared to SAPO-31, the increase in selectivity is limited. It is therefore likely that the better catalytic performance is related to a higher concentration of acid sites and presumably to the higher strength of these acid sites [28].

Metal-containing MeAPO-11 (Me = Mg, Mn, Co, Cr, Fe) molecular sieves have been reported to be active, selective, and stable catalysts for the selective isomerization of *n*-butene to isobutene [29–31]. The isobutene yields at 400 °C increases from 6% for $AlPO_4$-11 to around 40% for the framework-substituted analogs. At a reaction temperature of 80 °C, however, dimerization occurs to a large extent. This is also observed with MeAPO-5 catalysts, which are not very selective for isobutene formation ($S < 5\%$). It is argued that the active site is a weak acidic OH-group, possibly in the neighborhood of or in combination with a Lewis-acid site (Fig. 4). The proposed synergism between Brønsted- and Lewis-acid sites is further supported by a study of Arias et al. [30] showing that the skeletal isomerization efficiency during the transformation of 1-butene correlates well with the number of both Lewis- and Brønsted-acid sites of medium and high acid strength. It is known that Lewis sites alone cannot act as active centers for skeletal isomerization. For example, in the case of CrAPSO-11, oxidized and reduced samples have a different distribution of oxidation states of chromium that is not observed for the ion-exchanged Cr-APSO-11. An oxidative treatment of CrAPSO-11 results in an increase in the number of strong and medium acid sites as compared to those in the reduced and ion-exchanged samples [32]. The reduction process halves the number of strong Brønsted-acid sites and decreases the number of strong and medium Lewis-acid sites. The results of the physicochemical characterization are in line with the catalytic results that show that the selectivity for skeletal isomerization is

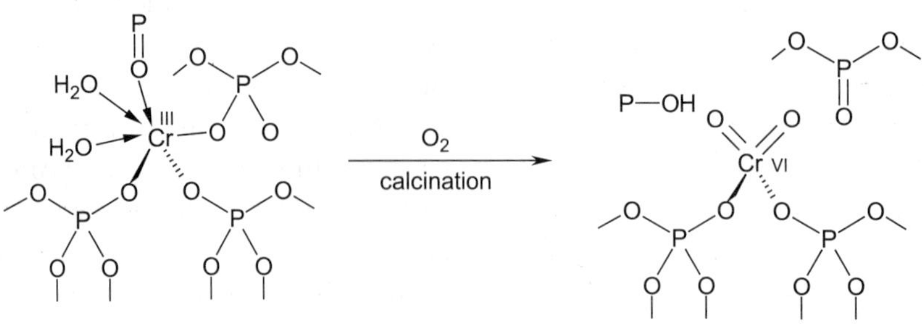

Figure 4 Model for the chromium site after reductive and oxidative treatment (after Ref. [33]). Metal ions act as Lewis acid sites interacting with Brønsted P-OH entities.

highest for the oxidized CrAPSO-11 sample. The authors conclude that their data are supporting a model for the chromium site in accordance with the one originally proposed by Chen and Sheldon [33] (Fig. 4).

The catalytic transformation of 1-butene over ZnAPSO-11, impregnated Zn/SAPO-11, and non-promoted SAPO-11 was compared by Escalante et al. [34]. The data indicate a higher skeletal isomerization selectivity for ZnAPSO-11, which is in agreement with the higher density of acid sites reported for this catalyst. The results were explained in terms of the model of Gielgens et al. [29], which suggests that the substitution of Al(III) by Zn(II) ions leads to partially unsaturated Zn(II) ions in the vicinity of structural P–OH groups. The mechanism of skeletal isomerization of 1-butene over CoAPO-11 was studied by Cejka et al. [35] using ^{13}C-labeled 1-butene. It is shown that a high selectivity to isobutene can be reached exclusively through the monomolecular reaction pathway. In this case, product formation is governed by restricted transition state selectivity, which leads to an inhibition of bulky bimolecular transition states. CoAPO-11 with channel dimensions of 0.39×0.63 nm is a prominent example for a domination of the monomolecular mechanism [35]. However, the bimolecular pathway with simultaneous formation of propene and pentenes and other by-products also contributes to the formation of isobutene. It can be assumed that the bimolecular reaction followed by isomerization mainly proceeds in topologies that are able to accommodate bulky dimers and codimers. The deactivation of solid acid catalysts for butene isomerization was reviewed by van Donk et al. [36], who discussed beneficial and harmful effects of carbonaceous deposits on microporous catalysts including $AlPO_4$.

MeAPOs are found to be selective catalysts for the isomerization of 1-pentene. Acidic –OH groups are believed to be the active and selective sites on the catalysts for the skeletal isomerization. The study by Höchtl et al. [37] on the isomerization of pentene over SAPO-5, SAPO-11, CoAPO-5, and CoAPO-11 reveals that in case of catalysts with low acidity (density and/or strength of the acid sites), the activity of the catalyst is not sufficient to reach the equilibrium distribution between linear and branched pentenes. If the acidity is too high, low selectivity for branched isomers is found provided that the structure allows the free formation of oligomers. Consequently, catalysts possessing acid sites with medium strength are the most favorable ones for skeletal isomerization. Molecular sieves with the AFI structure exhibit a low selectivity for skeletal isomerization due to rapid coking. Yang et al. [38] reported for the isomerization of 1-pentene over MeAPOs that the yield of the three pentene isomers, namely, 3-methyl-1-butene, 2-methyl-1-butene, and 2-methyl-2-butene, amounts to 60%.

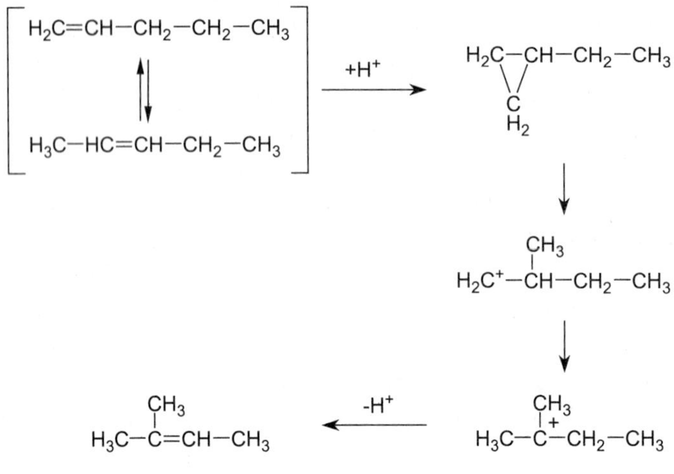

Figure 5 Monomolecular pathway for the production of 2-methyl-2-butene.

The favored formation of the latter isomer (2-methyl-2-butene) can be rationalized by a reaction mechanism through a protonated cyclopropane ring with subsequent C–C bond cleavage (Fig. 5).

The catalytic transformation of camphene over catalysts with AFI topology has been studied by Elangovan et al. and others [39–41]. The products formed are tricyclene, bornylene, and monocyclic terpenes (Fig. 6). The product distribution is influenced by the number and strength of the acid sites and the inverse of the weight hourly space velocity [$(\text{WHSV})^{-1}$]. Tricyclene is formed over catalysts possessing weak acid sites when a low $(\text{WHSV})^{-1}$ is used, whereas monocyclics such as dipentene, terpinolene, α- and γ-terpinenes, *p*-menthene, and *p*-cymene are formed over strong acid sites. With increasing reaction temperature, camphene conversion increases, which is accompanied by a decrease in tricyclene selectivity. Eswaramoorthy et al. [42] studied the transformation of 3-carene over CrAPO-5 and CrAPO-11 (Fig. 7). The (bifunctional) catalytic transformation of 3-carene involves its initial isomerization to *p*- and *m*-menthadienes over acid sites that are subsequently dehydrogenated to *p*- and *m*-cymenes, respectively, over the metal sites (e.g., Cr). The activity of the Cr-containing catalyst is further increased when Cr is reduced in flowing hydrogen before the reaction. The higher activity of the reduced catalyst as compared to the oxidized one is due to the reduction of extra-framework Cr^{6+} to Cr^{3+}. The presence of higher valent chromium ions associated with strong acid sites in the oxidized samples is responsible for the lower *p*/*m* ratio since a part of *p*-cymene is cracked to toluene over the strong acid sites.

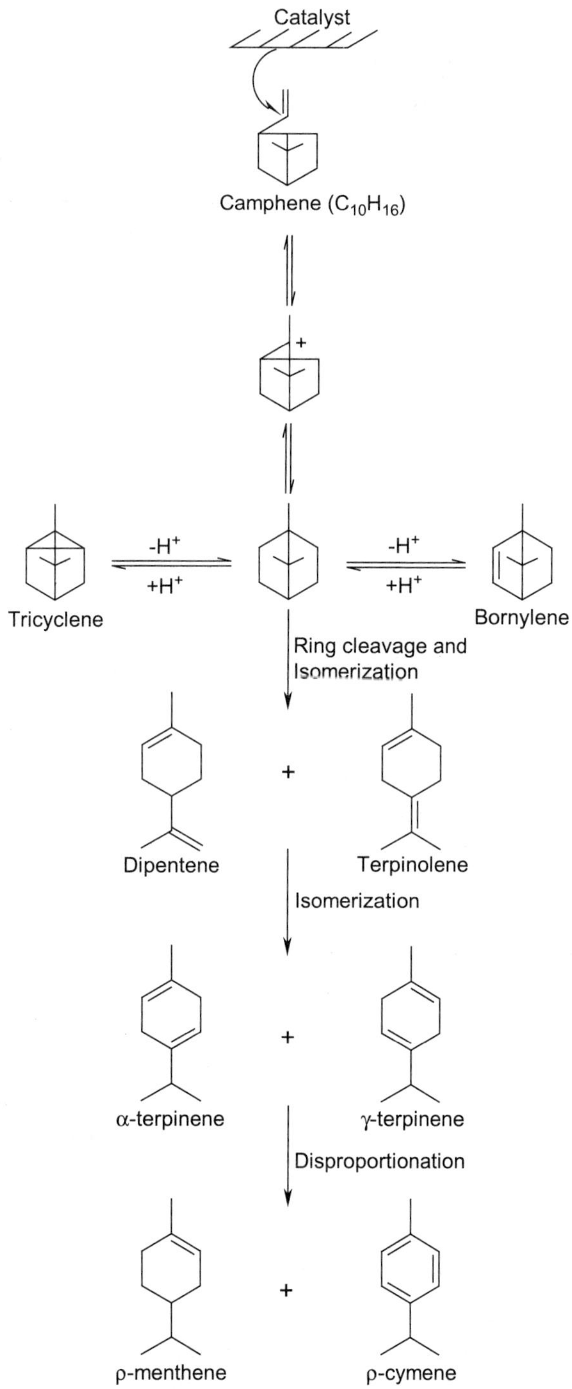

Figure 6 Reaction scheme of camphene conversion over acid $AlPO_4$.

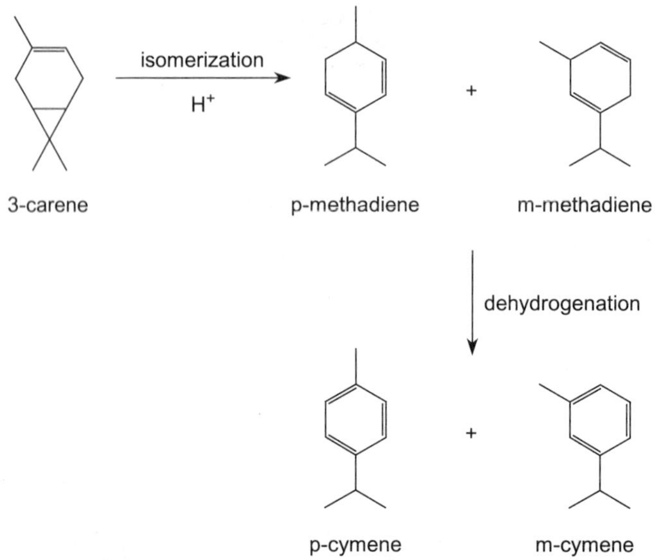

Figure 7 Transformation of 3-carene to monoterpenes.

Hydration of ethene in the gas phase over nickel- and zinc-containing $AlPO_4$ with AEL and AFI structure yields isobutanol, ethanol, and *n*-butanol [43,44]. With increasing partial pressure of water, the selectivity to ethanol and *n*-butanol increases at the expense of the isobutanol selectivity. Over Brønsted-acid sites (most likely generated by the introduction of nickel or zinc into the $AlPO_4$) ethanol is formed, which reacts further on Lewis-acid sites to yield isobutanol. With increasing partial pressure of water, these Lewis-acid sites are blocked (by water), which prevents the conversion of ethanol to isobutanol.

Light olefins such as ethene and propene are important raw materials for the production of polymers and alkylbenzenes (e.g., ethylbenzene and cumene). The modified fluid catalytic cracking (FCC) process is a major source of ethene and propene. Besides the necessity to obtain high yields of C_2 and C_3 olefins in the FCC units, it is required to upgrade low-value refinery petrochemical streams containing C_4 to C_6 olefins into ethene and propene in a separate unit. Lurgi has developed a fixed-bed catalytic process (Propylur), which is capable of converting C_4 to C_5 olefins from the steam cracker to ethene and propene with high yield [45]. In this context, the cracking of pentenes to C_2–C_4 light olefins over CoAPO-11 and SAPO-11 was studied by Bortnovsky et al. [46] in comparison with several zeolites including ZSM-5 and ZSM-11. Despite the high concentration of Brønsted-acid sites in SAPO-11, both SAPO-11 and CoAPO-11 exhibit low

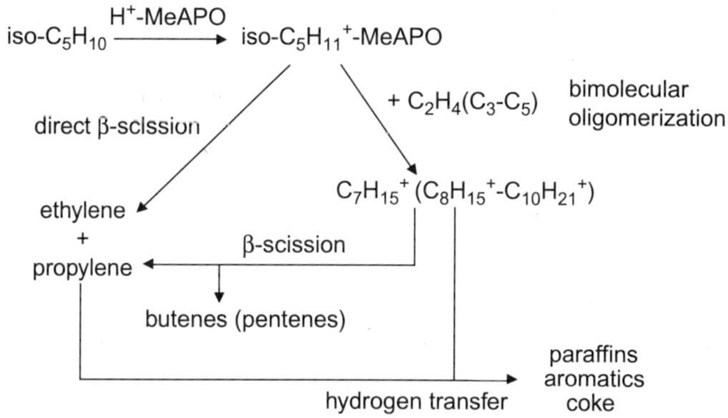

Figure 8 Scheme of isopentene cracking over acid catalysts (after Ref. [46]).

conversion in the cracking of 2-methyl-2-butene, which was used as a model feed. The weight ratio of propene and ethene was close to that at thermodynamic equilibrium (2.95) for SAPO-11 (2.65) and lower than thermodynamic equilibrium for CoAPO-11 (1.83). The propene/butenes ratio amounted to 2.39 for CoAPO-11, which implies that cracking of branched pentenes proceeds mainly through direct β-scission in the 10-membered ring pores of the AEL topology and not through oligomeric carbenium ion intermediates (Fig. 8). This conclusion is in line with the reported skeletal isomerization of butanes to isobutene, where a monomolecular mechanism is reported to occur on CoAPO-11, which results in a isobutene selectivity close to 100% at high conversion ($X = 46\%$) [35].

2.2 Conversion of alcohols

Methane transformation to more useful higher hydrocarbons is one of the most important topics of natural gas utilization. Although methane activation and its direct conversion to valuable compounds are still drawing increasing attention, the successful processes currently still use an indirect pathway (Fig. 9). Methane is transferred to synthesis gas (syngas) by partial oxidation or reforming, followed by conversion to higher carbons using the Fischer–Tropsch technology. In an alternative process, syngas is first converted to methanol employing a $Cu/ZnO/Al_2O_3$ catalyst and then to hydrocarbons using the methanol-to-gasoline (MTG), methanol-to-olefins (MTO) or methanol-to-propylene (MTP) pathway. Olah et al. [47] reported a three-step process for the conversion of methane to hydrocarbons through methyl halides. Here, methanol is directly converted to chloromethane by reaction with chlorine or hydrogen chloride and oxygen. The produced chloromethane is then converted to hydrocarbons over microporous (zeolite)

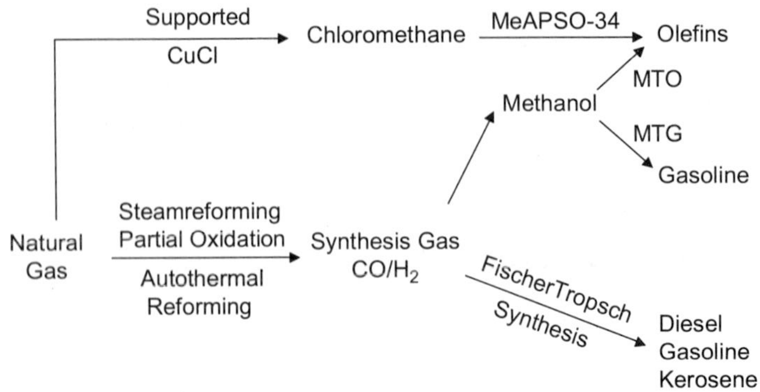

Figure 9 Natural gas conversion to higher hydrocarbons.

catalysts. Although ZSM-5 is the catalyst of choice for MTG and MTP, substituted $AlPO_4$ have been considered for MTO and the conversion of chloromethane.

The demand for olefinic feedstocks has increased rapidly in the past few years due to increased needs for synthetic fibers, plastics, and petrochemicals. Projected growth rates for light olefins are expected to remain above worldwide gross domestic product (GDP) growth rates, and there is an increasing need to apply production technologies that favor higher propene/ethene ratios. Growing global demand for crude oil will have a significant impact on the availability and pricing of traditional feedstocks for light olefin production. Consequently, other raw materials including natural gas, coal, and biomass are explored as feedstocks for the production of petrochemicals. Although a direct conversion of methane is not yet feasible, technology for the production of methanol from synthesis gas, which can be produced from any hydrocarbon feedstock including natural gas and biomass, is available and practiced today. The combination of methanol production with the MTG and the MTO processes provides an alternative route to gasoline and ethene and propene, respectively, which could be economically attractive under certain circumstances. The progress in MTG and MTO technology, in particular with respect to catalysts and the reaction mechanism, has been reviewed by Stöcker [48] some years ago. The molecular sieve–based technology for the production of lower olefins is often termed methanol-to-olefins (MTO), which can be considered as an intermediate step of the Mobil MTG process. The product distribution of methanol conversion over ZSM-5 depends, among others, on the contact time, as shown in Fig. 10. It is evident that C_2 to C_5 olefins are important intermediate products. In the reaction pathway, methanol is first dehydrated to dimethylether (DME).

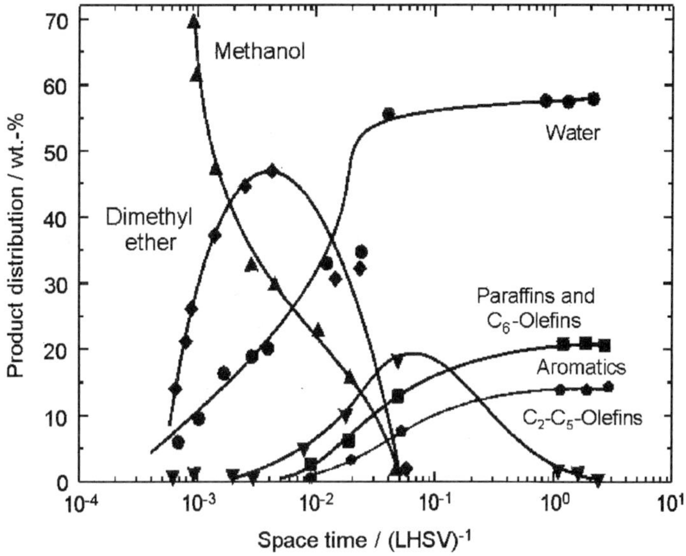

Figure 10 Product scheme for methanol conversion to hydrocarbons (adapted from Ref. [49]).

The equilibrium mixture formed is then converted to light olefins. In the last step, the light olefins react to form paraffins, aromatics, naphthenes, and higher olefins (Fig. 10). By judicious choice of the process conditions, namely, temperature, space velocity, total pressure, methanol partial pressure and catalyst used, the product distribution can be maximized toward the production of light olefins.

The detailed mechanism involved in the MTO process, however, is still a matter of debate. Two major issues are still discussed: (i) Are surface methoxy groups, which can be formed from methanol at bridging OH groups on acidic molecular sieves, active intermediates in the dehydration of methanol to DME? (ii) What is the mechanism for the first C–C bond formation? A large number of possible mechanisms have been described in the literature; some of them are based on surface-bound alkoxy species, oxonium ylides, carbenes, carbocations, or free radicals as key intermediates [48]. Several authors have suggested a hydrocarbon-pool mechanism to explain the formation of light olefins [50–56]. According to this proposal, large carbonaceous species are formed during the induction period of the reaction, to which can be further added reactants and/or split-off products in the steady state of the reaction (Fig. 11). The hydrocarbon-pool species were proposed to possess many characteristics of ordinary coke that may be described as $(CH_x)_n$ with $0 < x < 2$. Recent studies employing isotope

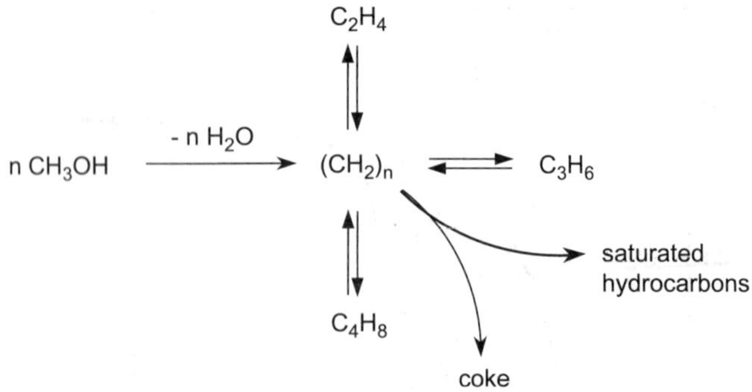

Figure 11 Schematic representation of the hydrocarbon pool in the methanol-to-hydrocarbon reaction.

labeling or *in situ* MAS NMR spectroscopy have verified that the MTO process is dominated by this hydrocarbon-pool route: Methanol is added to reactive organic species such as polymethylbenzenes [57], large olefins, cyclic carbenium ions [58], and probably methylbenzenium cations [59], while light olefins are formed by the elimination of alkyl chains from these organic intermediates. These mechanistic investigations have been reviewed in 2005 by Kolboe et al. [60], Haw and Marcus [61], and Hunger et al. [62].

The existence and possible role of surface-bound species, for example, surface methoxy groups, in the formation of methanol has received significant support from both experimental and theoretical studies as reviewed in references [60–62]. For example, surface methoxy groups, which exhibit a chemical shift of 56 ppm, have successfully been detected *in situ* or *ex situ* by ^{13}C MAS NMR spectroscopy during methanol conversion on H-SAPO-34 [63–67]. MAS NMR experiments also support the indirect route for DME formation, which involves the adsorption of methanol molecules on bridging OH groups to form methoxy groups ($ZOCH_3$, Z stands for the framework of the molecular sieve):

$$CH_3OH + ZOH \rightarrow (CH_3^{\delta+} \cdots OH_2^{\delta+} \cdots ZO^{\delta-})^{\#} \rightarrow ZOCH_3 + H_2O \tag{1}$$

Subsequent reaction with another methanol molecule results in the formation of DME:

$$\begin{aligned} ZOCH_3 + CH_3OH &\rightarrow (CH_3^{\delta+} \cdots CH_3OH^{\delta+} \cdots ZO^{\delta-})^{\#} \\ &\rightarrow ZOH + CH_3OCH_3 \end{aligned} \tag{2}$$

However, from the experiments conducted so far, the direct route, where two methanol molecules react with each other on one Brønsted-acid site cannot be ruled out. This pathway involves the simultaneous adsorption of two methanol molecules followed by reaction to DME and water in a single step:

$$\begin{aligned} 2\,CH_3OH + ZOH &\rightarrow (CH_3OH^{\delta+} \cdots CH_3^{\delta+} \cdots OH_2^{\delta+} \cdots ZO^{\delta-})^{\#} \\ &\rightarrow ZOH + CH_3OCH_3 + H_2O \end{aligned} \tag{3}$$

Several studies on the kinetics [68] and mechanism [69,70] of the methanol transformation have been published. Fig. 12 exhibits the proposed mechanism of methanol conversion to olefins [71].

The MTO technology has been extensively demonstrated in a demo plant owned by Hydro in Norway. The process (Fig. 13) converts methanol to ethene and propene at about 75–80% carbon selectivity [72]. The construction of the first commercial world-scale MTO unit is being underway in Lekki, Lagos State, Nigeria. The complex is based around a 2.5×10^6 tons year^{-1} methanol plant, which will provide the feedstock to a MTO unit that will subsequently produce 400,000 tons year^{-1} polypropylene (PP) and 400,000 tons year^{-1} high-density polyethylene (HDPE). The complex was scheduled to become operational by mid-2007 [73].

H-SAPO-34 is acknowledged to be a powerful catalyst with attractive performance in converting methanol to light alkenes, namely, ethene, propene, and butenes (MTO process) [74]. The structure of SAPO-34 (Fig. 14) together with the small size of the target molecules is the key to the MTO process. The small pore size (ca. 0.4 nm) restricts the diffusion of heavy and branched hydrocarbons resulting in a high selectivity to the desired small olefins. Zeolite ZSM-5, which is used in the MTG and MTP processes, produces a lower light olefin yield due to the larger pore openings (approximately 0.55 nm) (Fig. 14). Another key feature of the SAPO-34 molecular sieve is its optimized acidity relative to the aluminosilicate-based zeolitic materials. The somewhat lower acid strength of SAPO-34 as compared to ZSM-5 leads to much lower paraffinic by-product formation due to hydride transfer reactions. The MTO process utilizes a fluidized bed reactor (Fig. 13). Constant catalyst activity and product composition is maintained by continuous regeneration of a portion of the used catalyst with air. The propene/ethene ratio can be varied between 0.5 and 1.5, depending on the reaction conditions [72].

SAPO-18 (AEI) is a CHA-related structure and comparable to SAPO-34 in terms of activity and selectivity [75]. The conversion of methanol to

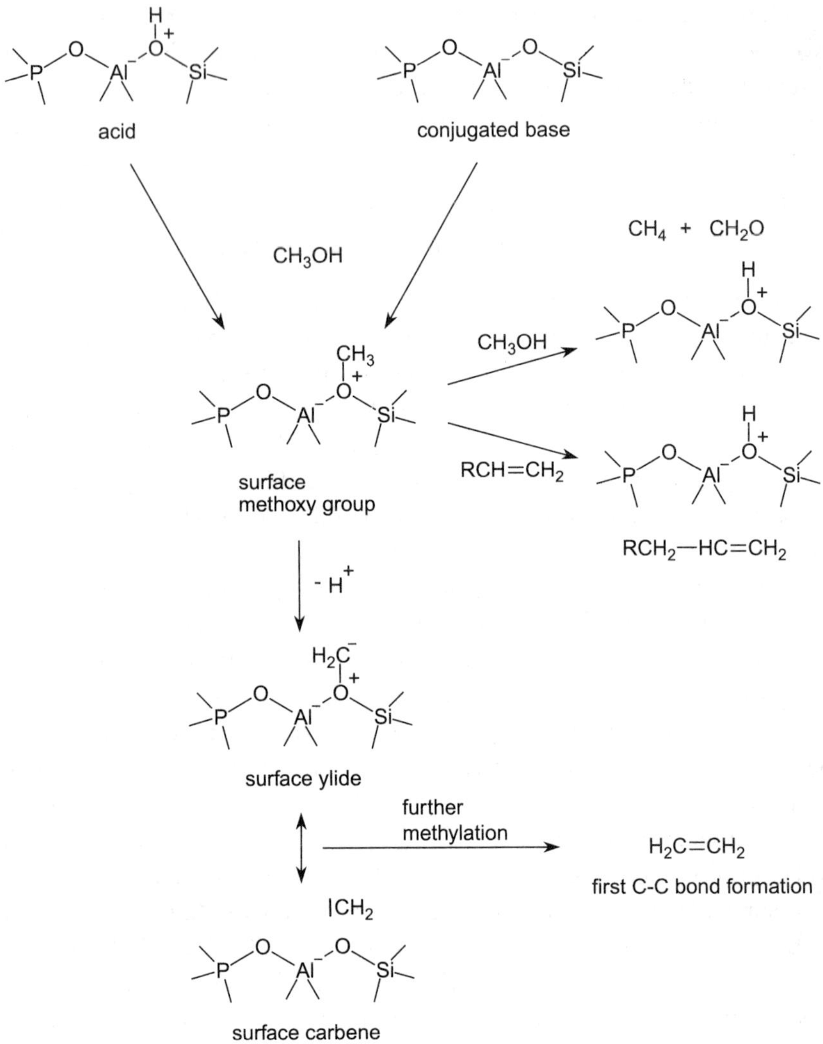

Figure 12 Proposed mechanism of the synthesis of olefins from methanol through the MTO reaction.

hydrocarbons was reported to reach 100% and the selectivity to ethene and propene in the hydrocarbon product may reach 80%. The catalytic activity of SAPO-34, which has roughly twice as many Brønsted-acid sites as good quality SAPO-18, drops off more rapidly than the later [75]. The kinetic behavior of SAPO-18 in the transformation of methanol to olefins was studied by Gayubo et al. [76–78]. The results are in agreement with the hydrocarbon-pool mechanism and show that the process is conditioned by

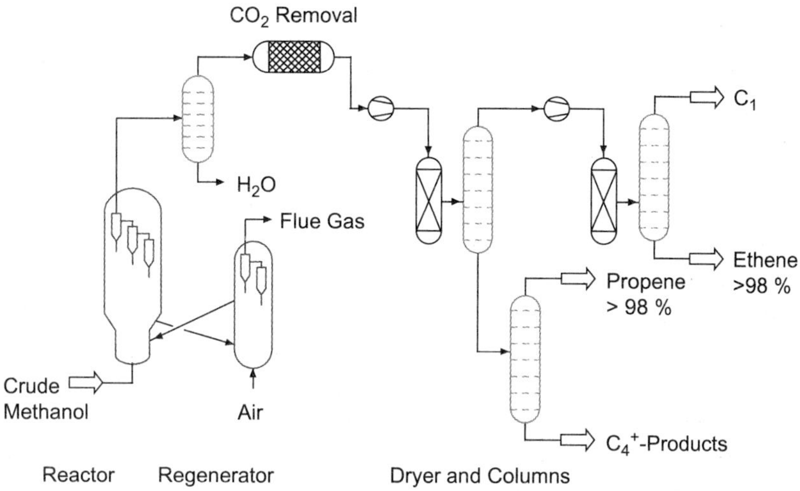

Figure 13 Flow chart of the UOP/Hydro MTO process.

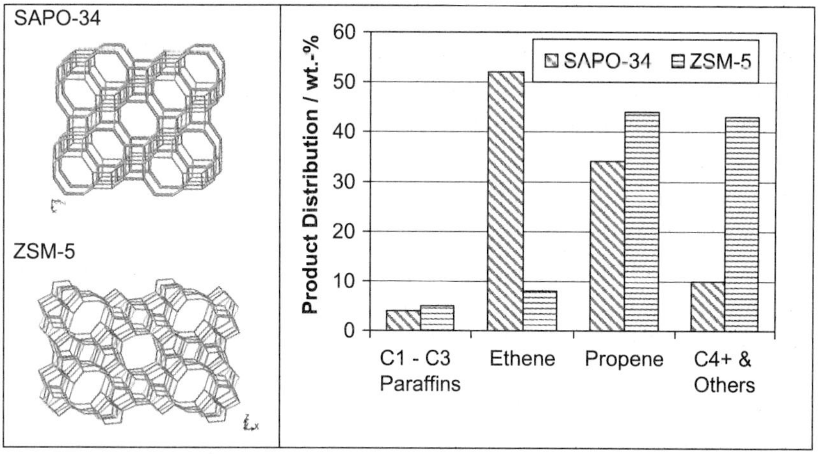

Figure 14 Comparison of the performance of SAPO-34 (CHA) and ZSM-5 (MFI) at maximum ethylene mode (after Ref. [72]).

an initial step in which the reactive intermediates for the production of olefins are generated. The influence of topology, total acidity, and acid site strength on the transformation of methanol to light olefins has been compared employing SAPO-11, SAPO-18, and SAPO-34 as catalysts [79]. In agreement with earlier studies [75], it is found that SAPO-18 has a lower deactivation rate than SAPO-34 due to a slightly lower acid strength and a lower density of strong acid sites on the surface, whereas SAPO-11 shows lower olefin selectivity and lower conversion [79]. It is further pointed out that the synthesis of SAPO-18 is more economical as compared to SAPO-34 due to the lower cost (and lower amount) of the organic template *N*,*N*-diisopropyl ethyl amine used in the synthesis of SAPO-18 in comparison with tetraethyl ammonium hydroxide (TEAOH) employed in the synthesis of SAPO-34.

For further improvements in the MTO process, transition metal substituted $AlPO_4$ with the CHA and AEI framework topology (MeAPSO-34 and MeAPO-18) have been tested by several groups [80–85]. Inui et al. [80] and Thomas et al. [81] reported independently that nickel-containing SAPO-34 (NiAPSO-34) is one of the best catalysts for the acid-catalyzed MTO conversion yielding close to 90% ethene at almost 100% conversion at a reaction temperature of 450 °C. The catalytic performance was maintained at this temperature for 13 h; during this time, no significant change in methanol conversion or ethene selectivity was observed [80]. Modification of the acidity by simultaneous introduction of silicon and a TMI is a convenient way of controlling or altering the performance of the catalyst in the MTO reaction. Among the various metal-modified small-pore molecular sieves, NiAPSO-34 has received the most attention. Several characterization results including ESR spectroscopy suggested that Ni ions are located in tetrahedral framework sites [81,86]. It was established by X-ray absorption spectroscopy (EXAFS and XANES) that Ni was exclusively located in tetrahedrally coordinated framework sites [80]. Loosely bound extra-framework protons were thought to be key components of these catalysts. In contrast, a clear effect of reaction temperature on the product distribution is reported by Thomas et al. [81] for a NiAPSO-34-type catalyst after 2 h on stream: a temperature increase from 250 to 450 °C results in an increase of methanol conversion from 90.5% to 95.5%, whereas the ethene selectivity drops from 94.5% to 60.5%. The conversion of methanol to olefins over MeAPSO-34 (Me = Co, Ni) has been studied by van Niekerk et al. [87]. The catalytic performance of the catalyst was found to be closely related to the number of strong acid sites determined by TPD of ammonia. The almost complete absence of C_5 and larger olefins was ascribed to the cage size of the CHA topology, which imposes a restriction on the formation of those compounds. Further treatments such as steaming, silanation, and poisoning of the strong acid sites all reduced the number of strong acid sites and, hence, the catalytic activity.

By controlling the uniform distribution of moisture in the gel, Inui and Kang synthesized NiAPSO-34 crystals with Si/Ni ratios of 40 and 100 and a sharp crystal size distribution around 0.85 μm [88,89]. With increasing nickel concentration, the acid strength measured by ammonia TPD decreases not only for the strong acid sites but also for the weak acid sites, which in turn results in increased ethene selectivity in the order SAPO-34 < NiAPSO-34 (100) < NiAPSO-34 (40) at high methanol conversion [90]. To improve ethene selectivity and mitigate coke formation, the acid sites located on the external surface of NiAPSO-34 crystals were selectively neutralized mechanochemically with basic oxides, for example, BaO supported on microspherical non-porous silica [91]. For methanol conversion, the modified NiAPSO-34 catalyst exhibited a higher selectivity to ethene and a longer catalyst lifetime than the parent material. This was rationalized with a decrease of coke formation on acid sites located on the external surface of the crystals. Impregnation of SAPO-34 with metal ions including K^+, Cs^+, Pt^{2+}, Ag^+, and Ce^{3+} was observed to reduce methane formation at higher temperatures thereby increasing the lower olefin selectivity [92]. It is suggested by the authors that DME is adsorbed at the extra-framework metal cations and that the thermal decomposition of DME to methane and carbon dioxide, which is thermodynamically possible at high temperature (e.g., 500 °C), is suppressed. It should, however, be noted that the unmodified catalyst SAPO-34 still exhibits a higher selectivity of lower olefins at 400 °C as compared to the modified samples [92].

To obtain NiAPSO-34 crystals containing larger quantities of nickel, nickel formate was used as the nickel source [93]. However, not only the amount of nickel incorporated in the framework but also the concentration of extra-framework nickel was enhanced. As a consequence, the selectivity for methane formation is increased. Sulfidation of extra-framework nickel by H_2S suppresses methane formation and restores high ethene selectivity [93]. Steam addition to the methanol feed also results in a higher ethylene selectivity and higher catalytic activity for methanol conversion with time-on-stream [94]. The presence of extra-framework Ni(0), as evident from magnetization studies, results in higher selectivity to (unwanted) methane [95,96]. Loosely bound extra-framework protons were thought to be important for the highly selective conversion of methanol to ethene.

Few reports are available on methanol conversion over CoAPSO-44 and MnAPSO-34 [82,85,87,95]. Although these catalysts exhibit activities and C_2–C_4 selectivities comparable to SAPO-34, MnAPSO-34 showed the most resistance towards deactivation [95]. Other small-pore $AlPO_4$ such as SAPO-17 (ERI), SAPO-18 (AEI), and SAPO-35 (LEV), as well as SAPO-11 doped with Ni or Cr have also been tested for methanol conversion in comparison with SAPO-34-based catalysts [97,98]. NiAPSO-34 proved to have the best performance with respect to both ethene selectivity and catalyst lifetime. MeAPSO-*n* samples appeared to be generally better catalysts than Me-SAPO-*n* materials. Incorporation of Cr into SAPO-34 also increases ethene selectivity and catalyst lifetime, most probably due to a reduction of the number of strong and moderate acid sites. In contrast, NiAPO-18, which has essentially the same framework structure, is found be one of the poorest catalysts for the MTO reaction [83]. ZnAPO-18, CoAPO-18, and MgAPO-18 are all very active at temperatures above 350 °C [84]. The lower ethene selectivity and higher propene and butene selectivity observed with Mg-, Co-, and ZnAPO-18 catalysts are also obtained for catalysts containing the same transition metals and having CHA structure [84]. The authors, therefore, suggest that the nature of the substituting element is more important than the difference in pore geometry between molecular sieves with AEI and CHA structure. Several authors tried to establish a correlation between Brønsted acidity and methanol conversion activity [99,100]. Besides selecting a molecular sieve with a different topology, the most straightforward way to change the strength and the concentration of acid sites is the isomorphous substitution of elements having different atomic electronegativities. Moreover, the concentration and the distribution of the TMIs in the framework of the molecular sieve may be changed. Hocevar et al. [82,101,102] investigated MeAPSO-34 and MeAPSO-44 molecular sieves with Me = Co, Mn, or Cr. They showed that introduction of TMIs influences the acidic strength of MeAPSO-34 and MeAPSO-44 and consequently the selectivity toward ethene formation, which follows the stability order of transition metal–ligand complexes suggested by Irving and Williams. This order is the consequence of the extra-stability of complexes due to contributions of the crystal field stabilization energy (CFSE) of atomic d-orbitals to the ligation energy. If the ligation energy of ethene (π-complex) with a transition metal is higher, it will be less prone to subsequent oligomerization, and consequently, such a catalyst is more selective for ethene. Assuming that this model is valid, Ni-containing samples should be the most selective ones for ethylene, since Ni-complexes have the highest CFSE. This was at least observed for NiAPSO-34 [74,81].

Kaiser tested MeAPOs and MeAPSOs to form light olefins from a 70:30 wt.% water–methanol mixture [103]. MgAPO-34, CoAPSO-34, and MnAPO-34 can completely convert methanol, but methane and carbon dioxide selectivities are high. For CoAPO-34, the following product distribution at a reaction temperature of 425 °C has been reported: ethene 45.3 mol.%, ethane 0.8%, propene 27.1%, butenes 8.3%, methane 6.2%, and CO_2 10.0%. Under comparable conditions, over a SAPO-34 catalyst the ethene selectivity amounts to 51.4%, while methane and CO_2 selectivities of 4.3% and 5.5%, respectively, have been observed [104].

Some topologies having larger channel diameters, namely, AFI and ATS, have been tested by Lischke et al. [105] and Akolekar [106]. The best catalytic results with respect to the formation of lower olefins are achieved with NiAPO-5, which contains a sufficiently high density of Brønsted sites of moderate strength and a comparatively small portion of strong Lewis sites. However, for molecular sieves possessing 12-membered ring pores such as NiAPO-5 and ZnAPO-36, where aromatics formation takes place to a considerable extent, ethene and propene selectivity is low as compared to materials with CHA topology [105,106]. Methanol conversion has been proposed as a test reaction for the elucidation of framework cobalt in CoAPO-5 and CoAPSO-5 after pre-treatment in air or hydrogen at different temperatures [107].

The conversion of ethanol was studied over MAPO-5, MnAPO-11, MAPO-36, and MAPO-46. At conversions around 50%, the selectivity for aromatics is relatively low ($<4.5\%$) and predominantely C_2 aliphatics are formed [108–110]. Similar results were obtained for $AlPO_4$ with ATS topology containing Zn, Co, and Mn [111]. The conversion of 2-methyl-3-butyn-2-ol and isopropanol were also performed as test reactions over CrAPO-5 to assess the acidic and basic sites of the catalysts [112]. However, the conversion of ethanol (prepared from biomass) to ETBE is a more interesting reaction to work on since up to 15 vol.% ETBE can be added to gasoline. Until now, this reaction has only been explored using acid zeolites [113]. Recently, the catalytic dehydration of ethanol to ethene has been reported over SAPO-34 and NiAPSO-34 as catalysts [114]. At a reaction temperature of 350 °C, the ethene yield amounts to 86% and 92.3% for SAPO-34 and NiAPSO-34, respectively. For ZSM-5, an ethene yield of 93.7% was observed at a reaction temperature of 300 °C. However, the $AlPO_4$-based catalysts exhibit a higher stability as compared to the zeolite.

A cyclic two-step process for the conversion of methane to gasoline with chloromethane as intermediate has been proposed by Taylor et al. in 1988 [115]. In the first stage, chloromethane is produced through oxydehydrochlorination of methane over a supported CuCl catalyst. Over a shape-selective zeolite, for example, ZSM-5 or ZSM-11, chloromethane is transformed to hydrocarbons in the gasoline range and hydrogen chloride, which is recycled to the first stage [116]. The use of SAPOs and MeAPOs in the conversion of chloromethane or bromomethane [117] has been reported recently. The conversion amounts to 20% and 51% over SAPO-5 and SAPO-34, respectively, which is low as compared to ZSM-5

($X_{chloromethane}$ = 99.9%) and ZSM-34 ($X_{chloromethane}$ = 81%) [118]. Chloromethane is mainly transferred to light alkenes over SAPO-5 and SAPO-34. Although over SAPO-34, ethene (S = 25%), propene (S = 31%), and butenes (S = 15%) are predominantly formed, in particular the ethene selectivity is somewhat lower over SAPO-5 (S = 9%). In an accompanying study from the same groups, chloromethane conversion over SAPO-34 and MnAPSO-34 is compared [119]. A decrease in conversion from 96.5 to 81.1 is observed within 65 min for SAPO-34, whereas the conversion declines from 99.6 to 84.2 over MnAPSO-34. Moreover, the light alkene selectivity is slightly higher on the latter catalyst, showing that MnAPSO-34 exhibits a better performance compared to SAPO-34. The development of new MeAPSO-34 (Me = Mg, Mn, Co, Fe) catalysts has been reported recently [120,121]. The novel catalysts exhibit a more stable catalytic activity and improved light olefins production. Ethene production is favored over MgAPSO-34 and CoAPSO-34, whereas propene generation is more dominant over Mn- or Fe-containing catalysts.

The pinacol rearrangement is an acid-catalyzed dehydration of vicinal diols yielding carbonyl compounds. Pinacol (2,3-dimethyl-2,3-butanediol) rearrangement mainly yielding pinacolone (3,3-dimethyl-2-butanone) and 2,3-dimethyl-1,4-butadiene (Fig. 15) was found to proceed at relatively mild temperatures over MeAPOs [122]. However, a clear correlation between metal content, number, and/or strength of acid sites and the observed conversion and selectivities was not found. Hsien et al. [123] studied pinacol-type rearrangement reactions with 10 different vicinal diols as reactants over iron-substituted molecular sieves. FeAPO-5 exhibits higher catalytic activity as compared to FeZSM-5 and FeMCM-41 in pinacol

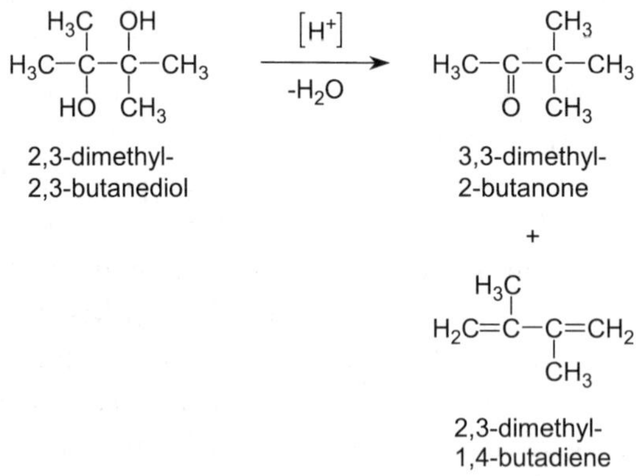

Figure 15 Pinacol-pinacolone rearrangement.

rearrangement of 2,3-dimethyl-2,3-butanediol mainly due to the facile removal of coordinated water molecules from Fe^{3+} centers of $AlPO_4$-5.

The synthesis of 2-adamantane derivatives from 1-adamantol (1-AdOH) was explored over SAPO-5 and CoAPO-5, the pore size of which is close to the kinetic diameter of the reactant and product molecules [124]. In the presence of chloroacetic acid as solvent, 1-AdOH forms 1-adamantyl acetate (1-AdOAc) and equilibrium is established between 1-AdOH and formed 1-AdOAc. This results in the reduction of polymerization of 1-AdOH to some extent. The 1,2-hydride shift of 1-adamantyl cation leads to the formation of the 2-adamantyl cation, resulting in the products of 2-adamantane derivatives, namely, 2-adamantanol (2-AdOH), 2-adamantanone (2-AdO), and 2-adamantyl acetate (2-AdOAc). Adamantane (AdH) is formed in the product ascribed to the hydride shift of 1-adamantyl and 1-hydroxy-adamantane. 2-AdO is produced through disproportionation of formed 2-AdOH over solid acid catalysts. The addition of AdH along with the reactant restricts the formation of AdH and some of the polymerized products from 1-AdOH and, in turn, increases the selective formation of 2-derivatives.

2.3 Conversion of paraffins

The catalytic properties of MeAPO-*n* molecular sieves were assessed with *n*-butane cracking as a test reaction for Brønsted-acid catalysts [125]. The materials exhibit activities that are both structure and metal dependent. Metal incorporation into the AFI framework imparts low activity for *n*-butane cracking, whereas the incorporation of magnesium into the CHA and ATS topology results in moderate to high activity [4]. The results of *n*-butane and *n*-hexane cracking on MgAPO-5 and CoAPO-5 suggest the presence of isolated strong acid sites in these materials [126]. For *n*-hexane, primary cracking, dehydrogenation followed by cracking, as well as secondary cracking were observed. The catalytic activity was found to be three times higher for MgAPO-5 and CoAPO-5 as compared to SAPO-5. The higher activity of the TMI-containing catalysts is ascribed to a higher contribution of the bimolecular cracking mechanism and not to higher acid site strength. The cracking of isobutane has been investigated over MgDAF-1, MAPO-5, and MAPO-36 [127]. The catalytic activity increases in the order MAPO-5 $<$ MgDAF-1 $<$ MAPO-36. In contrast to MAPO-36, the product selectivities observed with MAPO-5 vary little with temperature and mainly butenes and propene are formed ($Y = 50$–70%). The *n*-butene yield, however, never exceeds 11%; the observed product selectivities are ascribed to the presence of mild acid sites.

The conversion of a series of hydrocarbons including cyclohexane, cycloheptane, cyclooctane, *n*-octane, *n*-hexane, 3-methylpentane, and isooctane over MeAPOs has been reported [108–110,128,129]. The observed order of catalytic activity and aromatics selectivity follows the same trend as the number of strong acid sites determined for these catalysts. Moreover, the catalytic activity decreases in the order MAPO-*n* > CoAPO-*n* > ZnAPO-*n* > MnAPO-*n*. This order is also observed for other acid-catalyzed reactions. A detailed description of the observed activities and selectivities for aromatics formation is beyond the scope of this review.

The direct transformation of *n*-butane to isobutane is an important reaction from an industrial and an academic point of view [130]. However, only very few studies in the open literature are devoted to the use of SAPOs for the isomerization of *n*-butane [131,132]. Although pure SAPO-5 is inactive in this reaction, probably due to the low density of weak acid sites in this material, CrAPSO-5 and FeAPSO-5 exhibit some activity, but the conversion does not exceed 1%.

2.4 Conversion of aromatics

Alkylation of aromatics is gaining importance because of the wide application of its products such as ethylbenzene, cumene, and *p*-diethylbenzene as intermediates for the production of petrochemicals, polymers, and fine chemicals. Owing to stricter environmental regulations, the trend is to replace existing homogeneous catalysts with heterogeneous alternatives that cause less waste disposal problems. Hence, alkylation processes are developed that allow replacing the conventional mineral acid catalyst (namely, HF and H_2SO_4) with environmentally benign solid acids based on zeolites or $AlPO_4$ molecular sieves.

Aniline methylation has been studied over a series of TMI-containing AlPOs with AEL and AFI topology [133,134]. Alkylation of aniline with methanol ($n_{methanol}/n_{aniline} = 2$) gives mainly *N*-methylaniline and to a smaller extent *N*,*N*-dimethylaniline over $AlPO_4$-11 and $AlPO_4$-5 with yields of 9% and 12%, respectively. Over CoAPO-11 and ZAPO-11, the yields increase to 35% and 40%, respectively, again with predominant formation of *N*-methylaniline (yield $Y = 24\%$ and 26%, respectively). Over CoAPO-5 and ZAPO-5, the conversion increases an additional 10% compared to CoAPO-11 and ZnAPO-11 and the formation of *N*-methyltoluidine is observed. A similar trend is found for VAPO-5 and VAPO-11. The number of acid sites correlates nicely with the observed increase in aniline conversion.

The isomerization of *m*- or *o*-xylene is often used to characterize the acid sites in TMI-containing AlPOs and SAPOs [108–110,135–137]. Over large-pore molecular sieves, besides the formation of the other xylene isomers, unwanted transalkylation forming toluene and trimethylbenzenes also occurs. The xylene loss (which reflects the selectivity for transalkylation) decreases in the order MAPO-46 (40%) > MAPO-36 (19.7%) > MAPO-5 (10.3%) > MAPO-11 (5%), which probably reflects the differences in acidity rather than different channel geometries. This is further confirmed by the observed decrease in *o*-xylene conversion (MAPO-36 > CoAPO-36 > ZAPO-36), which resembles closely the number of strong acid sites determined by TPD of pyridine [128]. However, further increase in the magnesium content results in reduction of the activity of MAPO-36 [123].

Consequently, the same catalysts have also been tested for toluene disproportionation yielding benzene and xylenes [110,129,138]. Toluene conversion at 400 °C increases in the order MAPO-5 < MAPO-46 < MAPO-36. The acidity of the catalyst was also modified by substituing different metals resulting in the following conversion order: MAPO-36 ($X = 53\%$) > CoAPO-36 ($X = 44\%$) > ZnAPO-36 ($X = 37\%$) > MnAPO-36 ($X = 10\%$) [129].

Alkylation of toluene with ethanol has been studied over different MeAPOs with AEL, AFI, and AFS topologies [138–141]. The products formed are benzene, styrene, diethylether, and *m*-ethyltoluene, which is the commercially desired product. In a subsequent step, *m*-ethyltoluene is dehydrogenated to methylstyrene, which is the monomer for the production of polymethylstyrene. Maximum conversion is achieved at a reaction temperature of 350 °C and a toluene/ethylene ratio of 2. Toluene conversion and *m*-ethyltoluene selectivity increase with catalyst acidity, which depends on the structure and the nature of the isomorphously substituted metal ion (Table 2).

Alkylation of toluene with methanol was studied over SAPO-5 and SAPO-11 in comparison with ZSM-5 [142]. *p*-Xylene is the primary product, which undergoes subsequent isomerization to *o*- and *m*-xylene. The selective alkylation (methylation and ethylation) of toluene, ethylbenzene, *n*-propylbenzene, and cumene over MnAPO-11 and MnAPO-5 was reported by Yeom and Shim [143]. Methylation of alkyl aromatics was efficiently catalyzed by MnAPO-11, but ethylation was almost negligible. The difference of the activity is attributed to the room required by the transition state for the alkylation that is largely influenced by the size of the alcohol molecules. Moreover, it is found that the selectivity to the *o*-isomer in the methylation of alkyl aromatic compounds is also dependent on the size of the alkyl groups [143].

Table 2 Ethylation of toluene over different catalysts [139–141]

Catalyst	$X_{toluene}$ (%)	$Y_{m\text{-}ethyltoluene}$ (%)
NiAPO-11	10	–
ZnAPO-11	11	–
NiAPO-5	20	4.1
ZnAPO-5	28	4.2
$AlPO_4$-31	2.2	1.0
MnAPO-31	27.2	23.2
NiAPO-31	10.7	9.1
ZnAPO-31	21.7	18.7
MnAPO-46	55	19.3
NiAPO-46	26	10.0
ZnAPO-46	47	18.3

Reaction conditions: $T_R = 350\,°C$, WHSV $= 5\,h^{-1}$; $n_{toluene}/n_{ethene} = 1/2$.

Aluminophosphates are better catalysts for benzene isopropylation to cumene in terms of environmental safety and operating costs as compared to liquid acids. SAPO-5 shows a better cumene yield and selectivity than SAPO-11. The acidity of the catalyst plays a vital role in achieving selective formation of cumene. Cumene selectivity is higher over SAPO-5 as compared to zeolite H-beta [144]. SAPO-5 loaded with a small quantity of platinum (0.005 wt.%) shows a promising behavior in the transalkylation of commercial diisopropylbenzene (DIPB, cumene column bottoms) with benzene [145]. Higher DIPB/benzene ratios and higher space velocities were found to give better cumene selectivity; the apparent activation energy amounts to approximately $130\,kJ\,mol^{-1}$.

Transalkylation of toluene with trimethylbenzene was investigated by Dumitriu et al. [146] using various $AlPO_4$-based catalysts with AFI framework structure. They reported that catalysts with AFI structure are able to catalyze the transalkylation reaction of toluene with trimethylbenzenes and exhibit a good resistance to coking. The conversion varies with catalyst composition and especially with the nature of the metal. 1,2,4-Trimethylbenzene was found to be more reactive than 1,3,5-trimethylbenzene. Disproportionation and isomerization of the formed xylenes and trimethylbenzenes were observed as secondary reactions. 1,2,4-Trimethylbenzene disproportionates further, whereas 1,3,5-trimethylbenzene predominantly isomerizes. The product distributions reflect the acidity of the catalyst; it is suggested that transalkylation and disproportionation reactions occur on medium and strong acid sites, whereas isomerization is even catalyzed by weakly acidic sites.

Alkylation of benzene with a mixture of linear olefins (C_{10}–C_{13}) aiming at the production of linear alkylbenzenes was studied by Lian et al. [147] over a boron- and zirconium-containing BSiZrAPO-5 catalyst, which shows a high catalytic activity and a high selectivity to the corresponding 2-phenylalkane and 3-phenylalkane products in the alkylation of benzene with linear olefins. The authors propose that this environmentally benign catalyst may replace the currently used HF in the production of linear alkylbenzenes.

Butylation of phenol with *tert*-butanol as alkylating agent (Fig. 16) was reported on three medium-pore AlPO$_4$, namely, AlPO$_4$-11, AlPO$_4$-31, and AlPO$_4$-41 by Satyanarayana et al. [148]. In this study, the activity of the aforementioned molecular sieves was compared with those of their silicon-substituted counterparts. It is reported that the butylation of the highly reactive phenol is even catalyzed by the only weakly acidic terminal hydroxyl groups of the AlPO$_4$. It is observed that a reaction temperature of 140 °C is

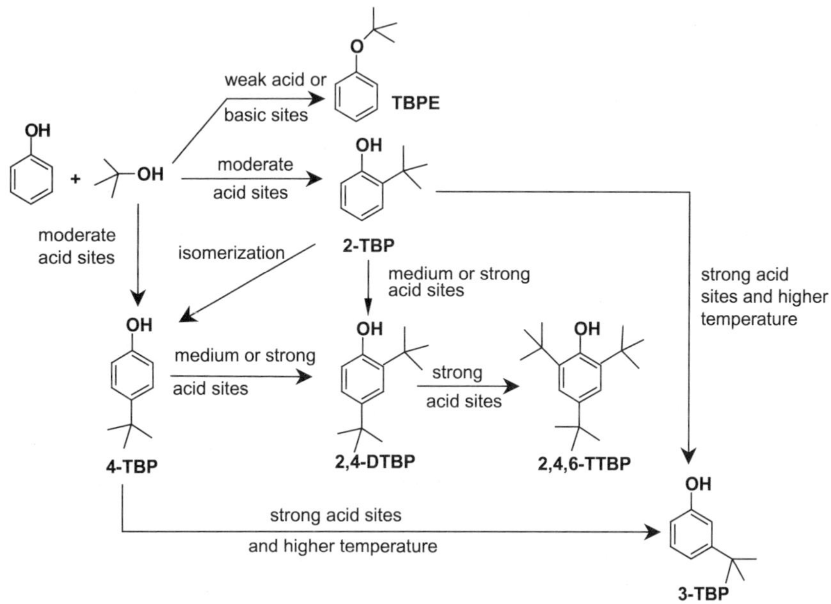

Figure 16 Effect of acid sites on the alkylation of phenol using *tert*-butanol.

sufficient for a considerable butylation of phenol ($X_{phenol} \approx 10\%$). However, the selectivities to the industrially important products 4-*tert*-butylphenol (4-TBP) and 2,4-di-*tert*-butylphenol (2,4-DTBP) are enhanced only at higher temperatures. An equimolar concentration of phenol and *tert*-butanol as well as low space velocities are required to increase the yield for 4-TBP and 2,4-DTBP. Although SAPO-11, SAPO-31, and SAPO-41 exhibit high catalytic activity as expected, it is interesting to note that the corresponding AlPOs are also active toward this reaction. The acid strength of the terminal hydroxyl groups obviously is sufficient for the dehydration of C_4 alcohols and subsequent alkylation. The ethylation of ethylbenzene with ethanol over $AlPO_4$-41, CoAPO-41, CoAPSO-41, and MnAPSO-41 was reported by Rajesh et al. [149]. Their main observation is that 1,4-diethylbenzene occurs as the major product at low temperatures while the meta isomer (1,3-diethybenzene) predominates at higher temperature. Raj and Vijayaraghavan [150] studied the same reaction over the bimetal-substituted $AlPO_4$ molecular sieves MgMnAPO-5 and MnZnAPO-5. MnZnAPO-5 has been found to be more active in benzene ethylation than MgMnAPO-5, whereas MnAPO-5 is found to be more active as compared to ZnAPO-5 although the degree of isomorphic substitution is the same. The authors suggest that this is most likely a consequence of the unpaired electron in the d-subshell of Mn [151]. A similar trend was reported recently for the isopropylation of ethylbenzene with 2-propanol [152]. Vapor-phase and liquid-phase alkylation of benzene, toluene, and ethylbenzene with different alcohols, namely, ethanol, 2-propanol, methanol, and *tert*-butanol, is compared over MnAPO-5 and MnAPO-11 [153]. An increase in the carbon chain length and the bulkiness of the alkylating agents result in rapid deactivation of the catalyst in vapor-phase alkylation reactions. The syntheses of mono- and dialkyl benzenes by alkylation of aromatics with C_1–C_3 alcohols are catalyzed by Brønsted-acid sites, which are generated by introduction of TMIs into the $AlPO_4$ framework. However, the formation of the desired products is accompanied by side reactions generating high molecular weight products resulting in catalyst coking. Thus, the observed strong deactivation of the catalysts might be due to (a) the microporous nature of the catalysts that offers diffusional constraints for the products formed and (b) the formation of dimers, trimers etc. of ethene and propene over the acid sites of the catalyts, contributing to the formation of bulky coke precursors [153].

The ion exchange of Lewis-acidic metal ions such as Fe^{3+}, Ga^{3+}, In^{3+}, La^{3+}, or Ce^{3+} into MeAPOs has been introduced as a new strategy for generating catalysts that exhibit both, Lewis and Brønsted acidity [154–157]. It is assumed that the actual ions used for ion exchange are $M(OH)_2^+$

based on the Hirschler–Planck mechanism, which are converted into MO^+ during calcination. The resulting Me-MAPO-36 catalysts were tested for toluene disproportionation [155] as well as in *tert*-butylation of phenol [154]. The ion-exchanged catalysts were found to be more active than the parent MAPO-36. Ion-exchanged ZAPO-5 catalysts were also tested in the synthesis of 7-hydroxy-4-methyl coumarin by liquid-phase (solvent-free) condensation of ethyl acetoacetate and resorcinol. The highest conversion was observed for GaZAPO-5 resulting in a 7-hydroxy-4-methyl coumarin yield of 89.9% [156]. The corresponding MgAPO-5 catalysts were tested in the liquid-phase condensation of salicylaldehyde and diethyl malonate yielding selectivities up to 95% for 3-ethyl coumarin carboxylate [157].

3. BIFUNCTIONAL CATALYSIS

Isomerization of *n*-paraffins plays an important role in the petroleum industry and several processes have been implemented using bifunctional catalysts uncluding the total isomerization process (TIP, UOP) and isodewaxing (Chevron). The branching of *n*-paraffins is required to increase the octane number of gasoline or to improve the low-temperature performance of diesel fuel. Bifunctional zeolite-based catalysts containing a Group VIII metal as a hydrogenation function and an acid function (Brønsted-acid sites) have long been known for their high efficiency in the isomerization of *n*-paraffins [158]. A major problem is that the isomerization reaction is always accompanied by unwanted hydrocracking that lowers the isomer yield and produces less valuable products. The pathways of isomerization and hydrocracking over bifunctional molecular sieve based catalysts are known in detail (Fig. 17a and b) [159]. It is well established that isomerization of *n*-alkanes occurs first and that cracking is a consecutive reaction which is faster for multibranched alkanes. Monobranched alkanes are less susceptible to cracking than multibranched alkanes [159]. Palladium- or platinum-containing $AlPO_4$ with AEL, AFI, ATO, and AFO topologies have been found to give high selectivities for the isomerization of wax or model feedstocks including *n*-decane, *n*-octane, *n*-heptane, and *n*-hexane (Table 3) [160]. As the hydrogenation function, palladium, platinum, or bimetallic Pd–Pt clusters were introduced. Acid sites are typically generated by isomorphous substitution of silicon, cobalt [161], or magnesium [162] into the $AlPO_4$ framework.

The isomerization of *n*-heptane was studied over a series of bifunctional SAPO-*n* based catalysts, which contained 0.1 wt.% palladium [169]. Best

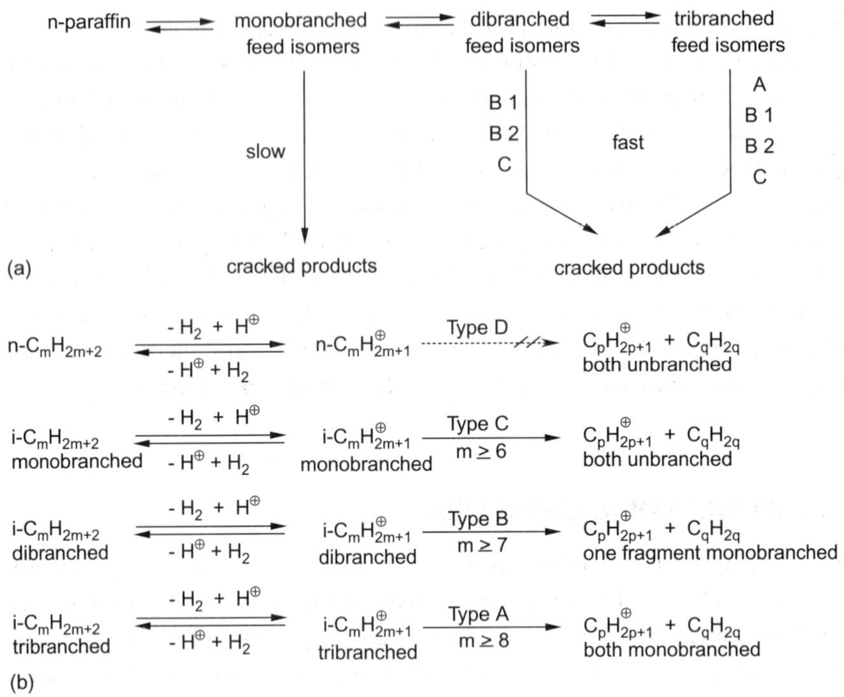

Figure 17 (a) Isomerization and hydrocracking pathways. (b) Reaction network for isomerization and hydrocracking of alkanes on bifunctional catalysts.

activities and selectivities for branched heptane isomers are achieved with SAPO-11 and SAPO-31. SAPO-17 and SAPO-5 show substantially lower activities. Over SAPO-5, the selectivity to cracked products is high, which is assumed to be caused by the reduced accessibility of parts of the bridging hydroxyl groups within the molecular sieve framework. Different locations of these acid sites are evidenced by IR spectroscopy recorded after the adsorption of *n*-heptane [169].

A series of papers was published by Campelo et al. on hydrocracking and isomerization of *n*-alkanes on bifunctional Pt/SAPO-5 and Pt/SAPO-11 catalysts [170–173]. These materials were prepared by impregnation of the SAPOs with $Pt(NH_3)_4Cl_2$ and subsequent reduction. The platinum loading was adjusted to 0.5 wt.%. In studies with *n*-hexane [171], *n*-heptane [172,174], and *n*-octane [170], it was found that the size of the pores of the SAPO support can determine, to a large part, the catalytic performance. The reactivity of the *n*-alkanes increases with chain length because of the involvement of more stable carbenium ion intermediates. The differences in selectivity between Pt/SAPO-5 and Pt/SAPO-11 have been explained by

Table 3 Survey of catalysts tested for the isomerization of *n*-alkanes

Feed	Catalyst	Metal content (wt.%)	Isomer yield (%)	Conversion $X_{n\text{-alkane}}$ (%) (T_R, °C)	Reference
n-Pentane	Pt/SAPO-11	0.5	2.5	≈ 5 (350)	[163]
n-Hexane	Pt/SAPO-5	0.5	62.5	68.8 (375)	[171]
	Pt/SAPO-11	0.5	26	48.4 (425)	[171]
n-Heptane	Pd/SAPO-11	0.7	17.6	38.9 (400)	[174]
	Pt/SAPO-11	1.1	2.0	7.3 (400)	[174]
	Pt/SAPO-5	0.5	38	60 (400)	[172]
	Pt/SAPO-11	0.5	48	70 (400)	[172]
	Pt/SAPO-5	0.5	26.2	40.6 (350)	[164]
	Pd/SAPO-5	0.1	3.6	23.5	[169]
	Pd/SAPO-11	0.1	55.6	82.6	[169]
	Pd/SAPO-17	0.1	15.0	19.2	[169]
	Pd/SAPO-31	0.1	67.6	72.1	[169]
	Pd/SAPO-5	1.0	4.6	≈ 5 (300)	[161]
	Pd/CoAPO-5	1.0	4.1	≈ 5 (300)	[161]

Table 3 *(Continued)*

Feed	Catalyst	Metal content (wt.%)	Isomer yield (%)	Conversion $X_{n\text{-alkane}}$ (%) (T_R, °C)	Reference
	Pd/SAPO-11	1.0	4.7	≈ 5 (300)	[161]
	Pd/CoAPO-11	1.0	4.7	≈ 5 (300)	[161]
n-Octane	Pt/SAPO-11	0.5	39	67	[170]
	Pt/SAPO-5	0.5	28	56	[170]
	Pt/SAPO-11	1.0	30	47 (300)	[177]
	Pt/SAPO-5	1.0	20	51 (300)	[177]
	Pt/SAPO-31	1.0	43	76 (340)	[177]
	Pt/SAPO-41	1.0	54	70 (300)	[177]
	Pt/SAPO-11	1.0	28.4	30 (331)	[190]
	Pd/SAPO-11	1.0	28.5	30 (345)	[190]
	Pd/SAPO-5	1.0	5.7	30 (341)	[190]
	Pt-Pd/SAPO-11	1.0	67.4	78.5 (250)	[178]
	Pt-Pd/SAPO-41	1.0	6.8	6.8	[180]
	Pt/SAPO-5	0.5	26.8	91.6	[181]

	Pt/SAPO-11	0.5	70.8	83.2	[181]
	Pt/SAPO-31	0.5	69.7	89.5	[181]
	Pt/SAPO-41	0.5	68.5	87.9	[181]
	Pt/SAPO-34	0.5	2.72	4.5	[181]
n-Decane	Pt/SAPO-41	0.5	78.9	89.5 (320)	[183]
	Pt/SAPO-41	0.5	64.1	93.8 (290)	[182]
	Pt/MgAPO-5	0.5	8	35 (260)	[162]
	Pt/MgAPO-11	0.5	62	92 (290)	[162]
	Pt/MgAPO-41	0.5	34	69 (320)	[162]
n-Dodecane	Pt/MgAPO-11 Pt/SAPO-11	0.5 0.5	73.0 70.7	86.5 (330) 83. 9 (335)	[185] [185]
n-Tetradecane	Pt/SAPO-11	0.37	86.2	91.5 (320)	[165]
n-Hexadecane	Pt/SAPO-11 Pt/SAPO-11	1.0 0.5	79.6 94.4	94 (340) 83.4 (350)	[190] [166]
	Pt/SAPO-11(nanosized)	0.5	64.6	76 (340)	[167]
	Pt/SAPO-11	05	48	82 (340)	[168]

the slower migration of the alkane intermediates in the smaller channels of SAPO-11 and by steric constraints at the pore mouths. The selectivity for isomerization decreases with chain length for *n*-alkenes in Pt/SAPO-5 and increases in Pt/SAPO-11. In a subsequent study by the same group, *n*-octane and isooctane hydroconversion have been investigated [175]. It was found that the reaction schemes for *n*-octane and isooctane transformation are not the same due to the different pore structure of the SAPOs used. The size of the pores as well as the shape of the space available near the acid sites is believed to determine the selectivity of *n*-octane and isooctane isomerization and cracking over platinum-containing bifunctional catalysts [175].

The conversion of *n*-heptane and *n*-hexane over Pt/SAPO-11 was investigated by Butt et al. [174,176]. The reaction pathways were found to be similar to other bifunctional catalysts on SAPO-11-based materials. With incorporation of sodium, a pronounced formation of aromatics is detected. The bifunctional nature of the catalyst is reflected in the difference in activity decay versus constant selectivity for cracking and isomerization. For isomerization over palladium-impregnated catalysts with AEL and AFI topoplogies, it was observed that the higher acidity of CoAPO compared with SAPO material is reflected by a higher activity and a slightly higher selectivity for dibranched isomers [162].

High selectivities for *n*-octane isomerization have been observed with medium-pore Pt/SAPO-11-, Pt/SAPO-31-, and Pt/SAPO-41-type materials, whereas preferential hydrocracking has been observed over the large-pore Pt/SAPO-5 [177]. Isomerization products almost exclusively consist of monobranched isomers with a negligible amount of dibranched species, which also accounts for the low yield of cracked products. The isomerization selectivity decreases in the order SAPO-41 > SAPO-31 > SAPO-11. The differences in selectivity are explained by diffusional restrictions and steric constraints. In subsequent papers from the same group, *n*-octane isomerization was studied over bifunctional Pt–Pd/SAPO-11 and Pt–Pd/SAPO-11 catalysts [178–180]. Approximately 90% selectivity toward isomerization (mainly monobranched isomers) was observed even at higher conversion ($X_{n\text{-octane}} = 79\%$). SAPO-41 was found to be slightly more selective than SAPO-11. Furthermore, small grains (0.1–0.2 μm) were more active than large grains (1–1.5 μm), but no significant effect on the product distribution was observed. The authors concluded that restricted transition state selectivity is responsible for the product distribution observed. The preferential formation of the monobranched 2-methylheptane can be explained through a protonated cyclopropane-type (PCP) transition state. The terminal PCP transition state from *n*-octane should be almost linear and thus would be accommodated more easily in the SAPO-*n* ($n = 11, 41$) pores than an internal C_8 transition state. This hypothesis is nicely supported by the large amounts of 2-methylheptane and *n*-octane formed from 3-methylheptane. Whether the concept of "pore mouth" catalysis is really necessary to explain the observed results is not yet decided. Remarkably, the medium-pore SAPO-11 and SAPO-41 molecular sieves exhibit a high *n*-octane isomerization selectivity while retaining some transition state selectivities as evident from the low yield of 3-methylheptane and 2,6-dimethylhexane [180]. In a recent paper, SAPOs with AFI, AEL, ATO, AFO, and CHA framework topology were compared to the zeolites ZSM-5, ZSM-22, and ZSM-23 with respect to their *n*-octane isomerization selectivity [181]. The results indicate that the selectivity to isomerization is highly influenced by the channel structure. It has to be pointed out, however, that due to the higher acid site density and strength of the zeolite-based catalysts, the maximum isomer yield is observed at a lower reaction temperature.

Pt/SAPO-41 has also been evaluated as a catalyst for the isomerization of *n*-decane. An isomer selectivity of approximately 90% is achieved at a conversion of approximately 88% [182,183]. The excellent performance of the catalysts based on SAPO-41 is ascribed to the pore dimensions (0.43×0.70 nm), which are slightly larger as compared to SAPO-11 (0.39×0.63 nm), and the small grain size of 0.2–0.5 μm.

The isomerization and hydrocracking of *n*-decane was studied over a composite catalyst using MgAPO-5 or MgAPO-11 as acid functions along with Pt/MCM-41 as the metal function [184]. The MgAPO-11-based bifunctional catalyst exhibited a higher isomerization yield that is ascribed to the medium acid strength and to the elliptical pore system of MgAPO-11, which prevents the formation of tribranched isomers resulting in a reduced cracking yield and, hence, higher isomerization yields. Hartmann and Elangovan [162] have compared Pt/MgAPO-*n* ($n = 5, 11, 41$) catalysts in the isomerization and hydrocracking of *n*-decane. It was observed that hydrocracking predominates over Pt/MgAPO-5, while isomerization is prevailing with Pt/MgAPO-41 and Pt/MgAPO-11 (Fig. 18). The influence of Mg concentration, which determines the number of acid sites, and Pt loading on the performance of the latter catalysts in the isomerization of *n*-dodecane was studied by Yang et al. [185]. The isomer yield reaches a maximum for a platinum loading of 0.5 wt.%, while isomer yield and conversion are maximized at a magnesium content of 0.4 wt.%. These results clearly show that a proper balance between the acidic and the metallic function is required for optimal performance. A lower Mg-content

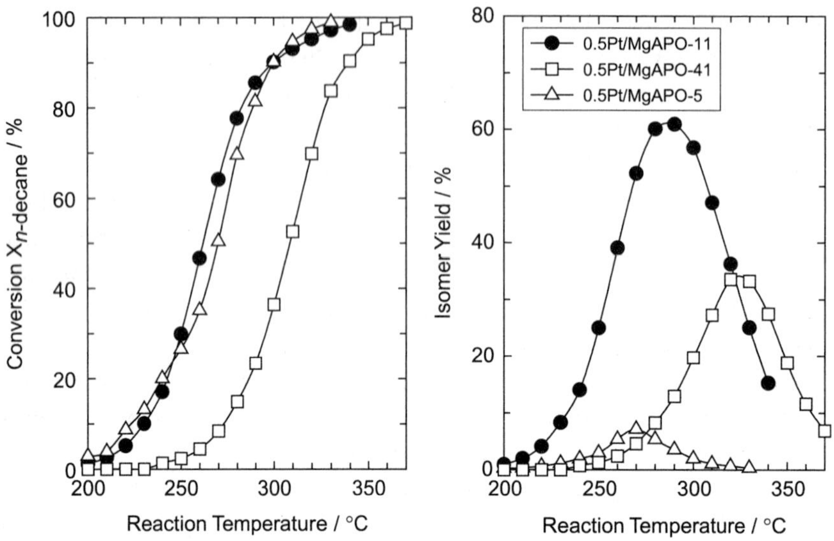

Figure 18 Isomerization of *n*-decane over Pt/MgAPO-*n* molecular sieves [162].

resulted in an inadequately low acidity, whereas higher Mg contents weakened the hydrogenation/dehydrogenation function of Pt.

Recently, Huang et al. [186] studied the hydroisomerization of *n*-dodecane over Pt and Pd supported on SAPO-11. The support was prepared from a H_2O–CTAB–butanol mixed-solvent system and has then been compared with respect to the catalytic properties to SAPO-11 synthesized by the conventional hydrothermal route. The largest isomer yield of 92% with a selectivity of 99% was observed over the support prepared in the H_2O–CTAB–butanol system. The results are ascribed to the smaller crystallite size and the presence of acid sites with medium strength. The preparation of SAPO-11 in non-aqueous media (ethylene glycol or glycerol) is reported to yield more active *n*-dodecane isomerization catalysts [187]. A conversion of 90% is reached at a significantly lower reaction temperature (250 vs. 350 °C). Moreover, a higher isomerization selectivity is observed (91% vs. 77% for the conventionally prepared SAPO-11). The authors ascribe the improved performance to a higher external surface area and a larger number of acid sites. The isomerization of *n*-dodecane over Pt/MeAPO-11 (Me = Mg, Mn, Co, or Zn) is reported by Yang et al. [188]. The acidic strength of the MeAPO-11 determines the activity of the bifunctional catalyst, while the isomerization selectivity is influenced by the structure and the metal function.

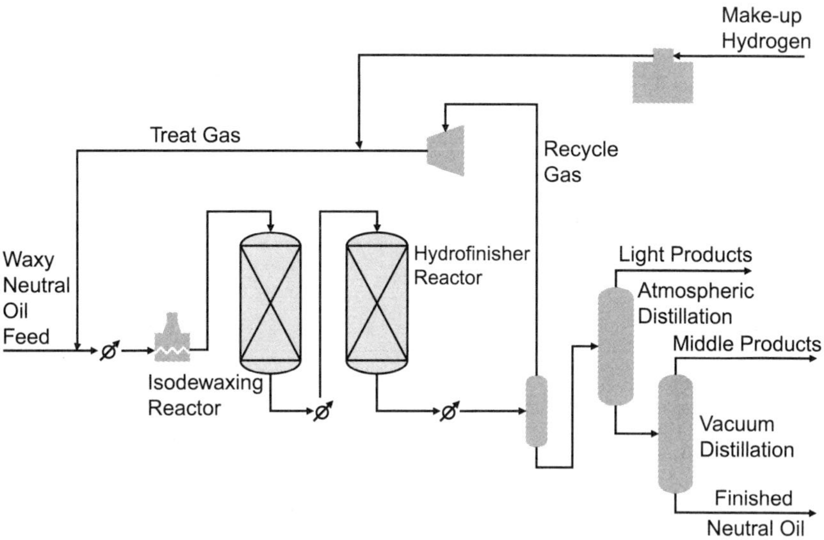

Figure 19 Process flow diagram of isodewaxing and hydrofinishing.

Chevron has developed a so-called isodewaxing process using a Pt/SAPO-11 molecular sieve [189,190]. This process is used to produce high yields of high viscosity index (VI) and low pour point lube oils from waxy feedstocks. It was first commercialized in Chevron's Richmond refinery in 1993. In the process (Fig. 19), waxy neutral oil feed from a hydrocracker/hydrotreater process step are preheated together with treat gas and fed to the isodewaxing reactor. Some of the normal paraffins are isomerized to high VI isoparaffins that lowers the pour point. Other paraffins are cracked to highly saturated light products such as high smoke point jet fuel and high cetane number diesel. The effluent from the isodewaxing reactor is then sent to the hydrofinishing reactor. Usually, the feed needs to be hydrotreated before the isodewaxing step to lower nitrogen and sulfur contents and, thus, ensure high product yields and long cycle length. Miller carried out a detailed analysis aimed at unraveling the factors related to the shape-selective isodewaxing catalyst [189,191]. The catalytic performance of Pt/SAPO-11 for *n*-hexadecane isomerization, which is taken as a model feed for long-chain paraffins present in the waxy feed stock, is compared to an amorphous Pt/silica-alumina catalyst (Table 4). A higher selectivity to monobranched isomers is achieved over the Pt/SAPO-11 catalyst. Moreover, this catalyst favors branching at the center of the hydrocarbon chain as seen from the high selectivity to 7- and 8-methyl-C_{15} resulting in a low pour point. Thus, the Pt/SAPO-11 catalysts have the advantage of

Table 4 Isomerization of hexadecane ($p = 7$ MPa, WHSV = 3.1 h^{-1}, $n_{H_2}/n_{HC} = 30$, $X_{n\text{-hexadecane}} = 96\%$) [189,190]

Catalyst	Pt/SAPO-11	Pt/SiO_2–Al_2O_3
Temperature	340	360
Isomerization selectivity (wt.%)	85	64
C_{16} product composition (wt.%)		
7-Methyl-C_{15}+8-methyl-C_{15}	22.9	6.1
6-Methyl-C_{15}		2.7
5-Methyl-C_{15}	7.1	3.2
4-Methyl-C_{15}	7.5	3.1
3-Methyl-C_{15}	8.1	3.2
2-Methyl-C_{15}	7.7	3.3
n-C_{16}	4.7	6.0
Dimethyl-C_{14}	29.8	37.8
Other C_{16}	12.2	34.6
Pour point (°C)	−51	−28
Viscosity (cSt) at −20 °C	15.99	16.46

both low cracking activity and good isomerization activity while suppressing the formation of multibranched isomers, which would readily crack to undesired lighter products. Also, moderate acidity (as determined by the number and strength of the acid sites) and strong hydrogenation functions may play a role. The remarkable performance of Pt/SAPO-11 is primarily attributed to the one-dimensional nature of the 10-membered ring pores, which induces the desired transition state selectivity and thereby minimizes secondary cracking reactions. In addition, passivation of the outer surface of the catalyst will usually also be beneficial to the performance. In the patent literature, Chevron has claimed criteria which are considered to be important for high-performance dewaxing catalysts: (1) medium-pore molecular sieve with oval-shaped pores, (2) crystallite size below 0.5 μm to minimize product diffusion rate limitations, (3) sufficient acidity, and (4) a Group VIII metal as hydrogenation function. In addition to SAPO-11, zeolites ZSM-22 and ZSM-23 are also claimed to fit this criteria for high-performance isodewaxing catalysts [191,192]. In this context, Walendziewski and Pniak [193] tested the isomerization of a mixture of n-C_{13} to n-C_{20} hydrocarbons over Pt/SAPO-11. Application of Pt/SAPO-11 with an alumina binder at a weight ration of 1:1 enables isomerization of the hydrocarbon mixture with a maximum isomerization selectivity of 84%. The relatively low isomerization selectivity in comparison with other reports in the literature is ascribed by the authors to the high amount of alumina binder in the catalyst.

The Boreskov Institute of Catalysis has developed a catalyst based on SAPO-31 for the isomerization of diesel feedstock. The properties of Pt/SAPO-31 (20 wt.% Al_2O_3 as binder) were tested in a pilot plant (Table 5) [194]. Depending on the reaction temperature, the obtained reduction in pour point is 25–46 °C yielding more than 97% diesel fraction. When the sulfur content of the feed is increased from 2 to 50 ppm, a decrease in catalytic activity is observed (Table 5).

A new field for the use of Pt/SAPO-11 catalysts is the isomerization of pre-hydrogenated sunflower oils aiming at the production of diesel blending

Table 5 Hydroprocessing of diesel feedstock in a pilot plant containing Pt/SAPO-31 as catalyst

Reaction temperature (°C)	Yield of diesel fraction (wt.%)	Cloud point (°C)	Pour point (°C)
Feed containing 2 ppm of sulfur			
Feedstock	100	−8	−14
280	99	−14	−25
300	98	−24	−35
320	97	−35	−46
Feed containing 50 ppm of sulfur			
Feedstock	100	−5	−20
300	99	−8	−24
320	98	−14	−31
335	98	−23	−37
340	98	−39	−56

stocks. Under favorable reaction conditions, biodiesel yields exceeding 88% with an *iso-/n*-paraffin ratio between 3.2 and 6.8 can be achieved. Cetane numbers between 79 and 85 were obtained, while the cold filter plugging points vary between 19 and 16 °C [195].

Using metal-containing $AlPO_4$-based catalysts, research was also conducted to explore the possibility of producing isobutene from *n*-butane in a single step, thereby avoiding the costs of two separate reactors for the metal-catalyzed dehydrogenation and the acid-catalyzed skeletal isomerization, which are presently used for the production of isobutene. The direct transformation of *n*-butane into isobutene (dehydroisomerization) was tested over bifunctional Pt/SAPO-11, Pt/MnAPSO-11, and Pt/MnAPO-11 catalysts [196,197]. In general, catalysts based on the 10-membered ring molecular sieve SAPO-11 and metal-substituted $AlPO_4$-11 and SAPO-11 catalysts have been found to be active for this reaction. Isomerization is believed to proceed through a bimolecular dimerization–cracking mechanism, involving the formation of C_8 carbenium ions. Since the reaction proceeds through relatively large C_8 carbenium ions, the pore geometry and size of the channel openings will play an important role in this reaction. At a reaction temperature of 500 °C, Pt/SAPO-11 was the most active catalyst; however, mainly cracking products are formed through hydrogenolysis on the metal sites. The selectivity to isobutane is approximately 20%; isobutane is also a valuable feedstock and is used for the production of isooctane (formed by alkylation with *n*-butenes). However, the isobutene selectivity is only 7%. Over Pt/MnAPSO-11 and Pt/MnAPO-11, hydrogenolysis is largely suppressed resulting in higher selectivities for isobutene ($S = 18\%$) and isobutane ($S = 42\%$ over Pt/MnAPO-11; $S = 30\%$ over Pt/MnAPO-11). The low selectivities for cracked products approximately 11.5% over Pt/MnAPSO-11 ($X_{n\text{-butane}} = 43\%$) and 6.5% over Pt/MnAPO-11 ($X_{n\text{-butane}} = 31\%$) confirm that on these catalysts the acidic and metal functions are well balanced and bifunctional catalysis predominates [196,197]. Wei et al. [198] studied the dehydroisomerization of *n*-butane over palladium-modified SAPOs. The conversion of *n*-butane over Pd/SAPO-11 is higher as compared to Pd/SAPO-5 and Pd/SAPO-34, although the pore size of the AFI structure is larger as compared to the AEL structure. Moreover, the metal dispersion is found to be higher on SAPO-11 as compared to other supports. The addition of a second element such as Sn, Pb, Zn, and In to Pt/H-SAPO-11 retarded the hydrogenolysis activity of Pt resulting in a decreased methane selectivity [199]. Among the second elements, tin was the most effective additive for increasing the isobutene selectivity. The formation of the intermetallic compounds Pt_3Sn, PtSn, and

$PtSn_2$ as revealed by XRD is believed to be responsible for the observed changes in selectivity. Machado et al. [200] studied the transformation of *n*-butane over Ga/SAPO-11 catalysts with particular attention to the role of extra-framework gallium species. Controlled incorporation of extra-framework Ga between 0.5 and 1.5 wt.% into SAPO-11 resulted in an excellent bifunctional catalyst giving moderate isobutene yields. A Ga/SAPO-11 catalyst which contains 0.5 wt.% Ga yielded 15% of dehydrogenated products; 27% of these products are formed through skeletal isomerization. In addition, undesired (hydro)cracking and hydrogenolysis were strongly suppressed.

The transformation of *n*-butane to aromatic hydrocarbons over Pt/SAPO-11 was studied by Kumar et al. [201]. The reduced Pt/SAPO-11 catalyst exhibited a higher *n*-butane conversion than the oxidized Pt/SAPO-11 material. However, the selectivity to aromatics is lower over the reduced catalyst.

From the large body of data available for the isomerization of *n*-alkanes over $AlPO_4$, some trends can be deduced: The main feature of medium-pore SAPOs such as SAPO-11, SAPO-41, and SAPO-31 is some kind of isomerization shape selectivity that results in relatively low branched paraffin isomers. Moreover, the mild acidity of SAPOs as compared to zeolites is believed to be beneficial for suppressing hydrocracking. However, optimal performance of SAPO-based catalysts is achieved at significantly higher reaction temperatures as compared to zeolites such as ZSM-22 and ZSM-23, which exhibit comparable isomer selectivities. It is well accepted that the balance between acid sites and metal sites are important for an imrovement of the catalytic performance. Here, the optimization of the synthesis procedure, the isomorphous substitution of TMIs into the $AlPO_4$ framework to generate acid sites and the extended use of bimetallic hydrogenation functions could be explored further to optimize the catalyst performance.

4. REDOX CATALYSIS

In the continual effort to transform fine chemical production into more environmentally benign technologies, metal-containing AlPOs and SAPOs offer tremendous potential as heterogenous oxidation catalysts for the production of these chemicals. This is demonstrated by the significant advancement in the field of heterogeneous catalysis by molecular sieves following the success of titanium-containing silicalite-1 (TS-1). Substitution of metal cations with redox properties such as Fe, Co, Mn, and V into $AlPO_4$ is expected to provide novel catalysts for heterogeneously catalyzed oxidation reactions. Hydrocarbons that are oxidized at the terminal carbon atom, namely, at the α- or 1-position, are important feedstocks for the chemical and pharmaceutical industries. Nevertheless, selective oxidation of alkanes at their terminal methyl group is still a challenge in modern catalysis research. It is well known that some enzymes are capable of performing selective terminal oxidations. In principle, selective partial oxidation is easier to control when hydrogen peroxide or organic hydroperoxides such as *tert*-butylhydroperoxide (TBHP) are used as oxygen donors, although from an economic point of view, the use of molecular oxygen, air or O_2, is preferred [202,203].

4.1 Oxidation of hydrocarbons with oxygen or air

The selective oxidation of linear paraffins (*n*-pentane, *n*-hexane and *n*-octane) over Co- and Mn-containing catalysts with AEI or ATS topologies in the liquid phase using air as an oxidant has been reported by Thomas et al. [204]. The regioselectivity reported for the oxidation of *n*-hexane in CoAPO-18 and MnAPO-18 is remarkable. After a reaction time of 24 h, 60% of the oxidation products are oxyfunctionalized at the terminal carbon atom and an additional 36% are oxidized in the 2-position yielding hexan-2-ol and hexan-2-one. Interestingly, the corresponding acid is formed with a high selectivity of 54%. Consequently, the oxidation of *n*-hexane to adipic acid has been attempted over CoAPO-18 and CoAPO-34 [205]. Over CoAPO-18, after a reaction time of 24 h (conversion 9.3%), a selectivity toward adipic acid of 35% is obtained. A reduced selectivity to adipic acid of 20% was achieved over CoAPO-34, while the formation of adipic acid was not observed with CoAPO-36 as catalyst. The terminal oxidation of dodecane with air was found to be highly selective at C_1 and C_2 with MnAPO-18, while with molecular sieves having larger pore diameters, such as MnAPO-5, MnAPO-11, and MnAPO-36, oxidation is favored at the carbon atoms in 3- or 5-position [206]. However, some of those experiments require well-formulated catalysts and have been found difficult to reproduce.

Cobalt-containing $AlPO_4$ with AFI, AEL, AEI, ATS, and VFI topologies have been explored for the oxidation of cyclohexane with oxygen or air aiming at preferential formation of cyclohexanone and cyclohexanol [207–213]. The oxidation products are important intermediates in the production of adipic acid and ε-caprolactam, which are used for the production of Nylon-6 and Nylon-6.6, respectively. Over CoAPO-5, cyclohexylhydroperoxide (CHHP) is the main product (Fig. 20), which is easily converted into monofunctional oxidation products. According to this study [207,208], CoAPO-11 is catalytically more active than CoAPO-5 despite the smaller channel diameter of CoAPO-11, which is expected to hinder diffusion of the

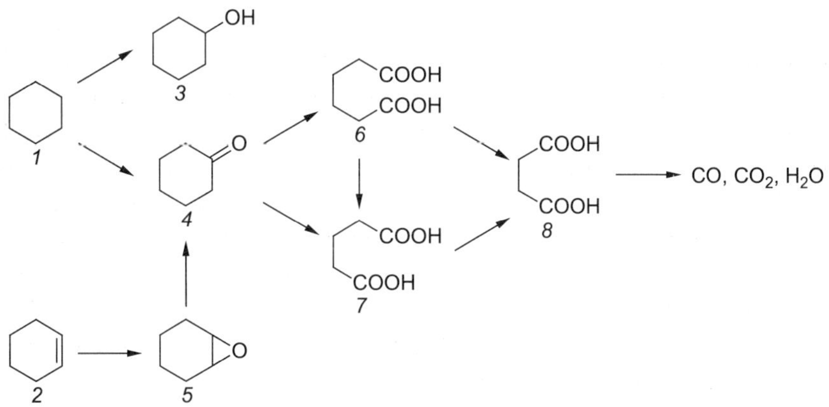

Figure 20 Main products and by-products of cyclohexane (*1*) and cyclohexene (*2*) oxidation: Cyclohexanol (*3*), cyclohexanone (*4*), cyclohexene oxide (*5*), adipic acid (*6*), glutaric acid (*7*), and succinic acid (*8*).

products. These catalytic results have been ascribed to a higher dispersion of cobalt in CoAPO-11. Over CoAPO-36, also minor quantities of adipic acid are formed [207,208]. Cyclohexane can be directly oxidized to adipic acid by air over FeAPO-5 and FeAPO-31 catalysts [214]. FAPO-5 mainly yields cyclohexanol and cyclohexanone as well as an adipic acid yield of roughly 15% (after 24 h at 373 K), while over FAPO-31 predominately adipic acid ($S = 65\%$) is formed. CrAPO-5 also predominately produces cyclohexanone [33]. Selectivities to adipic acid of around 8% are formed for CoAPO-5, MgVAPO-5, and CoVAPO-5, while for VAPO-5 the selectivity to adipic acid is only 3% [215]. The conversion after 2.5 h increases in the order MgVAPO-5 (2.8%) < CoVAPO-5 (5.8%) = VAPO-5 (5.7%) < CoAPO-5 (14.7%). The evaluation of MeAPSO-5 (Me = Fe, Co, Mn) catalysts containing more than one transition metal was reported by Zhou et al. It appears that the introduction of a suitable second or even third transition metal offers the possibility to increase the activity of the catalysts and also their selectivity to adipic acid [216,217].

The industrial production of adipic acid proceeds through a two-step liquid-phase oxidation of cyclohexane with air and nitric acid as oxidants. To increase the yield of adipic acid in an environmentally friendly one-step process, most studies seek reaction conditions that might favor the oxidation of cyclohexane and its intermediates [211]. CoAPO-5 has been found to be a potential catalyst for a one-step liquid-phase oxidation of cyclohexane to produce mainly adipic acid without any promoter being added and with no induction period [211]. This study shows that high cyclohexane concentration, high reaction temperature and high oxygen pressure will result in a 45% yield of adipic acid after a reaction time of 3 h ($X_{cyclohexane} = 50\%$) [218]. However, under these reaction conditions and in the presence of acetic acid, leaching of Co has been observed [219], which has claimed to be absent for solvent-free oxidations of cyclohexane on CoAPOs [210,211,220]. In a recent study by Iglesia et al. [221], it is shown that the formation rates of cyclohexanol and cyclohexanone in the oxidation of cyclohexane with air rise with increasing number of redox sites on CoAPO-5 and MnAPO-5. In a study by Tian et al. [222], MeAPOs (Me = Co, Mn, Cr, and Cu) with AFI, AEL, and CHA topology were tested in the oxidation of cyclohexane at 403 K using oxygen as the oxidant under solvent-free conditions. CoAPO-11 and MnAPO-11 exhibited the highest conversion (approximately 8% after 3 h); all MeAPO-11 catalysts showed a better activity than the corresponding MeAPO-5 and MeAPSO-34 materials. It is reported that the formation of the CHHP intermediate is the rate-determining step in cyclohexane oxidation [223]. Two pathways exist for the rapid decomposition of the intermediate CHHP: a heterolytic one (CHHP → cyclohexanone) and a homolytic one (CHHP → cyclohexanol). Since cyclohexanol is more reactive than cyclohexane under typical reaction conditions, it can be converted to cyclohexanone more easily. It was found that both pathways exist in the decomposition of CHHP over CoAPO-11 and CrAPO-5 [222]. From kinetic studies at conversions below 5%, Moden et al. [221] suggested a cyclohexane oxidation pathway over the Me sites in accordance with low-temperature oxidation reactions catalyzed by homogenous Mn and Co complexes:

An initial formation of CHHP (ROOH) is followed (at least when homogenous Me–ROOH complexes are involved) by the formation of ROO and RO radicals. More specifically, ROOH decomposition is proposed to involve Me–ROOH complexes and one-electron redox cycles (reactions 1–4 in Fig. 21). The scheme is adapted from the proposal by Black [224] by replacing Me ions in solution with active Me species in the $AlPO_4$. Thus, [Me^{2+}–O–H^{+}] and [Me^{3+}–O] represent reduced and oxidized species in reaction 2. It is likely that pathways resembling those in the scheme are involved on heterogeneous catalysts, but radical intermediates are likely to remain adsorbed on cationic sites. Thus, a description of this oxidation as free radical-like seems inappropriate [221]. Moreover, alternative pathways of surface-bound intermediates may offer opportunities for specific regioselectivity in oxygen insertion steps.

Recently, Ce/$AlPO_4$-5 (prepared by direct hydrothermal synthesis) has been suggested as an efficient catalyst for the oxidation of cyclohexane in a

Initiation

1. $[Me^{2+}\text{-O-}H^+] + ROOH \leftrightarrow [Me^{2+}\text{-O-}H^+]\text{-}ROOH$
2. $[Me^{2+}\text{-O-}H^+]\text{-}ROOH \rightarrow [Me^{3+}\text{-O}] + RO\cdot + H_2O$
3. $[Me^{3+}\text{-O}] + ROOH \leftrightarrow [Me^{3+}\text{-O}]\text{-}ROOH$
4. $[Me^{3+}\text{-O}]\text{-}ROOH \rightarrow [Me^{2+}\text{-O-}H^+] + ROO\cdot$

Propagation

5. $RO\cdot + RH \rightarrow ROH + R\cdot$
6. $R\cdot + O_2 \rightarrow ROO\cdot$
7. $ROO + RH \rightarrow ROOH + R\cdot$

Termination

8. $ROO\cdot + ROO\cdot \rightarrow ROH + R{=}O + O_2$

Figure 21 Oxidation pathway of ROOH (R = cyclohexyl) over Me sites (after Ref. [221]).

solvent-free system [225]. At 413 K and under an oxygen pressure of 0.5 MPa, cyclohexane conversion amounts to 13% (after a reaction time of 4 h) and a combined selectivity to cyclohexanol and cyclohexanone of approximately 92% has been observed, thus, exhibiting comparable performance to CoAPO-5 [215,221]. When solvents are used for the reaction, the conversion decreases in the following order: no solvent > acetonitrile > acetone > acetic acid > pyridine. Interestingly, adipic acid is the main product when acetic acid is employed as a solvent [225]. However, earlier studies have suggested that leaching is a major problem and that the homogenously catalyzed formation of adipic acid cannot be excluded.

It has been shown that TMI-containing catalysts with ATS and AEI framework topologies are also suitable for epoxidation of alkenes with air [203,226] and the Baeyer–Villiger oxidation of ketones to lactones [227]. However, many problems (leaching, long-term stability, pore blocking through products, structure, and uniformity of the active sites toward activity and selectivity) have to be solved before a commercial process can be realized [202,220].

p-Hydroxybenzaldehyde is an important chemical intermediate for the preparation of pharmaceutical, agricultural, and fragrance chemicals. Sheldon et al. [228,229] found that CoAPO-11 and CoAPO-5 are both effective, stable and recyclable solid catalysts for the facile oxidation of *p*-cresol to *p*-hydroxybenzaldehyde at 50 °C in NaOH/MeOH with conversion and

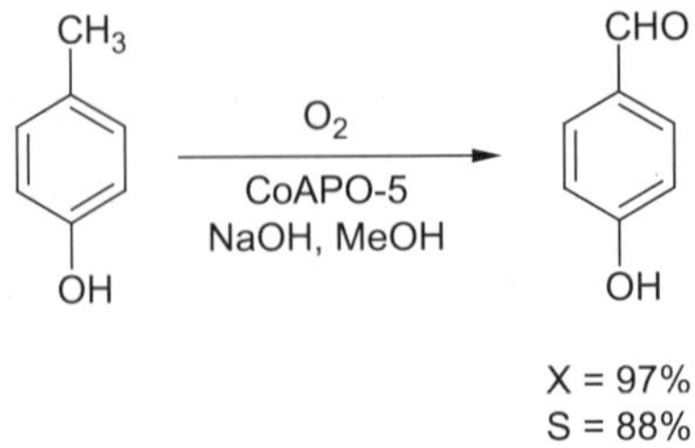

Figure 22 Oxidation of *p*-cresol to *p*-hydroxybenzaldehyde.

selectivity both reaching 90% (Fig. 22). The superior performance of these catalysts over cobalt salts used as homogeneous catalysts was attributed to the suppression of μ-hydroxo-bridged cobalt dimer formation over CoAPOs. However, it is reported that the catalyst is not stable under these conditions and dissolves slowly forming soluble cobalt compounds that serve at least partially as catalytically active species [211,220]. Moreover, the reaction is strongly governed by the solvent used. The overall performance followed the order methanol > ethanol > butanol. Small conversions were also observed with non-polar solvents such as chloroform or toluene [230].

Vanadium-containing aluminophosphates (VAPOs) with AEL, ATO, AFI, and ERI topologies were found to be active in the vapor-phase oxidation of toluene with molecular oxygen [231,232]. VAPO-11 and VAPO-31 exhibited high catalytic activity and high selectivity for benzaldehyde. The activity decreases in the order VAPO-11 > VAPO-31 > VAPO-5 ≈ VAPO-17, which is ascribed to the structural properties of these catalysts [231]. Recently, VAPO-5 and VAPSO-5 were synthesized with two different templates namely tripropylamine and hexamethylamine (HMA) by Venkatathri et al. [233] and tested in the oxidation of toluene with TBHP and the acetalization of benzaldehyde. The catalysts obtained with HMA as structure-directing agent (SDA) display a higher activity because the extent of vanadium incorporation is higher in this case.

The oxidative dehydrogenation (ODH) of alkanes is one of the important routes to obtain alkenes; thus, a strong research effort has been devoted to develop active and selective ODH catalysts. Although the production of propene from propane is a process that has been known for many years, it was for long not commercialized due to its unfavorable economics, in particular the difference in cost between propane and propene was too small. Recently, the demand and consequently the price for propene have increased and the production of propene from propane is gaining importance. There are currently seven plants in operation producing 2% of the world's propene demand for the manufacture of petrochemicals [45]. From thermodynamic considerations it is obvious that a catalytic approach is required as cracking of propane to ethene and methane is favored by thermodynamics. Moreover, propane conversion is limited and high reaction temperatures up to 600 °C are required. Above 600 °C, multiple dehydrogenation is an unwanted side reaction as the propadiene formed has to be completely removed from the polymer-grade propylene product.

Microporous aluminophosphates with AFI structure have been studied by Conception et al. [234–237] in the oxidation of paraffins. Pure $AlPO_4$-5 shows a low activity in the ODH of propane, while the catalytic performance is enhanced by the incorporation of vanadium ions [234]. A high selectivity to oxydehydrogenation products is achieved at a relative high O_2/alkane ratio and high alkane conversion ($X_{\mathrm{propene}} = 26\%$) resulting in an olefin yield of 15% at a reaction temperature of 540 °C [234]. The catalytic properties of VAPO-5 are superior as compared to vanadium oxide supported on $AlPO_4$-5. It has been suggested that site isolation is the key factor for the design of efficient catalysts for this reaction [238]. Isolated tetrahedrally coordinated V(V) species in the AFI framework structure are proposed to be active and selective sites for this reaction [234,237].

The manganese-substituted molecular sieve MnAPO-5 was reported to be an active catalyst for ethane dehydrogenation to produce ethene [239,240]. Good crystallinity was found to be an essential requirement for a good MnAPO-5 catalyst, because the occlusion of manganese oxide particles in the channels or a structural collapse during calcination results in decreased ethene selectivity [239]. At a reaction temperature of 600 °C, the selectivity to ethane amounts to 46.8% and 52.7% for VAPO-5 and CoAPO-5, respectively, at a conversion of 17% [241]. The presence of Mg(II) and V(V) species in MgVAPO-5 results in a more selective catalyst for ODH of ethane [237,242]. The authors relate the catalytic properties of MgVAPO-5 to the presence of acid sites [Mg(II)] near the redox sites [V(V)] in the framework. CoVAPO-5 has been shown to be even more active for ODH of ethane ($X = 44\%$) than MgVAPO-5 ($X = 8\%$), suggesting that the redox properties of cobalt are responsible for the observed activity [241]. It is found that the activity of different catalysts in the ODH of ethane decreases in the order CoVAPO-5 > VAPO-5 > CoAPO-5 > MgVAPO-5. Moreover, it has been observed that the olefin selectivity during the ODH of C_2–C_4 alkanes decreases with the length of the alkane [241]. The isomorphous substitution of Mg^{2+} or Co^{2+} into the $AlPO_4$ framework generates Brønsted-acid sites that modify the catalytic behavior of the catalyst, probably by changing the desorption rate of the olefinic intermediates. However, results of the spectroscopic characterization of VMgAPO-5 indicate that the vanadium ions are mainly located at extra-framework positions after the calcination step, which might account for the low activity observed [243].

Marchese et al. [244] showed that SAPO-34 is a good catalyst for the ODH of ethane. The high thermal stability of the CHA structure is essential when the reaction proceeds at temperatures between 600 and 700 °C. The reason for this catalyst being very selective for ethene in the ODH of ethane is that both ethene oxidation to carbon oxides and cracking reactions are significantly inhibited by the framework structure and/or the acid character of the catalysts. The introduction of variable amounts of Na^{+} and Ca^{2+} to balance the negative framework charge of the SAPO-34 structure allows the modulation of the acid properties (number and strength of acid sites) of the catalysts as well the access of ethane to the acid sites. It is commonly accepted that strong Brønsted-acid sites are undesirable for the ethane ODH reaction while the presence of transition metal cations enhances both selectivity and activity. Consequently, the catalytic behavior of the V- and/or Co-containing AlPO-18 (AEI topology) possessing a pore size similar to that of CHA was evaluated by Concepcion et al. [245,246]. The conversion of ethane as well as the ethene selectivity decreases in the order VCoAPO-18 > VAPO-18 > CoAPO-18 > VO_x/CoAPO-18 [246]. The influence of the framework structure in the ODH of ethane is shown in Fig. 23, which compares the ethene selectivity over vanadium and cobalt-containing $AlPO_4$ with AEI and AFI topology. It is found that Me-containing AlPOs with AEI topology exhibit higher ethylene selectivities as compared to the catalysts with AFI topology. The use of bifunctional catalysts and judicious choice of the catalyst structure appear to be interesting ways to optimize both activity and selectivity in this reaction. Finally, it is concluded that site isolation is a key factor in the design of efficient catalysts for ODH reactions. Besides this, the reducibility of the active sites and the acid–base properties of the catalysts are important factors in the enhancement of the catalytic performance by modifiying the rate of alkane conversion and the desorption rate of the reaction intermediates [246]. CoAPO-34 and MnAPO-34 (CHA framework topology) were also reported to be moderately active in the oxidation of CO with air [247].

The ammoxidation of toluene and benzyl alcohol with ammonia and air yielding benzonitrile was reported by Kulkarni et al. [248], who employed VAPOs and SAPOs as catalysts. Vanadium-containing SAPOs are also highly active and selective in the one-step ammoxidation of ethanol to acetonitrile [249]. The ammoxidation of 3- and 4-picolines over SAPOs

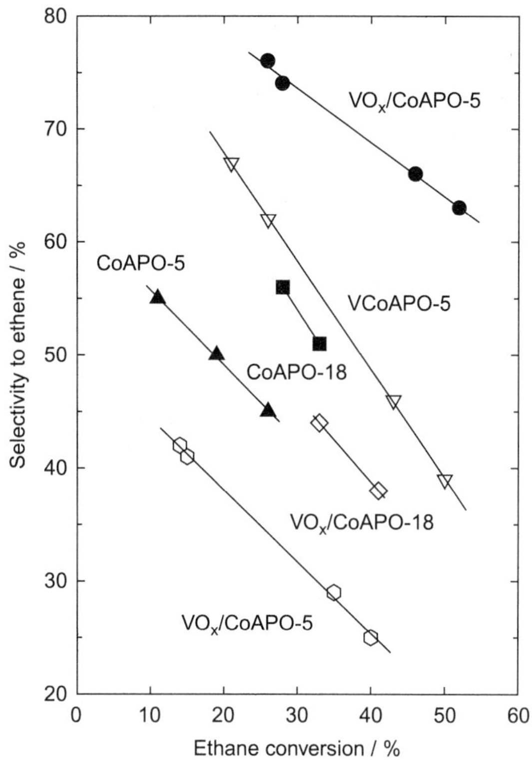

Figure 23 Influence of the framework structure and the substituted TMI on the selectivity in the ODH of ethane (after Ref. [246]).

with AFI topology containing V and combinations of V with Mo in comparison with modified SAPOs with AEL topology was studied by Srinivas et al. [250]. These novel materials were found to be very active and selective and are believed to have potential to replace the amorphous catalysts (V_2O_5 or α-$VOPO_4$ supported on Al_2O_3, SiO_2–Al_2O_3, or Cr_2O_3) presently used in the existing commercial process.

The ammoxidation of cyclohexanone to the corresponding oxime and further rearrangement to ε-caprolactam was studied by Raja et al. [251] over bimetallic catalysts of the type Me(II)Me(III)AlPO-36, where M = Co, Mn. The authors claim that these catalysts perform very well in converting cyclohexanone to the corresponding oxime and further to ε-caprolactam, because hydroxylamine is readily formed inside the pores from NH_3 and O_2 at the Me(III) active sites. The hydroxylamine then converts cyclohexanone to the corresponding oxime both inside and outside the pores of the catalyst. The formed oxime is isomerized over the

Brønsted-acid sites to ε-caprolactam inside the pores of the molecular sieve. It is interesting to note that oxygen yields hydroxylamine more efficiently than hydrogen peroxide or TBHP at the redox active sites.

As shown earlier, there have been some advances in the field of selective alkane functionalization using $AlPO_4$-based molecular sieves as catalysts. However, one important target that remained frustratingly elusive is selective oxidation at a terminal methyl position. A number of industrial applications might be on the horizon if linear terminal alcohols or acids can be produced from alkane feedstock rather than from olefins (through hydroformylation) as currently practiced. It is commonly accepted that selective methyl oxidation is among the major unsolved problems of organic chemistry. Although the work by Thomas et al. [204] does offer considerable encouragement, several questions still have to be answered, which might lead to the development of improved catalysts. Two questions were raised by Labinger [252]: (1) What grounds are there to believe that the radical-chain autoxidation mechanism is operative? (2) What would the active centers be to explain the observed results? Although trivalent Co(III) and Mn(III) do react with alkanes, typically more stringent conditions have to be employed. Moreover, a single-electron transfer would not be expected to give any terminal selectivity, but might well contribute to chain initiation (cf. Fig. 21). Metal oxo or peroxo species are possible, but there is little precedent for their formation directly from O_2. If the radical-chain autoxidation mechanism is operative, arguments based on the favored geometry become rather irrelevant, since the alkane does not encounter the active catalyst site. A more thorough discussion is beyond the scope of this review, but a detailed understanding of the underlying oxidation mechanism is a prerequisite for the development of improved catalyst for the desired terminal oxidation of alkanes.

4.2 Oxidation with peroxides

Direct hydroxylation of phenol yielding catechol (benzene-1,2-diol) and hydroquinone (benzene-1,4-diol) (Fig. 24) is observed over

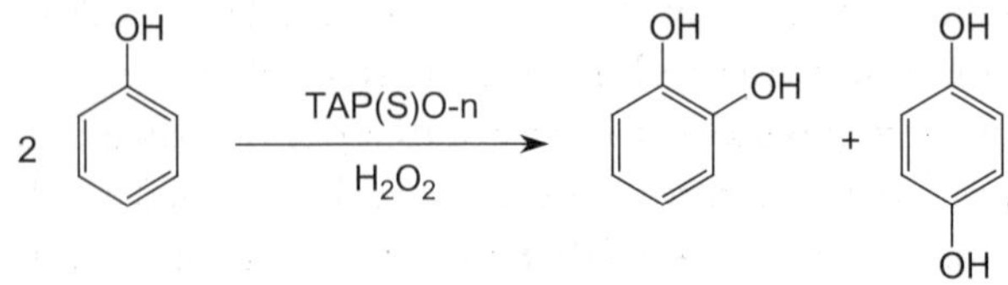

Figure 24 Hydroxylation of phenol yielding catechol and hydrochinone.

titanium-containing catalysts with AEL and AFI framework structure [253]. Hydroxylation occurs to the extent of 32.2% and 17.7% for TAPO-5 and TAPO-11, respectively, whereas in another study, TAPSO-5 and TiSiVPI-5 are found to efficiently catalyze the hydroxylation of phenol, especially under slow addition of H_2O_2 [254]. The conversion of phenol and the selectivities toward catechol and hydroquinone were found to be influenced by the polarity of the solvent, the Ti content, and the crystallinity of the catalyst. Solvents of medium polarity such as acetone are preferred because they can efficiently transfer the reactants to and products away from the catalyst surfaces. The catalytic activity of TAP(S)O-*n* molecular sieves is comparable to that of TS-1, with the catechol/hydroquinone ratios always higher than with TS-1. Some experimental results, however, suggest that the hydroxylation reaction proceeds mainly on the external surface of the catalysts due to complete pore filling of the catalysts by polar molecules from the reaction mixture [253,254].

The catalytic activities of MeAPO-11 (Me = Fe, Co, Mn) for the hydroxylation of phenol with hydrogen peroxide have been examined by Dai et al. [255]. They demonstrated that MeAPOs exhibit significant catalytic activities for this reaction and that introduction of these metals considerably improves the level of conversion. FeAPO-11 shows comparable performance to TS-1. The activity depends on the Al/Fe ratio and the level of acidity of the molecular sieve. MeAPOs of medium-pore size are more active than their counterparts with large pore diameters. However, the external surface of the catalyst was also found to play a significant role for their catalytic activity. Furthermore, it is an additional advantage to use $AlPO_4$ that are synthesized with a relatively cheaper template as compared to TS-1 were TPAOH is employed as SDA [256].

Chromium-containing catalysts with the AFI topology are found to catalyze alcohols to ketones with TBHP [220,257,258]. Carveol and 1-phenyl-1,2-ethanediol are oxidized chemoselectively (Fig. 25). These reactions presumably proceed through oxidation of alcohol through an oxochromium (VI) species and subsequent reduction of the chromium (VI) species by TBHP.

VAPOs with AFI and AEL topologies have been studied to a lesser extent as catalysts for the epoxidation of allylic alcohols such as geraniol and cinnamon alcohol using TBHP as an oxidant [259,260]. The stability of such catalysts against leaching, however, seems to be a problem. Haanepen et al. [261,262] studied the stability of VAPO-5 as catalysts for the epoxidation of 3-phenyl-2-propene-1-ol and the oxidation of 3-octanol by TBHP. They concluded that VAPOs are not suitable catalysts for oxidation

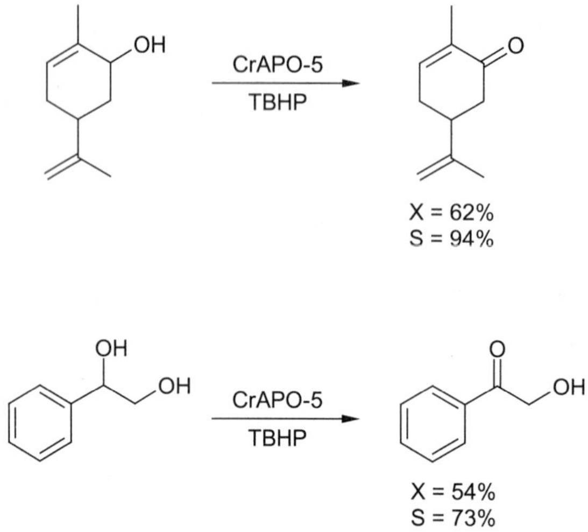

Figure 25 Oxidation of alcohols to ketones.

reactions in the liquid-phase where it is desired to improve the selectivity of the catalyst by means of its pore structure. In the early stage of the reaction, small amounts of vanadium are extracted from the framework. The amount depends on the substrate and the reaction conditions. The vanadium ions in solution contribute largely to the observed activity. Thereafter, the leaching of vanadium stops due to strong adsorption of polar molecules in the micropores of VAPO-5. This makes most vanadium sites inaccessible for reaction and leaching.

CrAPO-5 and CrAPO-11 catalyze various oxidation reactions using TBHP or O_2 as the oxidant (Fig. 26) [263]. CrAPO-5 catalyzes the oxidation of the side chains of aromatic substrates such as ethylbenzene, *p*-alkyltoluene (alkyl = ethyl, propyl, butyl), and *p*-ethylanisole with TBHP forming the corresponding ketones with high selectivity [257,264–266]. CrAPO-5 catalyzes the alkyloxidation of olefins with TBHP at 80 °C to the corresponding enones like α-pinene to verbenon (Fig. 27) [267]. However, it has also been demonstrated that in the presence of TBHP a small amount of chromium ions leaches into the solution, which might be responsible for the observed catalytic activity [264,268,269].

The oxidation of ethylbenzene with TBHP was studied over a series of MeAPO-11 catalysts (Me = Co, Mn, V) by Singh et al. [270]. It is reported that ethylbenzene conversion is higher over the catalysts prepared in the presence of fluoride ions, although the particle size of this materials is larger as compared to those prepared by conventional synthesis methods. The

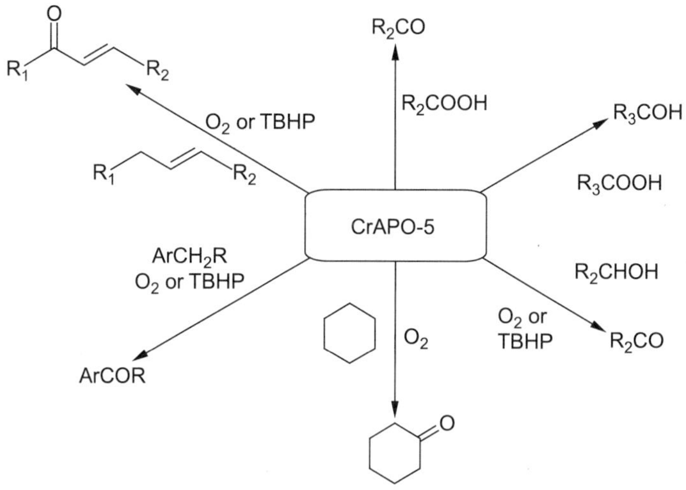

Figure 26 Oxidation reaction catalyzed by CrAPO-5.

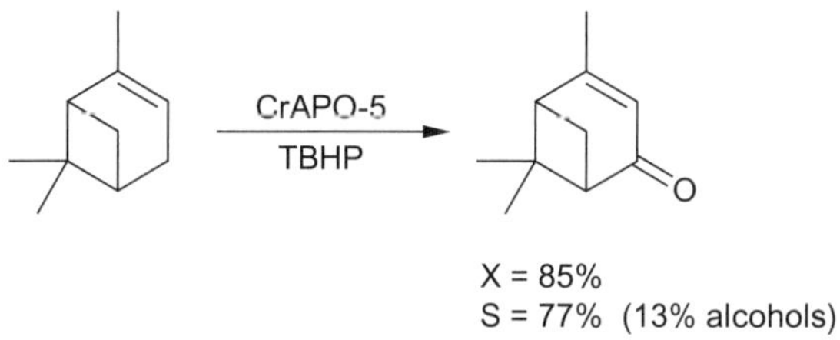

Figure 27 Oxidation of α-pinene to verbenon.

authors suggest that the catalytically active sites are the framework-substituted metals, which are more isolated and more stable as compared to the metal-impregnated samples [270]. The mechanism of ethylbenzene oxidation may involve a rapid and reversible complex formation between peroxide and catalyst without a change in the oxidation state of the metal as suggested earlier by Mimoun [271] and Huybrechts et al. [272] (Fig. 28[A]). Also, an intermolecular mechanism where redox metals are the catalytically active sites which change the oxidation state during the reaction may be involved (Fig. 28[B]) [273].

TAPSO-5 has been used as a catalyst for the liquid-phase epoxidation of cyclohexene [274,275]. When H_2O_2 is used as an oxidant, the formation of 1,2-cyclohexanediol is observed, which is a result of ring opening that may be catalyzed by the acid sites of the catalyst. When TBHP is used as the oxidant, the epoxide is formed very selectively. Only traces of other

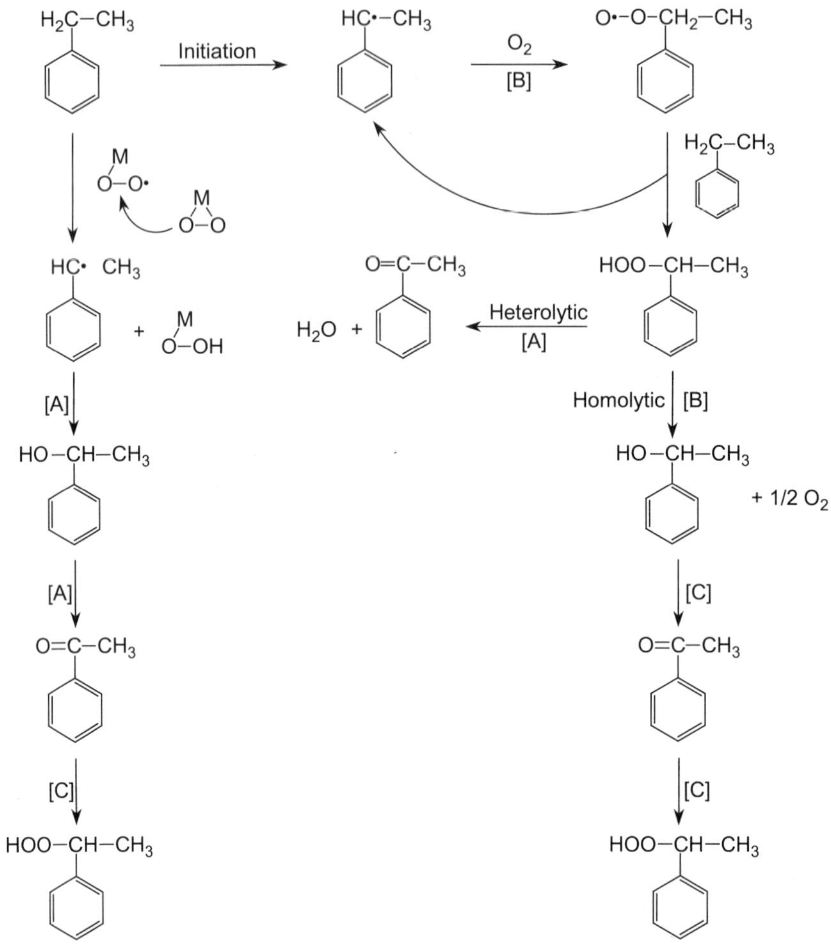

Figure 28 Proposed mechanism of ethylbenzene oxidation over MeAPO-11 (after Ref. [270]). [A] The reaction mechanism involves a rapid reversible complex formation between peroxide and catalyst without a change in the oxidation state, followed by heterolytic cleavage of the O–O peroxide bond. [B] Free radical mechanism involving decomposition of the alkyl peroxide through a heterolytic or a homolytic pathway. [C] Intermolecular mechanism, where the redox metal sites are the catalytically active sites, which change their oxidation state during the reaction.

products such as cyclohexanone, 1,2-cyclohexanediol and bicyclohexenyl are detected. A similar high selectivity for the epoxide has already been reported in an earlier paper by Rigutto and van Bekkum for the epoxidation of cyclohexene over VAPO-5 catalysts [276]. Titanium-containing AlPOs with AEI, AFI, and ATS topologies were found to be active for cyclohexane oxyfunctionalization and cyclohexanol and 2-hexanol oxidation in the presence of dilute H_2O_2 [277,278]. According to these studies, the oxidation activity is associated with framework-incorporated Ti atoms. The selectivity and H_2O_2 activity decrease in the order TAPO-5 > TAPO-11 > TAPO-11 with TAPO-5 being only slightly less active and less selective than TS-1 with similar titanium content. The substitution of Ti(IV) for Si in SAPO-37 may provide catalysts with interesting properties in oxidation reactions [279]. The titanium sites are largely available to H_2O_2 indicating that they are accessible and have the possibility to form the peroxo species needed in redox reactions [280]. However, it is reported that TAPSO-37 is sensitive to moisture even at ambient temperature, which limits its applicability to a dry atmosphere [279].

The oxidation of hydrocarbons such as benzene and cyclohexane with H_2O_2 and TBHP was investigated employing FeAPSO-37 and CrAPSO-37 as catalysts [281,282]. Only the chromium(III)-containing material is active as an oxidation catalyst, while over FeAPSO-37 no conversion was detected. Despite the framework stability under reaction conditions, leaching of small quantities of chromium occurs and the observed catalytic reaction is mainly due to chromium in solution. The oxidation of cyclohexene was studied over NbAPO-5 [283] and the redox activity was compared with NbS-1 and NbMCM-41. The differences in the reducibility and redox activity have been nicely correlated with ESR results showing that the reduction of Nb(V) to Nb(IV) is more difficult in NbAPO-5 than in NbS-1.

The oxidation of glycerol with H_2O_2 using a range of $AlPO_4$ catalysts has been investigated by McMorn et al. [284]. Unfortunately, the desired products glyceric acid and glycerol aldehyde were not formed. Instead, formic acid and monoformate esters of glycerol were observed as major products together with a complex mixture of acetals.

Iron-, cobalt-, and chromium-containing $AlPO_4$ with AFI topology were evaluated in the oxidative desulfurization of dibenzothiophene and 4,6-dimethyldibenzothiophene (as model sulfur contaminants in diesel fuel) in comparison with hexagonal mesoporous aluminophosphates (HMA) also containing the aforementioned TMIs [285]. The most active catalyst is CrHMA, which exhibits a significantly better performance than Cr-$AlPO_4$-5.

As in the case of hydrocarbon oxidation with air, the mechanistic pathway is largely determined by the catalyst used, and some general trends have been described by Sheldon et al. [286] for liquid-phase oxidations at metal ions in constrained environments. Metal-catalyzed oxidations with H_2O_2 or RO_2H can involve homolytic pathways through free radical intermediates ($HO-\cdot$, $HO_2-\cdot$, $RO-\cdot$, RO_2-) and/or heterolytic oxygen transfer processes. The latter can proceed through oxometal or peroxometal

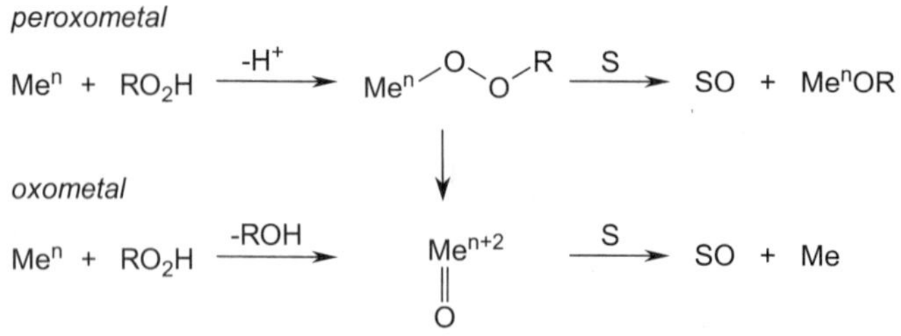

Figure 29 Hetorolytic pathways in catalytic oxidation reactions (S = substrate).

species as the active oxidant [287]. Heterolytic peroxometal pathways (Fig. 29) are favored when the metal in its highest oxidation state [e.g., Ti(IV), Mo(VI)] is both a Lewis acid and a weak oxidant. Catalysis is then due to the Lewis-acid character of the metal ion and its oxidation state does not change during the catalytic cycle. Strong (one-electron) oxidants such as the later and/or first row transition elements [e.g., Cr(VI), Mn(III), Co(III) and Fe(III)] favor oxometal pathways and/or homolytic decomposition of RO_2H. V(V) is both a strong Lewis acid and a relatively strong (one-electron) oxidant and, hence, exhibits all three types of activity. Typical examples of reactions involving peroxometal pathways are olefin epoxidation and oxidation of nitrogen compounds. In contrast, allylic and benzylic oxidations are typical for oxometal or free radical autoxidation pathways. Alcohol oxidations may involve peroxometal and oxometal pathways. Thus, it is suggested that titanium-containing molecular sieves operate through peroxometal routes and catalyze, for example, the oxidation of olefins. Consequently, one would not choose them for allylic or benzylic oxidation. Oxidation of alkanes with hydrogen peroxide in the presence of titanium- or vanadium-containing catalysts is unlikely to involve heterolytic oxygen transfer. The involvement of hydroxyl radicals formed by the metal-mediated homolytic decomposition of hydrogen peroxide seems to be more likely. Although some mechanistic understanding has been gained for oxidation reactions involving H_2O_2 or ROOH as oxidant, other problems including leaching of the active metal species under oxidizing conditions, blocking of active site by the polar reaction products and diffusional constraints currently prevent the use of metal-containing $AlPO_4$ as catalysts for selective oxidations on an industrial scale.

4.3 Alternative oxidants

Since the discovery of catalysts that use nitric oxide as oxidant as alternative to oxygen, the direct conversion of benzene to phenol has been a topic of interest. Solutia Inc. has developed a process for the direct oxidation of benzene to phenol employing FeZSM-5 as catalyst and N_2O as oxidant. In this context, Shiju et al. [288] have examined iron-substituted $AlPO_4$ with AFI structure as a potential catalyst for the hydroxylation of benzene. It is found that the phenol yield over FeAPO-5 amounts to 13.0% at a reaction temperature of 653 K, while around 23% are achieved for FeZSM-5. However, additional high-temperature heat treatments, which are required to obtain active FeZSM-5 catalysts, were not required for the iron-containing $AlPO_4$.

5. MISCELLANEOUS CATALYTIC APPLICATIONS

One of the important goals of catalysis research is to correlate the structural properties of the catalysts with their catalytic efficiency and to guide the development of new catalysts. Initial studies of ethene dimerization, which is catalyzed by nickel (I) and palladium (I) have been carried out to demonstrate this approach [289–291]. The reaction proceeds through two steps [292] involving the dimerization of ethylene to 1-butene and the subsequent isomerization of 1-butene to predominantly 2-butene on acidic catalysts [293]. Studies demonstrate that the process is indeed catalyzed by nickel (I) or palladium (I), which have been detected by ESR spectroscopy under catalytically relevant conditions [289–291].

ESR and ESEM spectroscopy enable a detailed investigation of the reaction on a molecular level. After the adsorption of ethylene, a π-bonded Ni(I)- or Pd(I)-ethylene complex is obtained (Fig. 30). In most cases, the active center is coordinated to one ethylene molecule at a distance of approximately 0.35 nm. Under reaction conditions, the ESR spectra change and new species are detected, which can be attributed to Ni(I)-butene

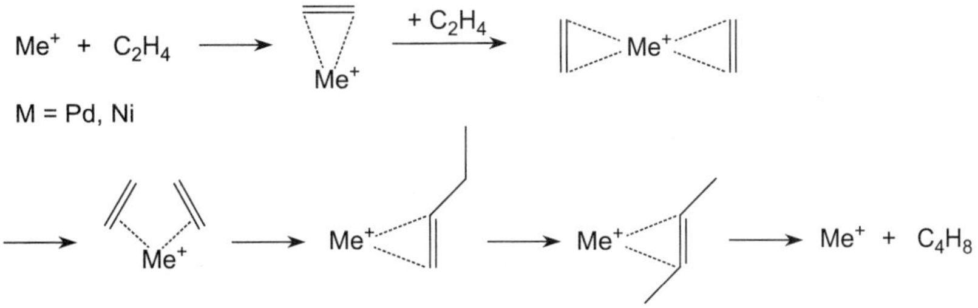

Figure 30 Proposed reaction mechanism for ethene dimerization over Ni(I) or Pd(I) containing catalysts.

complexes. The same complexes are obtained by *n*-butene adsorption on activated samples and therefore their assignment is unambiguous. Additionally, an ESEM analysis of the Ni(I)-butene complex shows an interaction with eight deuterium atoms at different distances, which is consistent with eight deuterium atoms located on one butane molecule. Butene complexes are not obtained with PdH-SAPO-34 and NiH-SAPO-34 as catalysts, which shows that due to the smaller channel size no reaction at temperatures below 100 °C occurs [290].

Nickel-containing $AlPO_4$-5 was prepared by direct synthesis and incipient wetness impregnation and tested in the amination of polyethyleneglycol (PEG) [294]. Polyetherpolyamines are used as curing agents for epoxy resins, plasticizers, cross-linking agents for textiles, defoamers, and drug carriers. It was found that the amine yield is higher for the catalysts prepared by the incipient wetness technique as compared to the catalyst prepared by direct hydrothermal synthesis. However, the nickel content of the latter catalysts is not reported; thus, the lower amine yield might be a consequence of the lower nickel content of the catalysts prepared by direct hydrothermal synthesis. The selectivity for primary and secondary amines is almost comparable for the catalysts prepared by impregnation, whereas primary amines are predominant over NiAPO-5. However, when tridymite is the prevailing phase, the selectivity to the secondary amine is almost 100% [294].

Fe/$AlPO_4$-5 catalysts prepared by impregnation in aqueous and organic media and subsequently activated by a reductive treatment were tested for CO hydrogenation in a stainless steel microreactor ($p = 1.1$ MPa, $T = 598$ K, WHSV $= 1200\,h^{-1}$, H_2/CO $= 1$). It was shown that Fe/$AlPO_4$-5 prepared through impregnation in an aqueous solution shows no activity, while the catalysts prepared by impregnation in organic media show conversions between 6% and 23%, depending on the type of organic solvent used [295]. MeAPO-5 (Me = Co, Fe) exhibits a low activity (approximately 3%) in synthesis gas (CO and H_2) conversion, whereas Co/$AlPO_4$-5 gives 42% CO conversion [296]. Significant differences in selectivity were observed for the different catalysts. The mixture of an iron-based Fischer–Tropsch catalyst with FAPO-*n* and FAPSO-*n* ($n = 5$, 11) molecular sieves, which provide acid sites to the catalyst, results in enhanced syngas conversion and improved selectivity for light hydrocarbons (C_2–C_4). It is anticipated that iron species associated with the molecular sieve framework participate in the reaction.

A one-step synthesis of methylisobutylketone (MIBK), which is the most important product derived from acetone, over palladium supported on $AlPO_4$-11 and SAPO-11 has been reported by Yang and Wu [297].

Their results show that a balance between condensation and hydrogenation is necessary to achieve a high selectivity for MIBK. The highest MIBK yield is obtained over 0.3 Pd/SAPO-11 at a reaction temperature of 200 °C, a H_2/acetone ratio of 1 and a WHSV of 0.7 h^{-1}. Basic sites are found to promote the activity of the catalyst in this reaction.

The interaction of nitric oxide with carbon monoxide on the surface of copper-containing $AlPO_4$ and SAPOs of the AFI structure has been studied by the transient response technique [298]. In the temperature range studied (60–300 °C), the catalysts start to interact first with the carbon monoxide above a definite temperature, which result in a reduction of the catalysts. Cu/$AlPO_4$-5 and Cu/SAPO-5 exhibit activity toward the conversion of nitric oxide to nitrogen above 200 °C, which is comparable to a CuO/Al_2O_3 catalyst. In contrast, MnAPO-5 und CoAPSO-11 do not interact with the components of the gas phase (NO, N_2O, and CO) and have no activity with respect to the interaction of nitric oxide and nitrous oxide with CO over the whole temperature range studied.

The catalytic activity of copper ion exchanged SAPO-5, SAPO-11, and SAPO-34 in the selective reduction of NO with C_3H_6 under an oxidizing atmosphere has been tested by Ishihara et al. [299]. Under the experimental conditions of their study, Cu-SAPO-34 exhibits higher activity for NO reduction than Cu-ZSM-5 and sustains its catalytic activity up to 600 °C. Although the catalytic activity of Cu-SAPO-11 is comparable to that of Cu-ZSM-5, the activity of Cu-SAPO-5 is significantly lower. CoAPSO-34 has been evaluated as a catalyst for NO conversion to N_2 in the presence of various hydrocarbons such as CH_4, C_3H_6, C_3H_8, C_8H_{18}, and $C_{16}H_{34}$ [300]. The NO conversion increases with the Co content of the sample up to $n_{Si}/n_{Co} = 5$. It was concluded that cobalt is mainly located in tetrahedral framework positions and causes an increase in the number of acid sites. A Ce-H-SAPO catalyst with undisclosed structure has been tested in the SCR of NO_x by acetylene in excess oxygen. However, the NO conversion to CO is low (6.6%) as compared to Ce-H-ZSM-5 (83%) and Ce-HY (75.4%) [301].

Copper-exchanged MgAPO-11 and ZnAPO-11 molecular sieves have been tested for NO decomposition by Dedecek et al. [302]. Both Cu-MgAPO-11 and Cu-ZnAPO-11 exhibited constant conversion for NO with time-on-stream. The turnover frequencies (TOF) per Cu atom at 770 K are comparable to those of Cu ions in cationic sites of Cu-ZSM-5.

Iron-containing FAPO-5 was evaluated in N_2O decomposition before and after steaming at 873 K [303]. In agreement with the results previously reported for FeZSM-5, steam treatment results in a significant increase of activity due to migration of framework iron to extra-framework positions and reduction of Fe(III) to Fe(II). However, the activity of steamed FAPO-5 is relatively low as compared to activity of isomorphously substituted FeZSM-5.

Photoionization of methylphenothiazine, which is of interest in solar energy utilization, was studied in microporous SAPOs with different pore sizes of CHA, AFI, AEL, and VFI topologies [304]. The photo yield of methylphenothiazine cation radicals was enhanced by an order of magnitude or more by the incorporation of the TMIs Ni, Co, and Cu as electron acceptors in framework or in ion-exchange sites of the microporous material. Ni(II) is the most efficient acceptor, whereas the addition of Co(II) into either ion-exchange or framework sites does not produce any significant effect on the photo yield. For Cu(II) as an electron acceptor, there is a distinct difference between copper located in framework against ion-exchange sites. It was also found that the photo yield increases with the SAPO pore size. Chromium-containing SAPO materials exhibited the highest photo yield among the TMIs studied [305]. The presence of chromium in the framework further increases the photo yield and the stability of the methylphenothiazine cation radical. It follows from these studies that SAPOs provide appropriate steric and electrostatic environments to retard electron back transfer and increase the lifetime of the photogenerated radical ions for many days or even months at room temperature.

Increasing attempts have been made to develop aligned molecular sieve crystals, especially in a membrane form, for sensors and nonlinear optical devices. A membrane of vertically aligned MeAPO-5 (Me = Co, V) crystals on anodic alumina as a support has been reported by Chao et al. [306]. The MeAPO crystals were found to be preferentially oriented along the *c*-axis. Thin films of MAPO-39 have been prepared through pulsed laser deposition (PLD) [307]. The MAPO-39 crystals grow with pores oriented primarily normal to the porous metal substrate. The separation of water/alcohol mixtures was also studied showing that selective permeation of water occurs. However, to the best of our knowledge, an evaluation of membrane reactors based on MeAPOs or MeAPSOs has so far not been reported.

SAPO-37 has been shown to be a promising catalyst for the degradation of HDPE [308]. The thermal degradation of HDPE produces a wide range of hydrocarbons from C_5 to C_{25}, whereas in the presence of SAPO-37, only C_1 to C_{12} hydrocarbons were detected.

NO_x abatement in gas emissions from both stationary and mobile sources is a relevant environmental issue, which may be tackled by means of catalytic processes. Thus, an increasing number of new materials are prepared and studied as possible catalysts for $DeNO_x$ reactions [309]. The catalytic properties of Co- and Cu-containing MeAPO-34 and MeAPSO-34 materials towards $DeNO_x$ activity was studied by Frache et al. [310]. It is revealed that only Cu-containing materials behave similar to Cu-ZSM-5 in NO and N_2O decomposition. The catalytic activity in N_2O decomposition of CuAPSO-34 is much higher as compared to CuAPO-34, most probably due to the presence of a larger fraction of monovalent copper ions in the former material. These conclusions are supported by FTIR spectroscopy of adsorbed NO and CO. CuAPO-34 shows a good activity even after treatment at 900 °C under dry conditions and at 400 °C in the presence of water vapor [311]. The same authors [312] have also studied the selective catalytic reduction of NO_x with hydrocarbons or CO over Co- and Cu-based $AlPO_4$. Moreover, CoAPO-34 exhibited the highest activity in the oxidation of NO to NO_2 [313]. The authors used a multi-technique approach employing various spectroscopic tools to understand the structure and morphology of the catalysts as well as the coordination and redox-state of the catalytic sites. They concluded that the protons are exchangeable sites for the introduction of redox centers such as Co(II)/Co(I) and Cu(II)/Cu(I), which leads to active catalysts for $DeNO_x$ reactions. Moreover, they claimed that when the TMIs are directly introduced into the synthesis gel, even more stable redox molecular sieves may be obtained. This is a necessary requirement when $DeNO_x$ reactions are run in the presence of water and/or at high temperatures.

Glycerol is a major by-product of biodiesel production from non-food crops. Although a small portion of the crude glycerol is purified for pharmaceutical and food grade applications, the majority is taken as waste. Efficient and effective use of such a material represents a great challenge. In this context, steam reforming of glycerol over Pt/SAPO-11 is evaluated for hydrogen production [314]. However, Pt/SAPO-11 exhibited remarkable deactivation and a lower catalytic activity as compared to Pt/USY and Pt/Al_2O_3.

To reduce the contents of sulfur compounds and olefins in FCC gasoline and at the same time preserve its RON, various hydro-upgrading processes that combine hydrodesulfurization (HDS) with hydrocracking and isomerization have been proposed. A novel ZSM-5/SAPO-11 composite catalyst has been synthesized through *in situ* crystallization of SAPO-11 on ZSM-5 [315]. The comparison of the catalytic performances of the mechanical mixture and the composite-based Ni–Mo catalysts for FCC gasoline hydro-upgrading showed that the composite catalysts exhibited improved gasoline RON, high yield of liquid products, good desulfurization activity, and slower coke formation as compared to the mechanical mixture.

6. CONCLUSIONS AND OUTLOOK

In the roughly 25 years since their discovery, $AlPO_4$ and SAPOs doped with TMIs have been tested for various heterogeneously catalyzed reactions including selective oxidations, acid-catalyzed reactions, and isomerization and hydrocracking over bifunctional catalysts. Although improvements have been made for some catalytic systems, to the best of our knowledge only two are close to be realized commercially. UOP and Norsk Hydro have built a pilot plant for the conversion of methanol into lower olefins using a SAPO-34-based catalyst and plant construction has been completed in Lagos, Nigeria, for the production of 40,000 tons $year^{-1}$ of high-density PP and polyethylene. Chevron is working on the realization of an isodewaxing unit using Pt/SAPO-11. With these two large-scale applications appearing at the horizon, additional new processes might emerge, if further improvements of the catalyst systems based on TMI-containing (silico)aluminophosphates are made.

The introduction of TMIs into (silico)aluminophosphates can be achieved by liquid-phase or solid-state ion exchange, through impregnation or through the addition of a transition-metal compound to the synthesis gel. The direct addition of TMIs into the synthesis gel might under favorable conditions result in isomorphous substitution of the TMIs into the $AlPO_4$ framework. To prove isomorphous substitution, several complimentary spectroscopic techniques including UV-Vis, IR, ESR, NMR, and Mössbauer spectroscopy have been employed, but the presentation of unambiguous evidence for true isomorphous substitution remains difficult.

It is generally accepted that isomorphous substitution of TMIs into $AlPO_4$ can be used in a number of cases to produce superior catalysts as compared to materials prepared by ion exchange or impregnation. It has been demonstrated in several studies that MeAPOs and MeAPSOs are powerful liquid-phase oxidation catalysts. However, the initial enthusiasm has been gradually reduced by the realization that many of these materials are not stable under oxidizing conditions in the liquid phase. Leaching seems to be a major problem, in particular with catalysts containing chromium, vanadium, or cobalt. Moreover, pore blocking by polar products severely reduces the lifetime of these catalysts, which, at present, prevents them from being used in large-scale commercial applications. Many of the studied liquid-phase oxidation reactions are quite selective. One ultimate goal, which is still far from being achieved, is the design of chiral redox molecular sieves. Such materials are required to catalyze asymmetric oxidations yielding the desired products with a high enantiomeric excess.

MeAPOs also represent an interesting alternative to SAPOs and zeolites in acid-catalyzed reactions. By judicious choice of the substituting element and a suitable framework structure, the number and the strength of the acid sites can be tuned. However, the MeAPOs employed in acid-catalyzed reactions typically contain acid sites ranging from mild to strong in the same catalyst. Therefore, the design of catalysts with uniform sites representing a narrow distribution of acid site strength is another challenge in modern catalysis research.

After more than 20 years of intensive research, the field of catalysis with microporous (silico)aluminophosphates containing TMIs has reached a certain maturity. With two large-scale processes around the corner, significant advances are now expected in the area of fine chemicals and intermediates production and more research work is expanding into this interesting field. The use of alternative reactor concepts including membrane and microstructure reactors in combination with optimized aluminophophate-based catalysts is still in its infancy. Moreover, the future use of *in situ* techniques such as NMR, IR, and UV-Vis spectroscopy might lead to an advanced understanding of aluminphosphate-based catalysts under "working conditions" and ultimately to improved tailor-made catalysts. The recent advances in the synthesis of microporous molecular sieves, namely, the use of ionic liquids [316] and the discovery of heteroatom-containing chiral $AlPO_4$ [317], are expected to revitalize research in this fascinating field of molecular sieve science in the future.

ACKNOWLEDGMENTS

The authors' work in this area was generously supported by Deutsche Forschungsgemeinschaft and Fonds der Chemischen Industrie.

REFERENCES

[1] S.T. Wilson, B.M. Lok, C.A. Messina, T.R. Cannan, E.M. Flanigen, J. Am. Chem. Soc. 104 (1982) 1146.

[2] B.M. Lok, C.A. Messina, R.L. Patton, R.T. Gajek, T.R. Cannan, E.M. Flanigen, J. Am. Chem. Soc. 106 (1984) 6092.

[3] E.M. Flanigen, B.M. Lok, R.L. Patton, S.T. Wilson, Pure Appl. Chem. 58 (1986) 1351.

[4] E.M. Flanigen, B.M. Lok, R.L. Patton, S.T. Wilson, in: Y. Murakami, A. Ijima, J.W. Ward (eds.), New Developments in Zeolite Science and Technology, Proceedings of the 7th International Zeolite Conference, Tokyo, Kodansha Ltd., Tokyo and Elsevier, Amsterdam, 1986, p. 103.

[5] S.T. Wilson, E.M. Flanigen, in: M.J. Occelli, H.E. Robson (eds.), Zeolite Synthesis, ACS Symp. Ser., vol. 398, ACS, Washington, DC, 1989, pp. 329–345.
[6] M. Hartmann, L. Kevan, Chem. Rev. 99 (1999) 635.
[7] B.M. Weckhuysen, R.R. Rao, J.A. Martens, R.A. Schoonheydt, Eur. J. Inorg. Chem. (1999) 565.
[8] S.T. Wilson, Stud. Surf. Sci. Catal. 168 (2007) 105.
[9] S.T. Wilson, US Patent 4,567,029 (assigned to Union Carbide Corp.), 1986.
[10] S.T. Wilson, Stud. Surf. Sci. Catal. 137 (2001) 229.
[11] J.A. Rabo, G.J. Gajda, Catal. Rev.-Sci. Eng. 31 (1989) 385.
[12] H. Nur, H. Hamdan, Mater. Res. Bull. 36 (2001) 315.
[13] C. De las Pozas, R. Lopez-Cordero, J.A. Gonzalez-Morales, N. Travieso, R. Roque-Malherbe, J. Mol. Catal. 83 (1993) 145.
[14] D. Stojakovic, N. Rajic, J. Porous Mater. 8 (2001) 239.
[15] I. Saadoune, F. Cora, C.R.A. Catlow, J. Phys. Chem. B 107 (2003) 3003.
[16] A. Chatterjee, J. Mol. Graphics Modell. 24 (2006) 262.
[17] M. Elanany, M. Koyama, M. Kubo, P. Selvam, A. Miyamoto, Microporous Mesoporous Mater. 71 (2004) 51.
[18] M. Elanany, D.P. Vercauteren, M. Kubo, A. Miyamoto, J. Mol. Catal. A 248 (2006) 181.
[19] Z. Zhu, Q. Chen, Z. Xie, W. Yang, C. Li, Microporous Mesoporous Mater. 88 (2006) 16.
[20] I. Saadoune, F. Cora, M. Alfredsson, C.R.A. Catlow, J. Phys. Chem. B 107 (2003) 3012.
[21] M. Hartmann, L. Kevan, Res. Chem. Intermed. 28 (2003) 625.
[22] H.O. Pastore, S. Coluccia, L. Marchese, Annu. Rev. Mater. Res. 35 (2005) 351.
[23] T. Kimura, Microporous Mesoporous Mater. 77 (2005) 97.
[24] A. Rhodes, Oil Gas J, April 5 (1999) 39.
[25] H.H. Mooiweer, K.P. de Jong, B. Kraushaar-Czarnetzki, W.H.J. Stork, B.H. Krutzen, Stud. Surf. Sci. Catal. 84 (1994) 2327.
[26] R.J. Pellet, P.K. Couglin, E.S. Shamshoum, J.A. Rabo, ACS Symp. Ser. 368, Ch. 13 (1988).
[27] J.A. Rabo, in: Zeolite Microporous Solids: Synthesis, Structure and Reactivity, NATO ASI Series, vol. 352, Kluwer, London, 1992, pp. 531–554.
[28] H.L. Zubowa, M., Richter, U. Roost, B. Parlitz, R. Fricke, Catal. Lett. 19 (1993) 67.
[29] L.H. Gielgens, I.H.E. Veenstra, V. Ponec, M.J. Haanepen, J.H.C. van Hooff, Catal. Lett. 32 (1995) 195.
[30] D. Arias, I. Campos, D. Escalante, J. Goldwasser, C.M. Lopez, F.J. Machado, B. Mendez, D. Moronta, M. Pinto, V. Sazo, M.M. Ramirez de Agudelo, J. Mol. Catal. A 122 (1997) 175.
[31] S.M. Yang, D.H. Guo, J.S. Lin, G.T. Wang, Stud. Surf. Sci. Catal. 84 (1994) 1677.
[32] D. Escalante, L. Giraldo, M. Pinto, C. Pfaff, V. Sazo, M. Matjushin, B. Mendez, C.M. Lopez, F.J. Machado, J. Goldwasser, M.M. Ramirez de Agudelo, J. Catal. 169 (1997) 176.
[33] J.D. Chen, R.A. Sheldon, J. Catal. 153 (1995) 1.
[34] D. Escalante, B. Méndez, G. Hernández, C.M. López, F.J. Machado, J. Goldwasser, M.M. Ramirez de Agudelo, Catal. Lett. 47 (1997) 229.
[35] J. Cejka, B. Wichterlova, P. Sarv, Appl. Catal. A 179 (1999) 217.
[36] S. van Donk, J.H. Bitter, K.P. de Jong, Appl. Catal. 212 (2001) 97.
[37] M. Höchtl, A. Jentys, H. Vinek, Appl. Catal. A 207 (2001) 397.
[38] S.M. Yang, S.Y. Liu, in: R. Van Ballmoos, J.B. Higgins, M.M.J. Treacy (eds.), Proceedings of the 9th International Zeolite Conference, Montreal 1992, Butterworth-Heinemann, Stoneham, MA, 1993, pp. 623–630.
[39] S.P. Elangovan, V. Krishnasamy, V. Murugesan, React. Kinet. Catal. Lett. 55 (1995) 153.

[40] S.P. Elangovan, B. Arabindoo, V. Krishnasamy, V. Murugesan, J. Chem. Soc., Faraday Trans. 91 (1995) 4471.
[41] C.M. Lopez, D. Arias, F.J. Machado, A. Tuel, React. Kinet. Catal. Lett. 74 (2001) 163.
[42] M. Eswaramoorthy, V. Mohan, V. Krishnasamy, Indian J. Chem. A 40 (2001) 605.
[43] C. Kannan, S.P. Elangovan, M. Palanichamy, V. Murugesan, Proc. Indian Acad. Sci. 110 (1998) 491.
[44] C. Kannan, S.P. Elangovan, M. Palanichamy, V. Murugesan, Indian J. Chem. A 37 (1998) 655.
[45] S. Ernst, S. Petzny, in: R. Dittmeyer, W. Keim, G. Kreysa, A. Oberholz (eds.), Winnacker/Küchler. Chemische Technik: Prozesse und Produkte, vol. 4, Wiley-VCH, Weinheim, 2005, pp. 523–639.
[46] O. Bortnovsky, P. Sazama, B. Wichterlova, Appl. Catal. A 287 (2005) 203.
[47] G.A. Olah, B. Gupta, M. Farina, J.D. Felberg, W.M. Ip, A. Husain, R. Karpeles, K. Lammertsma, A.K. Melhotra, N.J. Trivedi, J. Am. Chem. Soc. 1074 (1985) 7097.
[48] M. Stöcker, Microporous Mesoporous Mater. 29 (1999) 3.
[49] L.V. MacDougall, Catal. Today 8 (1991) 337.
[50] W.O. Haag, in: R.A. Bisio, D.G. Olson (eds.), Proceedings of the 6th International Zeolite Conference, Butterworth, Guildford, 1984, p. 466.
[51] W. Hölderich, H. Eichhorn, R. Lehnert, L. Marosi, W. Mross, R. Reinke, W. Ruppel, H. Schlimper, in: R.A. Bisio, D.H. Olson (eds.), Proceedings of the 6th International Zeolite Conference, Butterworth, Guildford, 1984, p. 545.
[52] R.M. Dessau, J. Catal. 99 (1986) 111.
[53] S. Kolboe, Acta Chem. Scand. A 40 (1986) 711.
[54] I.M. Dahl, S. Kolboe, Catal. Lett. 20 (1993) 329.
[55] I.M. Dahl, S. Kolboe, J. Catal. 149 (1994) 458.
[56] I.M. Dahl, S. Kolboe, J. Catal. 161 (1994) 304.
[57] W. Song, H. Fu, J.F. Haw, J. Phys. Chem. B 105 (2001) 12839.
[58] W. Song, J.B. Nicholas, J.F. Haw, J. Phys. Chem. B 105 (2001) 4317.
[59] H. Fu, W. Song, D.M. Marcus, J.F. Haw, J. Phys. Chem. B 106 (2002) 5648.
[60] U. Olsbye, M. Bjørgen, S. Svelle, K. Lillerud, S. Kolboe, Catal. Today 106 (2005) 108.
[61] J.F. Haw, D.M. Marcus, Top. Catal. 34 (2005) 41.
[62] W. Wang, Y. Jiang, M. Hunger, Catal. Today 117 (2006) 102.
[63] W. Song, J.F. Haw, J.B. Nicholas, K. Heneghan, J. Am. Chem. Soc. 122 (2000) 10726.
[64] W. Wang, A. Buchholz, M. Seiler, M. Hunger, J. Am. Chem. Soc. 125 (2003) 15260.
[65] F. Salehira, M.W. Anderson, J. Catal. 177 (1998) 189.
[66] A. Philippou, F. Salehira, D.P. Luigi, M.W. Anderson, J. Chem. Soc., Faraday Trans. 94 (1998) 2851.
[67] H. Fu, W. Song, D.M. Marcus, J.F. Haw, J. Phys. Chem. B 106 (2002) 5648.
[68] A.G. Gayubo, A.T. Aguayo, A.E. Sanchez del Campo, A.M. Tarrio, J. Bilbao, Ind. Eng. Chem. Res. 39 (2000) 292.
[69] F. Salehirad, M.W. Anderson, J. Chem. Soc., Farady Trans. 94 (1998) 2857.
[70] W. Song, J.F. Haw, J.B. Nicholas, C.S. Heneghan, J. Am. Chem. Soc. 122 (2000) 10726.
[71] P.E. Sinclair, C.R.A. Catlow, J. Chem. Soc., Faraday Trans. 93 (1997) 333.
[72] J.Q. Chen, A. Bozzano, B. Glover, T. Fuglerud, S. Kvisle, Catal. Today 106 (2003) 103.
[73] European Chemical News, October 14–20, 2002.
[74] T. Inui, H. Matsuda, H. Okaniwa, A. Miyamoto, Appl. Catal. 58 (1990) 155.
[75] J. Chen, P.A. Wright, J.M. Thomas, S. Natarajan, L. Marchese, S.M. Bradley, G. Sankar, C.R.A. Catlow, P.L. Gai-Boyes, R.A. Townsend, C.M. Lok, J. Phys. Chem. 98 (1994) 10216.

[76] A.G. Gayubo, R. Vivanco, A. Alonso, B. Valle, A.T. Aguayo, Ind. Eng. Chem. Res. 44 (2005) 6605.
[77] A.T. Aguayo, A.G. Gayubo, R. Vivanco, A. Alonso, J. Bilbao, Ind. Eng. Chem. Res. 44 (2005) 7279.
[78] A.G. Gayubo, A.T. Aguayo, A. Alonso, A. Atutxa, J. Bilbao, Catal. Today 106 (2005) 112.
[79] A.T. Aguayo, A.G. Gayubo, R. Vivanco, M. Olazar, J. Bilbao. Appl. Catal. A 283 (2005) 197.
[80] T. Inui, S. Phatanasri, H. Matsuda, J. Chem. Soc., Chem. Commun. (1990) 205.
[81] J.M. Thomas, Y. Xu, C.R.A. Catlow, J.W. Couves, Chem. Mater. 3 (1991) 667.
[82] S. Hocevar, J. Levec, J. Catal. 135 (1992) 518.
[83] J. Chen, J.M. Thomas, J. Chem. Soc., Chem. Commun. (1994) 603.
[84] R. Wendelbo, A. Akporiaye, I. Andersen, M. Dahl, H.B. Mostad, Appl. Catal. A 142 (1996) L197.
[85] M. Kang, J. Mol. Catal. A 160 (2000) 437.
[86] M.A. Djieugoue, A.M. Prakash, L. Kevan, J. Phys. Chem. B 103 (1999) 804.
[87] M.J. van Niekerk, J.C.Q. Fletscher, C.T. O'Connor, Appl. Catal. A 138 (1996) 135.
[88] T. Inui, Am. Chem. Soc., Div. Petr. Chem. Prepr. 35 (1990) 124.
[89] T. Inui, M. Kang, Appl. Catal. A 164 (1997) 211.
[90] T. Inui, in: Progress in Zeolite and Microporous Materials, Stud. Surf. Sci. Catal. 105 (1996).
[91] M. Kang, T. Inui, Catal. Lett. 53 (1998) 171.
[92] D.L. Obrzut, P.M. Adekkanattu, J. Thundimadathil, J. Liu, D.R. Dubois, J.A. Guin, React. Kinet. Catal. Lett. 80 (2003) 113.
[93] M. Kang, T. Inui, J. Mol. Catal. A 144 (1999) 329.
[94] M. Kang, T. Inui, J. Mol. Catal. A 140 (1999) 55.
[95] D.R. Dubois, D.L. Obrzut, J. Liu, J. Thundimadathil, P.M. Adekkanattu, J.A. Guin, A. Punnoose, M.S. Seehra, Fuel Proc. Technol. 83 (2003) 203.
[96] P. Dutta, A. Manivannan, M.S. Seehra, P.M. Adekkanattu, J.A. Guin, Catal. Lett. 94 (2004) 181.
[97] Z. Zhu, M. Hartmann, L. Kevan, Chem. Mater. 12 (2000) 2781.
[98] M.A. Djieugoue, A.M. Prakash, L. Kevan, J. Phys. Chem. B 104 (2000) 6452.
[99] L. Smith, A.K. Cheetham, L. Marchese, J.M. Thomas, P.A. Wright, J. Chen, E. Gianotti, Catal. Lett. 41 (1996) 13.
[100] J. Chen, P.A. Wright, S. Natarajan, J.M. Thomas, Stud. Surf. Sci. Catal. 84 (1994) 1731.
[101] S. Hocevar, J. Batista, V. Kaucic, J. Catal. 139 (1993) 351.
[102] N. Rajic, D. Stojakovic, S. Hocevar, V. Kaucic, Zeolites 13 (1993) 384.
[103] S.W. Kaiser, EP 249 915 (1987).
[104] S.W. Kaiser, Arab. J. Sci. Eng. 10 (1985) 361.
[105] G. Lischke, B. Parlitz, U. Lohse, E. Schreier, R. Fricke, Appl. Catal. A 166 (1998) 351.
[106] D.B. Akolekar, Appl. Catal. 112 (1994) 125.
[107] T. Tsoncheva, R. Dimitrova, Chr. Minchev, Appl. Catal. A 171 (1998) 241.
[108] D.B. Akolekar, J. Mol. Catal. A 104 (1995) 95.
[109] D.B. Akolekar, Zeolites 17 (1996) 283.
[110] D.B. Akolekar, S. Kaliguine, J. Chem. Soc., Faraday Trans. 89 (1993) 4141.
[111] D.B. Akolekar, J. Chem. Soc., Faraday Trans. 90 (1994) 1041.
[112] G. Zadrozna, E. Souvage, J. Kornatowski, J. Catal. 208 (2002) 270.
[113] F. Collignon, G. Poncelet, J. Catal. 202 (2001) 68.
[114] X. Zhang, R. Wang, X. Yang, F. Zhang, Microporous Mesoporous Mater. 116 (2008) 210.
[115] C.E. Taylor, R.P. Noceti, R.R. Schehl, Stud. Surf. Sci. Catal. 36 (1988) 483.

[116] C.E. Taylor, Stud. Surf. Sci. Catal. 130 (2000) 3633.
[117] S. Svelle, S. Aravinthan, M. Bjørgen, K.-P. Lillerud, S. Kolboe, I.M. Dahl, U. Olsbye, J. Catal. 241 (2006) 243.
[118] D. Zhang, Y. Wie, L. Xu, A. Du, F. Chang, B.-L. Su, Z. Liu, Catal. Lett. 109 (2006) 97.
[119] Y. Wie, Y. He, D. Zhang, L. Xu, S. Meng, Z. Liu, B.-L. Su, Microporous Mesoporous Mater. 90 (2006) 188.
[120] Y. Wei, D. Zhang, L. Xu, F. Chang, Y. He, S. Meng, B.L. Su, Z. Liu, Catal. Today 131 (2008) 262.
[121] D. Zhang, Y. Wei, L. Xu, F. Chang, Z. Liu, S. Meng, B.L. Su, Z. Liu, Microporous Mesoporous Mater. 116 (2008) 684.
[122] B.Y. Hsu, S. Cheng, Microporous Mesoporous Mater. 21 (1998) 505.
[123] M. Hsien, H.T. Sheu, T. Lee, S. Cheng, J.F. Lee, J. Mol. Catal. A 181 (2002) 189.
[124] S.P. Elangovan, K. Inoue, T. Okubo, A. Kojima, M. Ogura, Ind. Eng. Chem. Res. 46 (2007) 1039.
[125] H. Rastelli, B.M. Lok, J.A. Duisman, D.E. Earls, J.T. Mullhaupt, Can. J. Chem. Eng. 40 (1982) 44.
[126] J. Meusinger, H. Vinek, J. Lercher, J. Mol. Catal. 87 (1994) 263.
[127] P.A. Wright, C. Sayag, F. Rey, D.W. Lewis, J.D. Gale, S. Natarajan, J.M. Thomas, J. Chem. Soc., Faraday Trans. 91 (1995) 357.
[128] D.B. Akolekar, Catal. Lett. 28 (1994) 249.
[129] D.B. Akolekar, Appl. Catal. A 171 (1998) 261.
[130] M. Guisnet, Ph. Bichonn, N.S. Gnep, N. Essayem, Top. Catal. 11 (2000) 247.
[131] N. Kumar, J.I. Villegas, T. Salmi, D.Y. Murzin, T. Heikkila, Catal. Today 100 (2005) 355.
[132] A.C. Oliveira, N. Essayem, A. Tuel, J.M. Clacens, Y. Ben Taarit, Catal. Today 133–135 (2008) 56.
[133] S.P. Elangovan, C. Kannan, B. Arabindoo, V. Murugesan, Appl. Catal. A 174 (1998) 213.
[134] S.P. Elangovan, V. Krishnasamy, V. Murugesan, Indian J. Chem. A 34 (1995) 469.
[135] M.K. Dongare, D.P. Sabde, R.A. Shaikh, K.R. Kamble, S.G. Hegde, Catal. Today 49 (1999) 267.
[136] M.S. da Machado, J.P. Pariente, E. Sastre, D. Cardoso, M.V. Giotto, J.L.G. Fierro, V. Fornes, J. Catal. 205 (2002) 299.
[137] D.B. Akolekar, S. Bhargava, J. Mol. Catal. 122 (1997) 81.
[138] D.B. Akolekar, J. Catal. 144 (1993) 148.
[139] C. Kannan, S.P. Elangovan, M. Palanichamy, V. Murugesan, Indian J. Chem. Technol. 5 (1998) 65.
[140] V. Umamaheswari, C. Kannan, B. Arabindoo, M. Palanichamy, V. Murugesan, Proc. Indian Acad. Sci. 112 (2000) 439.
[141] K.K. Cheralathan, C. Kannan, M. Palanichamy, V. Murugesan, Indian J. Chem. A 39 (2000) 921.
[142] S. Basha, N. Subrahmanyam, J. Das, A.B. Halgeri, Indian Chem. Eng. 43 (2001) 75.
[143] Y.H. Yeom, S.C. Shim, J. Mol. Catal. A 180 (2002) 133.
[144] U. Sridevi, V.V. Bokade, C.V.V. Satyanarayana, B.S. Rao, N.C. Pradhan, B.K.B. Rao, J. Mol. Catal. A 181 (2002) 257.
[145] K. Kondamudi, S. Upadhyayula, J. Chem. Technol. Biotechnol. 83 (2008) 699.
[146] E. Dumitriu, C. Guimon, V. Hulea, D. Luctic, I. Fechete, Appl. Catal. A 237 (2002) 211.
[147] P. Lian, W. Gao, F. Yan, W. Liu, G. Li, H. Wang, G. Sun, Catal. Today 74 (2002) 137.
[148] C.V.V. Satyanarayana, U. Sridevi, B.S. Rao, Stud. Surf. Sci. Catal. 135 (2001) 3798.
[149] B. Rajesh, M. Palanichamy, V. Kazansky, V. Murugesan, J. Mol. Catal. A 187 (2002) 259.

[150] K.J. Antony Raj, V.R. Vijayaraghavan, Indian J. Chem. A 43 (2004) 254.
[151] V.R. Vijayaraghavan, K.J. Antony Raj, J. Mol. Catal. A 207 (2004) 41.
[152] K.J. Antony Raj, V.R. Vijayaraghavan, Catal. Lett. 96 (2004) 67.
[153] K.J. Antony Raj, E.J. Padma Malar, V.R. Vijayaraghavan, J. Mol. Catal. A 243 (2006) 99.
[154] S. Vishnu Priya, J. Herbert Mabel, S. Gopalakrishnan, M. Palanichamy, V. Murgesan, J. Mol. Catal. A 290 (2008) 60.
[155] S. Vishnu Priya, J. Herbert Mabel, S. Gopalakrishnan, M. Palanichamy, V. Murgesan, J. Porous Mater. DOI 10.1007/s10934-008-9214-y.
[156] S. Gopalakrishnan, K.R. Viswanathan, S. Vishnu Priya, J. Herbert Mabel, M. Palanichaamy, V. Murugesan, Microporous Mesoporous Mater. 118 (2009) 523.
[157] S. Gopalakrishnan, K.R. Viswanathan, S. Vishnu Priya, J. Herbert Mabel, M. Palanichaamy, V. Murugesan, Catal. Commun. 10 (2008) 23.
[158] J. Weitkamp, Ind. Eng. Chem. Proc. Res. Dev. 21 (1982) 550.
[159] J.A. Martens, P.A. Jacobs, J. Weitkamp, Appl. Catal. 20 (1986) 239.
[160] B. Kraushaar-Czarnetzki, R.J. Dogterom, W.H. Stork, K.A. Emeis, J.P. van Braam Houckgeest, J. Catal. 141 (1993) 140.
[161] M. Höchtl, A. Jentys, H. Vinek, Microporous Mesoporous Mater. 31 (1999) 271.
[162] M. Hartmann, S.P. Elangovan, Chem. Eng. Technol. 26 (2003) 1232.
[163] C.M. Lopez, Y. Guillen, L. Garcia, L. Gomez, A. Ramirez, Catal. Lett. 122 (2008) 267.
[164] R. Roldan, M. Sanchez-Sanchez, G. Sankar, F.J. Romero-Salguero, C. Jimenez-Sanchidrian, Microporous Mesoporous Mater. 99 (2007) 288.
[165] P. Liu, J. Ren, Y. Sun, Microporous Mesoporous Mater. 114 (2008) 365.
[166] S. Zhang, S.L. Chen, P. Dong, G. Yuan, K. Xu, Appl. Catal. A 332 (2007) 46.
[167] S. Zhang, S.L. Chen, P. Dong, Z. Ji, J. Zhao, K. Xu, Catal. Lett. 118 (2007) 109.
[168] T. Blasco, A. Chica, A. Corma, W.J. Murphy, J. Agundez-Rodriguez, J. Perez-Pariente, J. Catal. 242 (2006) 153.
[169] B. Parlitz, E. Schreier, H.L. Zubowa, R. Eckelt, E. Lieske, G. Lischke, R. Fricke, J. Catal. 155 (1995) 1.
[170] J.M. Campelo, F. Lafont, J.M. Marinas, J. Chem. Soc., Faraday Trans. 91 (1995) 1551.
[171] J.M. Campelo, F. Lafont, J.M. Marinas, Zeolites 15 (1995) 97.
[172] J.M. Campelo, F. Lafont, J.M. Marinas, Appl. Catal. A 152 (1997) 53.
[173] J.M. Campelo, F. Lafont, J.M. Marinas, J. Catal. 156 (1995) 11.
[174] M.A. Chaar, J.B. Butt, Appl. Catal. A 114 (1994) 287.
[175] J.M. Campelo, F. Lafont, J.M. Marinas, J. Chem. Soc., Faraday Trans. 91 (1995) 4171.
[176] M. Hoffmeister, J.B. Butt, Appl. Catal. A 82 (1992) 169.
[177] P. Mériaudeau, V.A. Tuan, V.T. Nghiem, S.Y. Lai, L.N. Hung, C. Naccache, J. Catal. 169 (1997) 55.
[178] P. Mériaudeau, V.A. Tuan, F. Lefebvre, V.T. Nghiem, C. Naccache, Microporous Mesoporous Mater. 22 (1998) 435.
[179] P. Mériaudeau, V.A. Tuan, F. Lefebvre, V.T. Nghiem, C. Naccache, Microporous Mesoporous Mater. 26 (1998) 161.
[180] P. Mériaudeau, V.A. Tuan, G. Sapaly, V.T. Nghiem, C. Naccache, Catal. Today 49 (1999) 285.
[181] H. Yunfeng, W. Xiangsheng, G. Xinwen, L. Silue, H. Sheng, S. Haibo, B. Liang, Catal. Lett. 100 (2005) 59.
[182] Y. Ma, N. Li, Y. Ren, S. Xiang, N. Guan, J. Mol. Catal. A 250 (2006) 9.
[183] X.T. Ren, N. Li, J.-Q. Cao, Z.Y. Wang, S.Y. Liu, S.H. Xiang, Appl. Catal. 298 (2006) 144.
[184] S.P. Elangovan, M. Hartmann, J. Catal. 217 (2003) 388.
[185] X. Yang, Z. Xu, Z. Tian, H. Ma, Y. Xu, W. Qu, L. Lin, Catal. Lett. 109 (2006) 139.

[186] X. Huang, L. Wang, L. Kong, Q. Li, Appl. Catal. A 253 (2003) 461.
[187] Z. Wang, Z. Tian, F. Teng, G. Wen, Y. Xu, Z. Xu, L. Lin, Catal. Lett. 103 (2005) 109.
[188] X. Yang, H. Ma, Z. Xu, Y. Xu, Z. Tian, L. Lin, Catal. Commun. 8 (2007) 1232.
[189] S.J. Miller, Stud. Surf. Sci. Catal. 84 (1994) 2319.
[190] S.J. Miller, Microporous Mater. 2 (1994) 439.
[191] S.J. Miller, US Patent 5,135,638 (assigned to Chevron Research and Technology), 1992.
[192] D.S. Santilli, M.M. Habib, T.V. Harris, S.I. Zones, Int. Pat. Appl. WO 92/01657 (assigned to Chevron Research and Technology), 1992.
[193] J. Walendziewski, B. Pniak, Appl. Catal. A 250 (2003) 39.
[194] O.V. Kikhtyanin, L.A. Vostrikova, G.A. Urzhuntsev, A.V. Toktarev, M.I. Tselyutina, I.D. Reznichenko, G.V. Echevsky, Stud. Surf. Sci. Catal. 174B (2008) 1227.
[195] J. Hancsok, M. Krar, Sz. Magyar, L. Boda, A. Hollo, D. Kallo, Stud. Surf. Sci. Catal. 170B (2007) 1605.
[196] A. Vieria, M.A. Tovar, C. Pfaff, B. Mendez, C.M. Lopez, F.J. Machado, J. Goldwasser, M.M. Ramirez de Agudelo, J. Catal. 177 (1998) 60.
[197] A. Vieria, M.A. Tovar, C. Pfaff, P. Betancourt, B. Mendez, C.M. Lopez, F.J. Machado, J. Goldwasser, M.M. Ramirez de Agudelo, J. Mol. Catal. A 144 (1999) 101.
[198] Y. Wei, G. Wang, Z. Liu, C. Sun, L. Xu, Stud. Surf. Sci. Catal. 135 (2001) 3822.
[199] T. Komatsu, H. Ikenaga, J. Catal. 241 (2006) 426.
[200] F.J. Machado, C.M. Lopez, Y. Campos, A. Bolivar, S. Yunes, Appl. Catal. A 226 (2002) 241.
[201] N. Kumar, R. Byggningsbacka, M. Korpi, L. Lindfors, T. Salmi, Appl. Catal. A 227 (2002) 97.
[202] M. Hartmann, S. Ernst, Angew. Chem. Int. Ed. 39 (2000) 888.
[203] J.M. Thomas, Angew. Chem. Int. Ed. 38 (1999) 3588.
[204] J.M. Thomas, R. Raja, G. Sankar, R.G. Bell, Nature 398 (1999) 227.
[205] R. Raja, G. Sankar, J.M. Thomas, Angew. Chem. Int. Ed. 39 (2000) 2313.
[206] R. Raja, J.M. Thomas, Chem. Commun. (1998) 1841.
[207] G. Sankar, R. Raja, J.M. Thomas, Catal. Lett. 55 (1998) 15.
[208] R. Raja, G. Sankar, J.M. Thomas, J. Am. Chem. Soc. 121 (1999) 11926.
[209] D.L. Vanoppen, D.E. De Vos, M.J. Genet, P.G. Rouxhet, P.A. Jacobs, Angew. Chem. Int. Ed. 34 (1995) 560.
[210] S.S. Lin, H.S. Wenig, J. Chem. Eng. Jpn. 27 (1994) 211.
[211] R. Kraushaar-Czarnetzki, W.G.M. Hoogervorst, W.J.H. Stork, Stud. Surf. Sci. Catal. 84 (1994) 1869.
[212] D.L. Vanoppen, P.A. Jacobs, Catal. Today 49 (1999) 177.
[213] F.J. Luna, S.E. Ukawa, M. Wallau, U. Schuchardt, J. Mol. Catal. 117 (1997) 405.
[214] M. Dugal, G. Sankar, R. Raja, J.M. Thomas, Angew. Chem. Int. Ed. 39 (2000) 2310.
[215] P. Concepción, A. Corma, J.M. López Nieto, J. Pérez-Pariente, Appl. Catal. A 143 (1996) 17.
[216] L. Zhou, J. Xu, H. Miao, X. Li, F. Wang, Catal. Lett. 99 (2005) 231.
[217] L. Zhou, J. Xu, C. Chen, F. Wang, X. Li, J. Porous Mater. 15 (2008) 7.
[218] S.S. Lin, H.S. Weng, Appl. Catal. A 105 (1993) 289.
[219] I. Belkhir, A. Germain, F. Fajula, E. Fache, J. Chem. Soc., Faraday Trans. 94 (1998) 1761.
[220] I.W.C.E. Arends, R.A. Sheldon, M. Wallau, U. Schuchardt, Angew. Chem. Int. Ed. 36 (1997) 1144.
[221] B. Moden, L. Oliviero, J. Dakka, J.G. Santiesteban, E. Iglesia, J. Phys. Chem. 108 (2004) 5552.
[222] P. Tian, Z. Liu, Z. Wu, L. Xu, Y. He, Catal. Today 93–95 (2004) 735.

[223] R. Pohorecki, J. Baldyga, W. Moniak, W. Podgorska, A. Zdrjkowski, P.T. Wierchowski, Chem. Eng. Sci. 56 (2001) 1285.
[224] J.F. Black, J. Am. Chem. Soc. 100 (1978) 527.
[225] R. Zhao, Y. Wang, Y. Guo, Y. Guo, X. Liu, Z. Thang, Y. Wang, W. Zhan, G. Lu, Green Chem. 8 (2006) 459.
[226] R. Raja, G. Sankar, J.M. Thomas, Chem. Commun. (1999) 829.
[227] R. Raja, J.M. Thomas, G. Sankar, Chem. Commun. (1999) 525.
[228] R.A. Sheldon, N. de Heij, in: W. Ando, Y. Mora-oka (eds.), The role of Oxygen in Chemistry and Biochemistry, Elsevier, Amsterdam, 1988, p. 243.
[229] J. Dakka, R.A. Sheldon, Netherlands Patent 9200968, 1992.
[230] O.S. Kim, S.H. Chang, W.S. Ahn, J. Mol. Catal. A 179 (2001) 175.
[231] N. Venkatathri, S.G. Hedge, S. Sivasankar, J. Chem. Soc., Chem. Commun. (1995) 151.
[232] M.H. Zahedi-Niaki, S.M.J. Zaidi, S. Kaliaguine, Appl. Catal. A 196 (2000) 9.
[233] N. Venkatathri, S.G. Hegde, S. Sivasanker, Indian J. Chem. A 42 (2003) 974.
[234] T. Blasco, P. Concepcion, J.M. Lopez Nieto, J. Perez-Pariente, J. Catal. 152 (1995) 1.
[235] P. Concepcion, J.M. Lopez Nieto, J. Perez-Pariente, Catal. Lett. 19 (1993) 333.
[236] P. Concepcion, J.M. Lopez Nieto, J. Perez-Pariente, Catal. Lett. 28 (1994) 9.
[237] P. Concepcion, J.M. Lopez Nieto, J. Perez-Pariente, J. Mol. Catal. 99 (1995) 173.
[238] J.M. Lopez Nieto, Top. Catal. 15 (2001) 189.
[239] B.Z. Wan, K. Huang, Appl. Catal. 73 (1991) 113.
[240] J.Y. Wu, S.H. Chien, B.Z. Wau, Ind. Eng. Chem. Res. 40 (2001) 94.
[241] P. Concepcion, A. Corma, J.M. Lopez Nieto, J. Perez-Pariente, Appl. Catal. A 143 (1996) 17.
[242] P. Concepcion, J.M. Lopez Nieto, A. Mifsud, J. Perez-Pariente, Appl. Catal. A 151 (1997) 373.
[243] T. Blasco, L. Fernandez, A. Martinez-Arias, M. Sanchez-Sanchez, P. Concepcion, J.M. Lopez Nieto, Microporous Mesoporous Mater. 39 (2000) 219.
[244] L. Marchese, A. Frache, G. Gatti, S. Coluccia, L. Lisi, G. Raoppolo, G. Russo, H.O. Pastore, J. Catal. 208 (2002) 479.
[245] P. Concepcion, L. Nieto, Catal. Commun. 2 (2001) 363.
[246] P. Concepcion, T. Blasco, J.M. Lopez Nieto, A. Vidal-Moya, A. Martinez-Arias, Microporous Mesoporous Mater. 67 (2004) 215.
[247] A. Ristic, G. Avgouropoulos, T. Ionnides, N.N. Tusai, V. Kaucic, Stud. Surf. Sci. Catal. 135 (2001) 4718.
[248] S.J. Kulkarni, R. Ramachandran Rao, M. Subrahmanyam, A.V. Ramarao, A. Sarkany, L. Guczi, Appl. Catal. A 139 (1996) 59.
[249] S.J. Kulkarni, R. Ramachandra Rao, A.V. Subrahmanyam, A.V. Rama Rao, J. Chem. Soc., Chem. Commun. (1994) 273.
[250] N. Srinivas, M. Radha Kishan, S.J. Kulkarni, K.V. Raghavan, Microporous Mesoporous Mater. 39 (2000) 125.
[251] R. Raja, G. Sankar, J.M. Thomas, J. Am. Chem. Soc. 123 (2001) 8153.
[252] J.A. Labinger, CATTECH 3 (1999) 18.
[253] N. Ulagappan, V. Krishnasamy, J. Chem. Soc., Chem. Commun. (1995) 373.
[254] B.Y. Hsu, S. Cheng, J.M. Chen, J. Mol. Catal. A 149 (1999) 7.
[255] P.S.E. Dai, R.H. Petty, C.W. Ingram, R. Szostak, Appl. Catal. A 143 (1996) 101.
[256] X. Meng, Y. Yu, L. Zhao, J. Sun, K. Lin, M. Yang, D. Ziang, S. Qiu, F.S. Xiao, Stud. Surf. Sci. Catal. 135 (2001) 4423.
[257] J.D. Chen, M.J. Haanepen, J.H.C. van Hooff, R.A. Sheldon, Stud. Surf. Sci. Catal. 84 (1994) 973.
[258] J.D. Chen, J. Dakka, E. Neeleman, R.A. Sheldon, J. Chem. Soc., Chem. Commun. (1993) 1379.
[259] M.S. Rigutto, J. van Bekkum, J. Mol. Catal. 81 (1993) 77.

[260] R. Joseph, M. Sasidharan, R. Kumar, A. Sudalai, T. Ravindranathan, J. Chem. Soc., Chem. Commun. (1995) 1341.

[261] M.J. Haanepen, J.H.C. van Hooff, Appl. Catal. 152 (1997) 183.

[262] M.J. Haanepen, A.M. Elemans-Mehring, J.H.C. van Hooff, Appl. Catal. 152 (1997) 203.

[263] R.A. Sheldon, M. Wallau, I.W.C.E. Arends, U. Schuchardt, Acc. Chem. Res. 31 (1998) 485.

[264] H.E.P. Lempers, R.A. Sheldon, Stud. Surf. Sci. Catal. 105 (1996) 1061.

[265] J.D. Chen, R.A. Sheldon, J. Catal. 153 (1995) 1.

[266] J.D. Chen, H.E.B. Lempers, R.A. Sheldon, J. Chem. Soc., Faraday Trans. 92 (1996) 1807.

[267] H.E.B. Lampers, R.A. Sheldon, Appl. Catal. A 143 (1996) 137.

[268] R.A. Sheldon, I.W.C.E. Arends, H.E.B. Lempers, Catal. Today 41 (1998) 387.

[269] H.E.B. Lempers, R.A. Sheldon, J. Catal. 175 (1998) 62.

[270] P.S. Singh, K. Kosuge, V. Ramaswamy, B.S. Rao, Appl. Catal. A 177 (1999) 149.

[271] H. Mimoun, Catal. Today 1 (1987) 281.

[272] D.R.C. Huybrechts, L. de Bruycker, P.A. Jacobs, Nature 345 (1990) 240.

[273] S.H. Jhung, Y.S. Uh, H. Chan, Appl. Catal. 62 (1990) 61.

[274] A. Tuel, Zeolites 15 (1995) 228.

[275] A. Tuel, Y. Ben Taarit, J. Chem. Soc., Chem. Commun. (1994) 1667.

[276] M.S. Rigutto, H. van Bekkum, Appl. Catal. 68 (1991) L1.

[277] M.H. Zahedi-Niaki, M.P. Kapoor, S. Kaliaguine, J. Catal. 177 (1998) 231.

[278] M.H. Zahedi-Niaki, M.P. Kapoor, S. Kaliaguine, in: M.M.J. Treacy, B.K. Marcus, M.E. Higgins, J.B. Higgins (eds.), Proceedings of the 12th International Zeolite Conference, Materials Research Society, Warrendale, PA, 1998, pp. 1221–1226.

[279] N. Jappar, Y. Tanaka, S. Nakata, T. Tatsumi, Microporous Mesoporous Mater. 23 (1998) 169.

[280] V.N. Shetti, P. Manikandan, D. Srinivas, P. Ratnasamy, J. Catal. 216 (2003) 461.

[281] E.V. Spinace, D. Cardoso, U. Schuchardt, Zeolites 19 (1997) 6.

[282] E.V. Spinace, U. Schuchart, D. Cardoso, Appl. Catal. A 185 (1999) L193.

[283] M. Hartmann, A.M. Prakash, L. Kevan, Catal. Today 78 (2003) 467.

[284] P. McMorn, G. Roberts, G.J. Hutchings, Catal. Lett. 63 (1999) 193.

[285] K.E. Jeong, H.J. Chae, U.C. Kim, S.Y. Jeong, W.S. Ahn, Diff. Def. Data Part B Sol. State Phen. 135 (2008) 89.

[286] R.A. Sheldon, I.W.C.E. Arends, H.E.B. Lempers, Catal. Today 41 (1998) 387.

[287] R.A. Sheldon, Top. Curr. Chem. 164 (1993) 21.

[288] N.R. Shiju, S. Fiddy, O. Sonntag, M. Stockenhuber, G. Sankar, Chem. Commun. (2006) 4955.

[289] M. Hartmann, L. Kevan, J. Chem. Soc., Faraday Trans. 92 (1996) 1429.

[290] M. Hartmann, L. Kevan, Stud. Surf. Sci. Catal. 105 (1996) 717.

[291] M. Hartmann, L. Kevan, J. Phys. Chem. 100 (1996) 4606.

[292] A. Alex, T. Clark, J. Am. Chem. Soc. 114 (1992) 506.

[293] P.A. Jacobs, D.J. Declerck, L.J. Vandamme, J.B. Uytterhoeven, J. Chem. Soc., Faraday Trans. I 71 (1975) 1545.

[294] C.M. Chen, J.M. Jehng, Catal. Lett. 93 (2004) 213.

[295] J. Ma, B. Fan, R. Li, J. Cao, Catal. Lett. 23 (1994) 189.

[296] M.L. Cubeiro, C.M. Lopez, A. Comenares, L. Texeira, M.R. Goldwasser, M.J. Perez-Zurita, F. Machado, F. Gonzalez Jimenez, Appl. Catal. A 167 (1998) 183.

[297] S.M. Yang, Y.M. Wu, Appl. Catal. A 192 (2000) 211.

[298] D. Panayotov, L. Dimitrov, M. Khristova, L. Petrov, D. Mehandjiev, Appl. Catal. B 6 (1995) 61.

[299] T. Ishihara, M. Kagawa, Y. Mizuhara, Y. Takita, Chem. Lett. (1992) 2119.

[300] M. Kang, J. Mol. Catal. A 161 (2000) 115.

[301] X. Wang, Y. Xu, S. Yu, C. Wang, Catal. Lett. 103 (2005) 101.
[302] J. Dedecek, J. Cejka, B. Wichterlova, Appl. Catal. B 15 (1998) 233.
[303] W. Wei, J.A. Moulijn, G. Mul, Microporous Mesoporous Mater. 112 (2008) 193.
[304] V. Kurshev, A.M. Prakash, R.M. Krishna, L. Kevan, Microporous Mesoporous Mater. 34 (2000) 9.
[305] K.T. Ranjit, Z. Chang, R.M. Krishna, A.M. Prakash, L. Kevan, J. Phys. Chem. B 104 (2000) 7981.
[306] K.J. Chao, C.N. Wu, H.C. Shih, T.G. Tsai, Y.H. Chiou, Stud. Surf. Sci. Catal. 105 (1996) 2187.
[307] L. Washmon-Kriel, K.J. Balkus, Microporous Mesoporous Mater. 38 (2000) 107.
[308] G.J.T. Fernandes, V.J. Fernandes, Jr., A.S. Araujo, Catal. Today 75 (2002) 233.
[309] Y. Traa, B. Burger, J. Weitkamp, Microporous Mesoporous Mater. 30 (1999) 3.
[310] A. Frache, B. Pallela, M. Cadoni, R. Pirone, P. Ciambelli, H.O. Pastore, L. Marchese, Catal. Today 75 (2002) 359.
[311] B.I. Pallela, M. Cadoni, A. Frache, H.O. Pastore, R. Pirone, G. Russo, S. Coluccia, L. Marchese, J. Catal. 217 (2003) 100.
[312] A. Frache, M. Cadoni, S. Coluccia, L. Marchese, Stud. Surf. Sci. Catal. 135 (2001) 5096.
[313] A. Frache, E. Gianotti, L. Marchese, Catal. Today 77 (2003) 371.
[314] G. Wen, Y. Xu, H. Ma, Z. Xu, Z. Tian, Int. J. Hydrogen Energy 33 (2008) 6657.
[315] Y. Fan, D. Lei, G. Shi, X. Bao, Catal. Today 114 (2006) 388.
[316] L. Han, Y. Wang, C. Li, S. Zhang, X. Lu, M. Cao, AIChE J. 54 (2008) 280.
[317] X. Song, Y. Lin, L. Gan, Z. Wang, J. Yu, R. Xu, Angew. Chem. Int. Ed. 47 (2008).

SUBJECT INDEX

有　奖　征　集　反　馈　意　见

尊敬的读者：

科学出版社科爱森蓝文化传播有限公司（简称“科爱传播”）立足国际合作，致力于为科技专业人士提供优质的信息服务。我们很想通过自己的努力最大限度地满足您的需求，您的哪怕是一点点的建议和意见，都将成为我们改进工作的重要依据。

科爱传播
KE'AI COMMUNICATIONS

我们将在每年的 6 月份、12 月份各一次从半年的参与者中抽取幸运者 10 名，幸运者可以从“科爱传播”的出版物中任选价值 1000 元的图书（10 册以内）作为奖品（全部出版物信息可在我们的网站上查到）。

1．**您所购买的图书书名：**《______________________________》

您于________年___月____日在（通过）________________________购买到此书。

您认为本书的定价：□偏高　□合适　□偏低

您认为本书的内容有约____%对您有用。

2．**您认为我们出版物的质量：**

内容质量（学术水平、写作水平）：□很好　□较好　□一般　□较差

译介质量（翻译水平、文字水平）：□很好　□较好　□一般　□较差

印制质量（印制、包装）：□很好　□较好　□一般　□较差

3．**您所在的专业领域：**________________________

4．**在您获取专业知识和专业信息的主要渠道中，排在前三位的是：**

1．________　2．________　3．________

A.网络　B.期刊　C.图书　D.报纸　E.电视　F.会议　G.内部交流　H.其他：________

5．**您还需要哪些类型的图书？**

□专著　□教材　□实验手册　□辞典工具书　□文集　□其他：________

□书摘（原版书摘编）　□刊摘（国外学术期刊重要文献摘编）

6．**您还希望我们从国外引进哪些专业方向的图书（或期刊）？**

7．您建议采用何种引进形式？

☐翻译　☐影印　☐摘编（只从原书中选部分内容引进）

☐导读（原文影印加少量中文介绍）　☐注解（原文影印加大量中文介绍）

8．您是否愿意与我们合作，参与编写、编译、翻译图书或其他科技信息？

9．请列举您近两年看过的，您认为最有参考价值、对您帮助最大的 1~2 本书：

书名	著作者	出版社	出版日期	定价

10．您还有什么别的意见、建议？（可另附纸）

● 请告诉我们您准确的地址和联系办法：

姓名：__________ 性别：______ 生日：______年__月__日

单位：____________________职务/职称：__________

地址 ：____________________

E-mail：__________________电话：____________

传真：______________ 手机：______________

回邮地址（也可以通过 E-mail 反馈）：

北京东黄城根北街 16 号　科学出版社 科爱传播中心 杨 琴（收）　邮编：100717

联系电话：010-64006871；传真：010-64034056

编辑部电话：010-64034507

投稿及读者反馈：editor@kbooks.cn, keai@mail.sciencep.com

（注：本反馈单复印有效，也可以在线下载：http://www.kbooks.cn）